Student Solutions Manual

Physics for Scientists and Engineers
Foundations and Connections
with Modern Physics

Volume 2

Eric Mandell
Bowling Green State University

Brian Utter
James Madison University

Debora M. Katz
United States Naval Academy

T0195270

CENGAGE
Learning·

Australia · Brazil · Mexico · Singapore · United Kingdom · United States

For product information and technology assistance, contact us at **Cengage Learning Customer & Sales Support, 1-800-354-9706**.

For permission to use material from this text or product, submit all requests online at **www.cengage.com/permissions** Further permissions questions can be emailed to **permissionrequest@cengage.com**.

ISBN: 978-0-534-46767-8

Cengage Learning
20 Channel Center Street
Boston, MA 02210
USA

Cengage Learning is a leading provider of customized learning solutions with office locations around the globe, including Singapore, the United Kingdom, Australia, Mexico, Brazil, and Japan. Locate your local office at: **www.cengage.com/global**.

Cengage Learning products are represented in Canada by Nelson Education, Ltd.

To learn more about Cengage Learning Solutions, visit **www.cengage.com**.

Purchase any of our products at your local college store or at our preferred online store **www.cengagebrain.com**.

Printed in the United States of America
Print Number: 01   Print Year: 2015

Preface

This *Student Solutions Manual* has been written to accompany the textbook *Physics for Scientists and Engineers: Foundations and Connections with Modern Physics, First Edition, Volume 2*, by Debora M. Katz, and provides solutions to selected end-of-chapter problems. It is hoped that providing detailed, stepped-out solutions for the problems selected will allow students further practice in methods of problem solving.

Every textbook chapter has a matching chapter in this book and very often reference is made to specific equations or figures in the textbook. Solutions are given for approximately 20 problems in each textbook chapter. Problems were selected to illustrate important concepts in each chapter. The solutions follow the two-column worked example format—INTERPRET and ANTICIPATE, SOLVE, and CHECK and THINK presented in the text.

An important note concerning significant figures: When the statement of a problem gives data, we should only keep the number of significant figures in our final answer as dictated by the number of significant figures in the data and the rules for dealing with those figures in calculations (see Chapter 1, pages 10–11). The last digit is uncertain; it can, for example, depend on the precision of the values assumed for physical constants and properties. When a calculation involves several steps, we carry out intermediate steps to many digits, but we write down only three. We "round off" only at the end of any chain of calculations, never anywhere in the middle.

ACKNOWLEDGMENTS

We take this opportunity to thank everyone who contributed to this edition of the *Student Solutions Manual to accompany Physics for Scientists and Engineers: Foundations and Connections with Modern Physics, First Edition, Volume 2.*

Special thanks for managing and directing this project go to Senior Product Manager Charles Hartford, Product Manager Rebecca Berardy Schwartz, Content Developers Susan Dust Pashos and Ed Dodd, Associate Content Developer Brandi Kirksey, and Product Assistant Brendan Killion.

Our appreciation goes our reviewers, Douglas Sherman and James D. Olsen. Their careful reading of the manuscript and checking the accuracy of the problem solutions contributed immensely to the quality of the final product. Any errors remaining in the manual are the responsibility of the authors.

Finally, we express our appreciation to our families for their inspiration, patience, and encouragement, during the production of this manual.

We sincerely hope that this *Student Solutions Manual* will be useful to you in reviewing the material presented in the text, and in improving your ability to solve problems and score well on exams. We welcome any comments or suggestions that could help improve the content of this study guide in future editions; and we wish you success in your study of physics.

Eric Mandell
Bowling Green, OH

Brian Utter
Harrisonburg, VA

Debora M. Katz
Annapolis, MD

Table of Contents

23

Electric Forces

3. (N) An object has a charge of 35 nC. How many excess protons does it have?

INTERPRET and ANTICIPATE

A positively charged object has lost electrons and therefore has an excess number of protons relative to the number of electrons. While 35 nC is a small charge, the charge per electron is *much* smaller, so we expect that this will correspond to a large number of electrons.

SOLVE

The charge is due to a certain number of excess protons each with a charge $+e$. Applying Equation 23.1, we solve for N.

$$N = \frac{q}{e} = \frac{35 \times 10^{-9}\,\text{C}}{1.6 \times 10^{-19}\,\text{C/proton}}$$

$$N = \boxed{2.2 \times 10^{11}\,\text{protons}}$$

CHECK and THINK

Indeed, there is a large number of excess protons (10^{11}), or equivalently, electrons lost, that correspond to the 35 nC charge.

---

6. (N) A sphere has a net charge of 8.05 nC, and a negatively charged rod has a charge of −6.03 nC. The sphere and rod undergo a process such that 5.00×10^9 electrons are transferred from the rod to the sphere. What are the charges of the sphere and the rod after this process?

INTERPRET and ANTICIPATE

The amount of charge transferred is found by multiplying the number of electrons transferred by the charge of a single electron. Also, note that this amount of charge will be subtracted from the rod and added to the sphere. Thus, we expect the net charge of the sphere to decrease.

SOLVE

First, find the charge that is transferred from the rod to the sphere.

$$\Delta q = \left(5.00 \times 10^9\right)\left(-1.60 \times 10^{-19}\,\text{C}\right) = -8.00 \times 10^{-10}\,\text{C}$$

Then, the remaining charge of the sphere is found by adding this charge to the sphere's initial charge.	$Q_{sphere} = 8.05 \times 10^{-9}$ C $+ \left(-8.00 \times 10^{-10} \text{ C}\right) = \boxed{7.25 \times 10^{-9} \text{ C}}$
Then, the remaining charge of the rod is found by subtracting the charge transferred from the rod's initial charge.	$Q_{rod} = -6.03 \times 10^{-9}$ C $- \left(-8.00 \times 10^{-10} \text{ C}\right)$ $Q_{rod} = \boxed{-5.23 \times 10^{-9} \text{ C}}$

CHECK and THINK

As expected, the rod became less negatively charged and the sphere became less positively charged. Though there are two possible signs for the electric charge transferred between two objects, the concept of adding the charge to one and subtracting it from the other is unchanged.

---

9. (N) A 50.0-g piece of aluminum has a net charge of +4.20 μC.

Aluminum has an atomic mass of 27.0 g/mol, and its atomic number is 13.

a. Calculate the number of electrons that were removed from the initially neutral aluminum to produce this charge.

b. Determine what fraction of the original number of electrons this removed number represents.

c. By how much did the original mass of the aluminum decrease after charging?

INTERPRET and ANTICIPATE

The total charge removed to leave the net positive charge is equal to the number of electrons times the charge per electron $-e$. We can then compare this to the total number of electrons by determining the number of atoms in 50 g of aluminum times the number of electrons per atom. Finally, we can determine the mass of the electrons determined in part (a). We expect that this will be a large number of electrons, but far less than the total number of electrons. Given the relatively small mass of electrons compared to protons and neutrons, the loss of electrons should be a *very* small fraction of the total mass.

SOLVE **a.** The positive net charge occurred by losing electrons. We can relate the total charge of electrons ($q = -4.20$ μC) to the number of electrons N using Equation 23.2.	$N = \dfrac{q}{-e} = \dfrac{-4.20 \times 10^{-6} \text{ C}}{-1.60 \times 10^{-19} \text{C/electron}}$ $N = \boxed{2.63 \times 10^{13} \text{ electrons}}$

b. First determine the number of atoms from the mass, molecular weight of aluminum, and the number of atoms per mole (Avogadro's number). Each atom of aluminum contains 13 electrons.

$$50.0 \text{ g Al}\left(\frac{\text{mol}}{27.0 \text{ g}}\right)\left(6.02\times10^{23}\frac{\text{atoms}}{\text{mol}}\right)\left(13\frac{\text{electrons}}{\text{atom}}\right)=1.45\times10^{25} \text{ electrons}$$

Equivalently, we can determine this result step by step:	$\dfrac{50.0 \text{ g}}{27.0 \text{ g/mol}}=1.85$ moles of aluminum
	$\left(1.85 \text{ mol}\right)\left(6.02\times10^{23}\dfrac{\text{atoms}}{\text{mol}}\right)=1.11\times10^{24} \text{ atoms}$
	$\left(1.11\times10^{24} \text{atoms}\right)\left(13\dfrac{\text{electrons}}{\text{atom}}\right)=1.45\times10^{25} \text{ electrons}$
Now, divide the answer to part (a) by this total number of electrons to get the desired fraction.	fraction of electrons lost $=\dfrac{2.63\times10^{13}}{1.45\times10^{25}}=\boxed{1.81\times10^{-12}}$
c. Multiply the number of electrons removed by the mass of the electron.	$m=\left(\dfrac{-4.20\times10^{-6} \text{ C}}{-1.60\times10^{-19} \text{C/electron}}\right)\left(9.11\times10^{-31}\dfrac{\text{kg}}{\text{electron}}\right)$
	$m=\boxed{2.39\times10^{-17} \text{ kg}}$ or 2.39×10^{-14} g

CHECK and THINK
As expected, the charge is composed of a huge number of electrons (10^{13}), but an extremely small fraction of the total number (10^{-12}) and an even smaller fraction of the mass (10^{-14} out of 50 g).

15. (N) A charge of -36.3 nC is transferred to a neutral copper ball of radius 4.35 cm. The ball is not grounded. The excess electrons spread uniformly on the surface of the ball. What is the number density (number of electrons per unit surface area) of excess electrons on the surface of the ball?

INTERPRET and ANTICIPATE
Given the charge, we can determine the number of electrons. With the surface area of the ball, it will be straightforward to calculate the number density (number per area).

SOLVE	
Using Eq. 23.2, the charge is related to the number of electrons times the charge per electron.	$N = \dfrac{q}{-e} = \dfrac{-36.3 \times 10^{-9} \text{ C}}{-1.60 \times 10^{-19} \text{ C/electron}}$ $N = 2.27 \times 10^{11} \text{ electrons}$
The surface area of a sphere is $4\pi r^2$.	$A = 4\pi r^2 = 4(3.14)(0.0435 \text{ m})^2 = 0.0238 \text{ m}^2$
The number density is the number of electrons per area.	$\dfrac{N}{A} = \dfrac{2.27 \times 10^{11}}{0.0238 \text{ m}^2} = \boxed{9.54 \times 10^{12} \text{ m}^{-2}}$ (electrons per square meter)

CHECK and THINK
The number density is about 10^{13} electrons per square meter. Not surprisingly, it's a very large number of electrons spread over the surface of the copper ball.

16. (N) Two identical conductors are brought into contact. Initially, one conductor has a charge of +30.0 μC. What is the charge of each conductor afterward? Does it matter how the contact is made?

Charges are free to move between conducting materials and because the two conductors are identical, the charges will be divided evenly between them. Therefore, each will have a final charge of +15.0 μC. Once they are in electrical contact, the charges will be transferred, regardless of how contact is made. If these were insulators, the charge would not be free to move and charge would not easily be transferred between the two objects. In that case, charges can be transferred by rubbing them together and so it does matter how contact is made.

20. (C) An electroscope is a device used to measure the (relative) charge on an object (Fig. P23.20). The electroscope consists of two metal rods held in an insulated stand. The bent rod is fixed, and the straight rod is attached to the bent rod by a pivot. The straight rod is free to rotate. When a positively charged object is brought close to the electroscope, the straight movable rod rotates as shown in Figure P23.20. Explain your answers to these questions:
a. Why does the rod rotate in Figure P23.20?
b. If the positively charged object is removed, what happens to the electroscope?
c. If a negatively charged object replaces the positively charged object in Figure P23.20, what happens to the electroscope?
d. If a charged object touches the top of the fixed conducting rod and is then removed,

what happens to the electroscope?

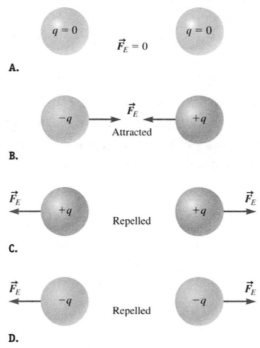

A. $q = 0$ $\vec{F}_E = 0$ $q = 0$

B. $-q$ → $\vec{F}_E$ ← $+q$ Attracted

C. $\vec{F}_E$ ← $+q$ Repelled $+q$ → $\vec{F}_E$

D. $\vec{F}_E$ ← $-q$ Repelled $-q$ → $\vec{F}_E$

Figure P23.20

---

a. When the positively charged object is brought close to the electroscope, it becomes polarized and negative charges move to the platform at the top while a net positive charge is left at the bottom. The two rods then both have a net positive charge and repel each other, causing the moveable rod to rotate away (to maximize the separation of the positive charges due to the repulsive force).

- - - - - Negative charge

Insulating stand

Fixed conducting rod

Movable conducting rod

Rotates

Positive charge

Figure P23.20aANS

b. Without the positive charge polarizing the electroscope, the charge on the platform returns to the bottom. The neutral electroscope is then uncharged and not polarized, as initially. Typically, the moveable rod is weighted slightly such that when the electroscope is uncharged, the moveable rod rotates back to a vertical position, so we would expect the rod to rotate back to this initial position.

c. When the negatively charged object is brought close, the electroscope polarizes in the opposite way as in part (a). A net positive charge will reside on the top platform leaving a net negative charge at the bottom. Since the two rods are negatively charged, they again repel and the moveable rod rotates away from the fixed one.

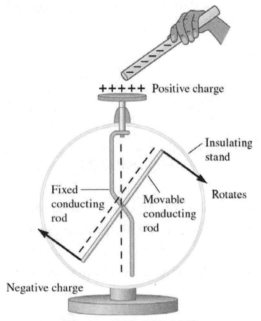

+++++ Positive charge

Insulating stand

Fixed conducting rod

Movable conducting rod

Rotates

Negative charge

Figure P23.20bANS

d. In this case, charge can be transferred to the electroscope causing it to have a net charge even when the object is removed. The rods will therefore be charged (with all parts of the electroscope having the same sign charge) and the rods will repel causing the moveable rod to rotate away from the fixed one.

23. (N) Two coins are placed on a horizontal insulating surface a distance of 1.5 m apart and given equal charges. They experience a repulsive force of 2.3 N. Calculate the magnitude of the charge on each coin.

INTERPRET and ANTICIPATE
Since typical coins are small compared to the 1.5 m separation, they can be treated as point charges. Coulomb's law then governs the force between them.

SOLVE The charge on each coin is the same, so we call it q and apply Coulomb's law (Equation 23.3) for the magnitude of the force.	$F_E = \dfrac{kq^2}{r^2}$
Solve for q and insert numerical values.	$q = \sqrt{\dfrac{F_E r^2}{k}} = \sqrt{\dfrac{(2.3)(1.5^2)}{8.99 \times 10^9}} = \boxed{2.4 \times 10^{-5} \text{ C}}$

CHECK and THINK
The charges on the coins could be either positive or both negative and the coins would repel. The magnitude is 24 μC.

29. (A) Two particles with charges q_1 and q_2 are separated by a distance d, and each exerts an electric force on the other with magnitude F_E.
a. In terms of these quantities, what separation distance would cause the magnitude of the electric force to be halved?
b. In terms of these quantities, what separation distance would cause the magnitude of the electric force to be doubled?

INTERPRET and ANTICIPATE
We will use Eq. 23.3 to express the initial electric force between the particles. We can then use the same equation written for an unknown distance r between the charges in parts (a) and (b) in order to solve for the distance when the electric force would be halved, or doubled. The particles should be further apart when the electric force is halved and closer together when the electric force is doubled.

SOLVE First, write Eq. 23.3 for the case where the particles are separated by the distance d.	$F_E = \dfrac{k\lvert q_1 q_2 \rvert}{d^2}$
a. Express the electric force using Eq. 23.3 when the particles are separated by a distance r, where the electric force has now been halved.	$F = \dfrac{1}{2} F_E$ $\dfrac{k\lvert q_1 q_2 \rvert}{r^2} = \dfrac{1}{2}\dfrac{k\lvert q_1 q_2 \rvert}{d^2}$
Solve for the distance r.	$\dfrac{1}{r^2} = \dfrac{1}{2d^2}$ $r = \boxed{d\sqrt{2}}$

b. Express the electric force using Eq. 23.3 when the particles are separated by a distance r, where the electric force has now been doubled.	$F = 2F_E$ $\dfrac{k\lvert q_1 q_2 \rvert}{r^2} = 2\dfrac{k\lvert q_1 q_2 \rvert}{d^2}$
Solve for the distance r.	$\dfrac{1}{r^2} = \dfrac{2}{d^2}$ $r = \boxed{d/\sqrt{2}}$

CHECK and THINK

As expected, the particles must be further apart for the electric force to decrease and the particles must be closer together for the electric force to increase. Note, however, that it is not doubling or halving the distance between the particles that accomplishes this feat.

31. (N) Two electrons in adjacent atomic shells are separated by a distance of 5.00×10^{-11} m.

a. What is the magnitude of the electrostatic force between the electrons?

b. What is the ratio of the electrostatic force to the gravitational force between the electrons?

INTERPRET and ANTICIPATE
Part (a) is a straightforward application of the Coulomb force law. We are then asked to compare the result to the gravitational force using Newton's law of gravitation. We expect that gravity is negligible for electrons in an atom, so the electric force should be *much* larger than the gravitational force.

SOLVE **a.** We can calculate the force using Equation 23.3. Each electron has a charge e and the separation distance is given.	$F_E = \dfrac{k\lvert q_1 q_2 \rvert}{r^2}$ $F_E = \dfrac{\left(8.99 \times 10^9 \ \text{N} \cdot \text{m}^2 / \text{C}^2\right)\left(\lvert -1.60 \times 10^{-19} \ \text{C}\rvert\right)^2}{\left(5.00 \times 10^{-11} \ \text{m}\right)^2}$ $F_E = \boxed{9.21 \times 10^{-8} \ \text{N}}$

b. Calculate the gravitational force using the mass of the electron and the separation distance given.	$F_g = \dfrac{Gm_1 m_2}{r^2}$ $F_g = \dfrac{\left(6.67\times10^{-11}\ \text{N}\cdot\text{m}^2/\text{C}^2\right)\left(9.11\times10^{-31}\ \text{kg}\right)^2}{\left(5.00\times10^{-11}\ \text{m}\right)^2}$ $F_g = 2.21\times10^{-50}\ \text{N}$
Taking the ratio, we see that the electric force is an astounding 10^{42} times larger!	$\dfrac{F_E}{F_g} = \dfrac{9.21\times10^{-8}\ \text{N}}{2.21\times10^{-50}\ \text{N}} = \boxed{4.16\times10^{42}}$

CHECK and THINK

As expected, the electrostatic force is _much_ larger than the gravitational force—10^{42} times larger!!

---

34. (N) One end of a light spring with force constant $k = 125$ N/m is attached to a wall, and the other end to a metal block with charge $q_A = 2.00\ \mu C$ on a horizontal, frictionless table (Fig. P23.34). A second block with charge $q_B = -3.60\ \mu C$ is brought close to the first block. The spring stretches as the blocks attract each other so that at equilibrium, the blocks are separated by a distance $d = 12.0$ cm. What is the displacement x of the spring?

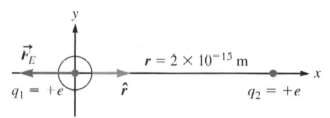

$\vec{F}_E$ $r = 2 \times 10^{-15}$ m

$q_1 = +e$ $\hat{r}$ $q_2 = +e$

Figure P23.34

INTERPRET and ANTICIPATE
As the second block is brought near the first block, there is an electrostatic attraction that pulls the first block to the right. The system comes to a static equilibrium, so the electrostatic attraction must be balanced by the spring force as the spring is stretched.

SOLVE The electric force on q_A (found using Eq. 23.3, Coulomb's law) is balanced by the spring force ($F = -Kx$), where K is the spring	$\dfrac{k\left|q_A q_B\right|}{d^2} = Kx$

constant and x is the displacement from equilibrium. Note that we are using K for the spring constant in order to avoid confusion with the constant k in Coulomb's law.	
Now, solve for x.	$x = \dfrac{k\|q_1 q_2\|}{Kd^2}$ $x = \dfrac{(8.99 \times 10^9 \,\text{N} \cdot \text{m}^2/\text{C}^2)(2.00 \times 10^{-6}\,\text{C})(3.60 \times 10^{-6}\,\text{C})}{(125 \,\text{N/m})(0.120 \,\text{m})^2}$ $x = \boxed{0.0360 \,\text{m}} = 3.60 \,\text{cm}$

CHECK and THINK
The spring has stretched 3.6 cm, at which point the spring force pulling to the left on the block is equal in magnitude to the Coulomb force pulling to the right due to the charge on block B.

38. (N) Given the arrangement of charged particles in Figure P23.38, find the net electrostatic force on the 5.65-μC charged particle.

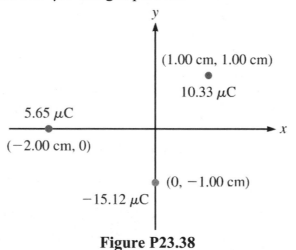

Figure P23.38

INTERPRET and ANTICIPATE
We will need to use Eq. 23.3 to write the magnitude of the electric force between the 5.65-μC charged particle and each of the other particles separately, interpret the vector direction of the electric force in each case, and then use the principle of superposition to perform a vector sum to find the net, or total, electric force on the 5.65-μC charged particle. It is somewhat difficult to anticipate the direction of the net electric force given

the arrangement of particles, but since the negatively charged particle is nearer and has a greater absolute value of charge, we expect its effect to dominate, and likely the 5.65-μC charged particle experiences a net electric force that points down and to the right.

SOLVE First, use the geometry of the figure to find the distance between the 5.65-μC particle and each of the others.	$d_{10.33} = \sqrt{(-0.0200 \text{ m} - 0.0100 \text{ m})^2 + (0.0100 \text{ m})^2}$ $d_{10.33}^2 = 1.00 \times 10^{-3} \text{ m}^2$ $d_{-15.12} = \sqrt{(-0.0200 \text{ m})^2 + (0.0100 \text{ m})^2}$ $d_{-15.12}^2 = 5.00 \times 10^{-4} \text{ m}^2$
Then, use Eq. 23.3 to find the magnitude of the electric force between each of the particles and the 5.65-μC particle.	$F_{10.33} = \dfrac{k\|q_1 q_2\|}{r^2}$ $F_{10.33} = (8.99 \times 10^9 \text{ N} \cdot \text{m}^2 / \text{C}^2) \dfrac{\|(10.33 \times 10^{-6} \text{ C})(5.65 \times 10^{-6} \text{ C})\|}{1.00 \times 10^{-3} \text{ m}^2}$ $F_{-15.12} = \dfrac{k\|q_1 q_2\|}{r^2}$ $F_{-15.12} = (8.99 \times 10^9 \text{ N} \cdot \text{m}^2 / \text{C}^2) \dfrac{\|(-15.12 \times 10^{-6} \text{ C})(5.65 \times 10^{-6} \text{ C})\|}{5.00 \times 10^{-4} \text{ m}^2}$
Next, determine the direction of the electric force in each case by finding the magnitude of the angle of the lines of action, with respect to the positive x axis, between each of the particles and the 5.65-μC particle. We retain extra digits for the application of the significant figure rules when later computing the net force.	$\tan\theta_{10.33} = \left\|\dfrac{0.0100 \text{ m}}{0.0300 \text{ m}}\right\|$ $\theta_{10.33} = 18.435°$ $\tan\theta_{-15.12} = \left\|\dfrac{0.0100 \text{ m}}{0.0200 \text{ m}}\right\|$ $\theta_{-15.12} = 26.565°$

We must now write each of the electric forces as a vector in component form. For the 10.33-μC particle, the force is repulsive and will result in a negative x and y component.	$$\vec{F}_{10.33} = \left(8.99 \times 10^9 \text{ N} \cdot \text{m}^2 / \text{C}^2\right) \frac{\left|\left(10.33 \times 10^{-6} \text{ C}\right)\left(5.65 \times 10^{-6} \text{ C}\right)\right|}{1.00 \times 10^{-3} \text{ m}^2}$$ $$\times \left[-\cos(18.435°)\hat{i} - \sin(18.435°)\hat{j}\right]$$ $$\vec{F}_{10.33} = \left(-498\hat{i} - 166\hat{j}\right)\text{N}$$
For the -15.12-μC particle, the force is attractive and will result in a positive x and negative y component. We retain extra digits for the application of the significant figure rules when later computing the net force.	$$\vec{F}_{-15.12} = \left(8.99 \times 10^9 \text{ N} \cdot \text{m}^2 / \text{C}^2\right) \frac{\left|\left(-15.12 \times 10^{-6} \text{ C}\right)\left(5.65 \times 10^{-6} \text{ C}\right)\right|}{5.00 \times 10^{-4} \text{ m}^2}$$ $$\times \left[\cos(26.565°)\hat{i} - \sin(26.565°)\hat{j}\right]$$ $$\vec{F}_{-15.12} = \left(1.374 \times 10^3 \hat{i} - 687\hat{j}\right)\text{N}$$
The net electric force is then found using the principle of superposition. We add the two force vectors together.	$$\vec{F}_{net} = \left(-498\hat{i} - 166\hat{j}\right)\text{N} + \left(1.374 \times 10^3 \hat{i} - 687\hat{j}\right)\text{N}$$ $$\boxed{\vec{F}_{net} = \left(876\hat{i} - 853\hat{j}\right)\text{N}}$$

CHECK and THINK

Though the geometry of the picture is a bit complicated, we are able to anticipate the negatively charged particle will have a greater effect because of its proximity to the subject particle.

42. Three identical conducting spheres are fixed along a single line. The middle sphere is equidistant from the other two so that the center-to-center distance between the middle sphere and either of the other two is 0.125 m. Initially, only the middle sphere is charged, with $q_{middle} = +35.6$ nC. The middle sphere is later connected by a conducting wire to the

12

sphere on the left. The wire is removed and then used to connect the middle sphere to the sphere on the right. The wire is again removed.

a. (C) What is the charge on each sphere?

When the center sphere is connected with the charge on the left, the charge equilibrates. Since they are identical conductors, they each end up with half the charge (35.6/2 = 17.8 nC). When the second wire is connected, the 17.8 nC charge that is now on the center conductor is redistributed so that the center and right conductor each end up carrying half the charge (17.8/2 = 8.9 nC), while the charge on the first one is unchanged. So, from left to right, they end up at 17.8 nC, 8.9 nC, and 8.9 nC.

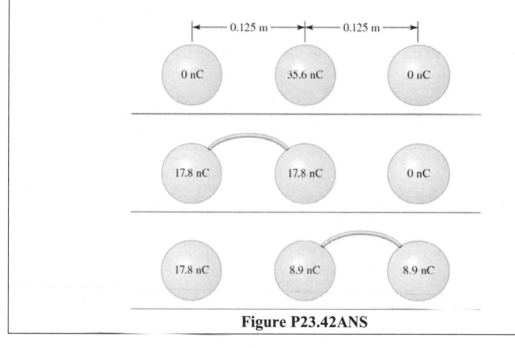

Figure P23.42ANS

b. (C) Which sphere experiences the greatest electrostatic force?

Considering the different pairs of conductor, the force between the left and center conductors is the largest because the total charge is greatest while the separation is small. The center charge experiences a force to the right due to the left conductor, but this is offset by the force to the left due to the charge on the right. The conductor on the left experiences two forces to the left which add together. These reinforce each other, so the conductor on the left experiences the largest net force. (The conductor on the right experiences two forces to the right, but it is further from the large charge on the left and therefore will have a smaller total magnitude force than the conductor on the left.)

c. (N) What is the magnitude of that force?

INTERPRET and ANTICIPATE Given the set-up in parts (a) and (b), we only need to apply Coulomb's law to find the forces due to the middle and right hand charge and add them together.	

SOLVE Apply Coulomb's law (Eq. 23.3). As discussed in part (b), since all charges are positive, both forces on the conductor on the left point to the left and therefore their magnitudes add.	$$F_E = \frac{k|q_1 q_2|}{r^2}$$

We calculate the forces of the middle charge on the charge on the left:

$$F_{ML} = \frac{\left(8.99 \times 10^9 \ \text{N} \cdot \text{m}^2 / \text{C}^2\right)\left(17.8 \times 10^{-9} \text{C}\right)\left(8.90 \times 10^{-9} \text{C}\right)}{\left(0.125 \ \text{m}\right)^2}$$

$$F_{ML} = 9.12 \times 10^{-5} \ \text{N}$$

And now the force due to the charge on the right on the charge on the left:

$$F_{RL} = \frac{\left(8.99 \times 10^9 \ \text{N} \cdot \text{m}^2 / \text{C}^2\right)\left(17.8 \times 10^{-9} \text{C}\right)\left(8.90 \times 10^{-9} \text{C}\right)}{\left(0.250 \ \text{m}\right)^2}$$

$$F_{RL} = 2.28 \times 10^{-5} \ \text{N}$$

Finally, add the two together.	$F_{tot} = F_{ML} + F_{RL}$ $F_{tot} = 9.12 \times 10^{-5} \ \text{N} + 2.28 \times 10^{-5} \ \text{N}$ $F_{tot} = \boxed{1.14 \times 10^{-4} \ \text{N}}$

CHECK and THINK
After thinking through parts (a) and (b), we are left only to apply Coulomb's law.

47. (N) A sphere of mass 5.00 g carries a positive charge of 30.0 nC and remains stationary when placed 5.00 cm directly above a second charged sphere that is fixed to a tabletop. What must be the charge of the second sphere?

INTERPRET and ANTICIPATE

For the sphere to remain stationary, the net force on it must be zero. That is, there must be a repulsive force due to the charge on the tabletop that is equal and opposite the force of gravity on the sphere. To find the electric force between the charges, we use Coulomb's law. Since charge of the upper ball is positive, the charge on the fixed lower ball must also be positive to create a repulsive force.

This is actually an *unstable* equilibrium. With any slight misalignment, the sphere will actually fall out to the side. However, based on the question, we assume that it remains directly above the charge on the tabletop and therefore only concern ourselves with its vertical position.

SOLVE	
We apply Newton's second law on the sphere. The acceleration of the subject must be zero since it is in static equilibrium, so the sum of the forces is zero.	$\sum \vec{F} = \vec{F}_1 + \vec{F}_2 = 0$ $\vec{F}_1 = -\vec{F}_2$
There are two forces on the upper ball: the gravitational force downward, $F_1 = mg$, where m is the mass of the upper ball, and the repulsive Coulomb force F_2 acting upward on the sphere. These forces have equal magnitudes and point in opposite directions.	$mg = k\dfrac{\lvert q_1 q_2 \rvert}{r^2}$ $m = 0.00500$ kg $r = 0.050$ m $q_1 = 3.0 \times 10^{-8}$ C $q_2 = ?$
Insert numerical values and solve. The charge on the fixed sphere is $+4.5 \times 10^{-7}$ C.	$q_2 = \dfrac{mgr^2}{q_1 k}$ $q_2 = \dfrac{\left(0.00500\ \text{kg}\right)\left(9.81\ \text{m/s}^2\right)\left(0.0500\ \text{m}\right)^2}{\left(3.00 \times 10^{-8}\text{C}\right)\left(8.99 \times 10^9\ \text{N} \cdot \text{m}^2 / \text{C}^2\right)}$ $q_2 = \boxed{4.55 \times 10^{-7}\ \text{C}}$

CHECK and THINK

The charge is positive such that the repulsive Coulomb force supports the weight of the sphere.

Chapter 23 – Electric Forces

52. (C) Four charged particles q, $-q$, q, and $-q$ are fixed at the corners of a square with side length L as shown in Figure P23.52. If another charged particle of magnitude Q is placed at the center of the square, will it be in static equilibrium? Does the sign of the charge Q matter? Explain.

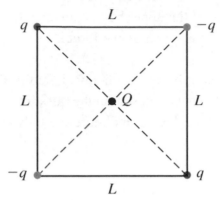

Figure P23.52

The problem has symmetry and using it will make solving the problem much easier. The forces between the added charge Q and the fixed charges are shown in the figure, assuming that Q is positive.

All four forces are equal in magnitude, $\left|\vec{F}_1\right|=\left|\vec{F}_2\right|=\left|\vec{F}_3\right|=\left|\vec{F}_4\right|=k\dfrac{qQ}{a^2}$, where a is the length of half the diagonal. As shown in figure, $\vec{F}_2$ cancels $\vec{F}_4$ and $\vec{F}_1$ cancels $\vec{F}_3$. Thus, the net force on charge Q is zero – so it will be in static equilibrium.

If the charge Q is negative, all four forces flip in the opposite directions, but the forces are still equal and opposite for each pair, and the charge is still in static equilibrium.

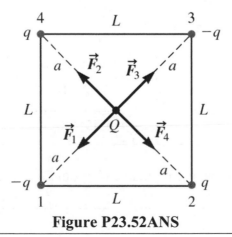

Figure P23.52ANS

55. (A) Three small metallic spheres with identical mass m and identical charge $+q$ are suspended by light strings from the same point (Fig. P23.55). The left-hand and right-hand strings have length L and make an angle θ with the vertical. What is the value of q in terms of k, g, m, L, and θ?

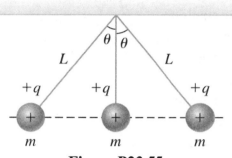

Figure P23.55

INTERPRET and ANTICIPATE
The three charges are in static equilibrium, so the net force on each is zero. We can take into account the forces due to gravity (weight), tension, and electric charge (Coulomb's law) on each sphere and set the net force equal to zero.

SOLVE We first sketch a free body diagram for the sphere on the right, which includes the electrostatic (Coulomb) force, weight, and tension.	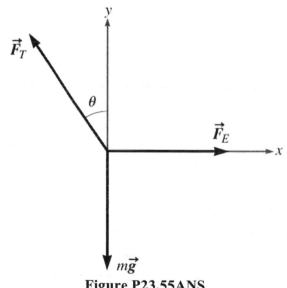 **Figure P23.55ANS**
The net force is zero. Consider the y component first.	$\Sigma F_y = 0:\quad F_T\cos\theta = mg \quad \rightarrow \quad F_T = \dfrac{mg}{\cos\theta}$

Now the x components.	$\Sigma F_x = 0:$ $F_E = F_T \sin\theta = \left(\dfrac{mg}{\cos\theta}\right)\sin\theta = mg\tan\theta \qquad (1)$
We can also calculate the Coulomb force (Eq.23.3) on this rightmost charge due to the charges on the left and the center. All three charges have the same magnitude q. The distance to the center charge r_1 can be found using geometry. Since the left hand sphere makes the same angle, the distance between the left and right spheres r_2 is just double the distance from the center to the right.	$r_1 = L\sin\theta$ $r_2 = 2r_1 = 2L\sin\theta$
Now apply Coulomb's law.	$F_E = \dfrac{kq^2}{r_1^2} + \dfrac{kq^2}{r_2^2}$ $F_E = \dfrac{kq^2}{\left(L\sin\theta\right)^2} + \dfrac{kq^2}{\left(2L\sin\theta\right)^2}$ $F_E = \dfrac{5kq^2}{4L^2\sin^2\theta} \qquad (2)$
Combine equations (1) and (2) and solve for q.	$\dfrac{5kq^2}{4L^2\sin^2\theta} = mg\tan\theta$ $\boxed{q = \sqrt{\dfrac{4L^2 mg\tan\theta\sin^2\theta}{5k}}}$

CHECK and THINK

This expression is not immediately transparent, but let's consider a few cases to see if it makes sense. If the length and angle are kept constant, then increasing the mass would lead to an increase in the charge. That is, if the mass was larger, the charge would have to be larger to balance the increased weight in order to keep the angle the same. Similarly, a larger angle (keeping the length and mass constant) leads to an increase on the right hand side and a larger charge. In other words, if the mass and length was unchanged, a larger charge (and larger repulsive force) is associated with a larger angle from center for the outside charges. The expression seems to make sense!

60. (N) Two otherwise identical, small conducting spheres have charges +5.0 μC and −2.0 μC. When placed a distance r apart, each experiences an attractive force of 3.0 N. The spheres are then touched together and moved back to a distance r apart. Find the magnitude of the new force on each sphere.

INTERPRET and ANTICIPATE

This problem will require applying Coulomb's law. First, knowing the force and the two charges, we can determine the separation. And then, knowing that identical conductors will equilibrate with the same charge after placed in contact, we can calculate the new force given the charges and the separation that we determined.

SOLVE Apply Coulomb's law, Equation 23.3.	$$F_E = k\frac{|q_1 q_2|}{r^2}$$
We can calculate the distance r given the charges and the magnitude of the force.	$$r = \sqrt{k\frac{|q_1 q_2|}{F_E}}$$ $$r = \sqrt{\left(8.99\times10^9\,\text{N}\cdot\text{m}^2/\text{C}^2\right)\frac{\left(5\times10^{-6}\,\text{C}\right)\left(2\times10^{-6}\,\text{C}\right)}{\left(3\,\text{N}\right)}}$$
After the conductors are brought into contact, the net charge of 5 μC − 2 μC = 3 μC is split evenly between them, so they each have 1.5 μC. We can now calculate the force.	$$F_E = k\frac{|q_1 q_2|}{r^2}$$ $$F_E = \frac{\left(8.99\times10^9\,\text{N}\cdot\text{m}^2/\text{C}^2\right)\left(1.5\times10^{-6}\,\text{C}\right)\left(1.5\times10^{-6}\,\text{C}\right)}{\left(\sqrt{\left(8.99\times10^9\,\text{N}\cdot\text{m}^2/\text{C}^2\right)\frac{\left(5\times10^{-6}\,\text{C}\right)\left(2\times10^{-6}\,\text{C}\right)}{\left(3\,\text{N}\right)}}\right)^2}$$ $$F_E = \boxed{0.675\,\text{N}}$$

CHECK and THINK

Perhaps not surprisingly, as the initial charges were opposite in sign and therefore the final charges have a lower magnitude than the initial (with the same separation), the magnitude of the force is somewhat smaller. Though we're not asked, the force is now repulsive rather than attractive.

65. (A) Figure P23.65 shows two identical conducting spheres each with charge q suspended from light strings of length L. If the equilibrium angle the strings make with the vertical is θ, what is the mass m of the spheres?

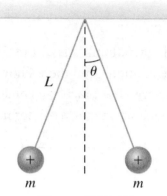

Figure P23.65

INTERPRET and ANTICIPATE

The gravitational force pulls the spheres downward while the electrostatic force pushes them apart. Since they are in static equilibrium, we will use the fact that the net force on each sphere is zero. We don't have values, so we'll derive a symbolic answer.

SOLVE

We first sketch one of the spheres (the one on the left). F_T is the tension in the string and F_E is the repulsive electrical force exerted by the other sphere.

Figure P23.65ANS

Since the sphere is in static equilibrium, the net force is zero. Consider the vertical component first.

$$\Sigma F_y = 0 \quad \rightarrow \quad F_T \cos\theta = mg$$

$$F_T = \frac{mg}{\cos\theta} \tag{1}$$

Now, solve for the horizontal component, using equation (1) to substitute for the tension.	$\sum F_x = 0 \qquad \rightarrow \qquad F_E = F_T \sin\theta$ $F_E = \dfrac{mg}{\cos\theta}\sin\theta = mg\tan\theta$
At equilibrium, the distance separating the two spheres is $r = 2L\sin\theta$. We can then use Coulomb's law (Equation 23.3) to express the electric force between the two charges q.	$F_E = k\dfrac{\lvert q_1 q_2 \rvert}{r^2}$ $\dfrac{kq^2}{\left(2L\sin\theta\right)^2} = mg\tan\theta$
Solve for m.	$m = \boxed{\dfrac{kq^2}{4L^2 g \tan\theta \sin^2\theta}}$

CHECK and THINK

The answer seems a bit opaque at first, but let's think about a couple cases. If the length and angle are kept constant, we see that if the charge q increases, we get a larger mass m. That is, if we had a larger charge, we would need a larger mass to keep the same angle. That sounds right at least. Similarly, if we had a constant charge and angle, a longer string length would correspond to a smaller mass. That's because with a longer string *and the same angle*, the spheres would be further apart. Therefore, the electrostatic force would be smaller and we need a smaller mass to balance that force. Okay, that seems to make sense as well.

68. (A) Two positively charged spheres with charges $4e$ and e are separated by a distance L and held motionless. A third charged sphere with charge Q is set between the two spheres and along the line joining them. The third sphere is in static equilibrium. What is the distance between the third charged sphere and the sphere that has charge $4e$?

INTERPRET and ANTICIPATE

The third sphere is in static equilibrium, so the forces due to the other charges are balanced. We expect that the third charged sphere must be closer to the sphere with the smaller charge (since a smaller charge that is closer can exert the same force as a larger charge that's further away).

SOLVE We sketch the situation described, assuming that Q is	

a positive charge. The two fixed charges are labeled $q_1 = 4e$ and $q_2 = e$. Only two electrostatic forces are exerted on the subject—$\vec{F}_1$, the force exerted by q_1 and $\vec{F}_2$, the force exerted by q_2.	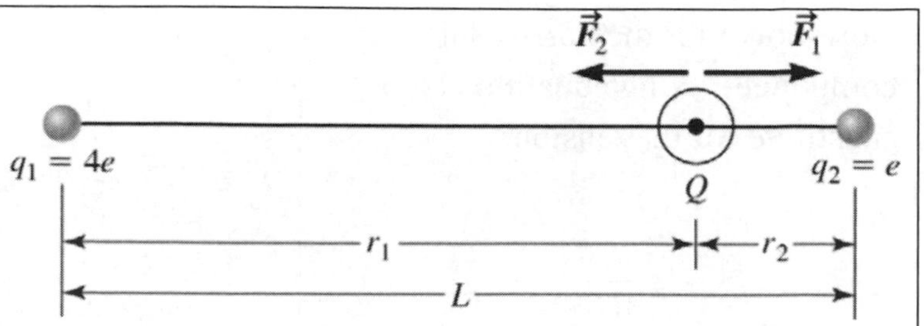 $\vec{F}_2$ $\vec{F}_1$ $q_1 = 4e$ Q $q_2 = e$ r_1 r_2 L **Figure P23.68ANS**
We apply Newton's second law. The third charge is in static equilibrium, so the sum of the forces is zero and the electrostatic forces must have equal magnitude and point in opposite directions.	$\sum \vec{F} = \vec{F}_1 + \vec{F}_2 = 0$ $\vec{F}_1 = -\vec{F}_2$ $F_1 = F_2$
The electrostatic force is given by Coulomb's law (Equation 23.3). We know that both charge q_1 and q_2 are positive.	$k\dfrac{\|q_1 Q\|}{r_1^2} = k\dfrac{\|q_2 Q\|}{r_2^2}$ $\dfrac{q_1}{r_1^2} = \dfrac{q_2}{r_2^2}$
We need the center-to-center distances r_1 and r_2 based on the figure above.	$r_1 = x_0 \quad$ and $\quad r_2 = L - x_0$
Substitute r_1 and r_2 and solve for x_0.	$\dfrac{q_1}{\left(x_0\right)^2} = \dfrac{q_2}{\left(L - x_0\right)^2} \quad \rightarrow \quad \left(\dfrac{L - x_0}{x_0}\right)^2 = \dfrac{q_2}{q_1}$ $\left(\dfrac{L - x_0}{x_0}\right) = \sqrt{\dfrac{q_2}{q_1}} \quad \rightarrow \quad L - x_0 = x_0\sqrt{\dfrac{q_2}{q_1}}$ $L = x_0\left(\sqrt{\dfrac{q_2}{q_1}} + 1\right) \quad \rightarrow \quad x_0 = L\left(\sqrt{\dfrac{q_2}{q_1}} + 1\right)^{-1}$

| Finally insert $q_1 = 4e$ and $q_2 = e$ and simplify the expression for x_0. | $$x_0 = L\left(\sqrt{\frac{e}{4e}} + 1\right)^{-1}$$ $$x_0 = \boxed{\frac{2}{3}L}$$ |

CHECK and THINK

The third charge is indeed closer to the smaller charge as we predicted. Since Q cancels from our expression, neither its magnitude nor sign actually matters. In other words, we cannot tell whether Q is positive or negative and we cannot determine its magnitude, only it's position between our known charges matters!

72. (N) Three particles with charges of 1.0 μC, −1.0 μC, and 0.50 μC are placed at the corners A, B, and C of an equilateral triangle with side length 0.10 m as shown in Figure P23.72. Find the net force on the charge at point C.

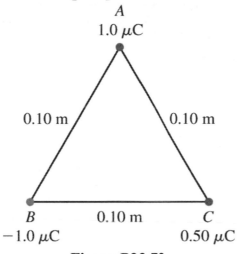

Figure P23.72

INTERPRET and ANTICIPATE

We calculate the two forces due to charges at corners at A and B on the charge at C using Coulomb's law and add the two vector forces. The electrostatic force between charges at the corners A and C is repulsive and the force between charges at the corners B and C is attractive, as shown, so we expect the final forces to be down and to the left.

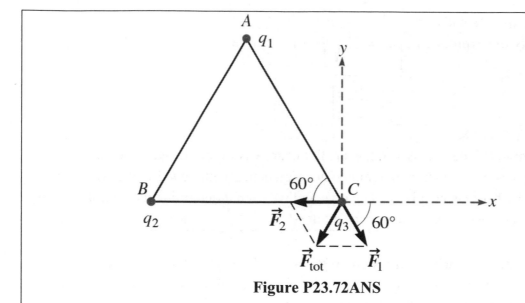

Figure P23.72ANS

SOLVE First, we find the magnitude of the repulsive force F_1 on charge $q_3 = 0.50\ \mu C$ due to charge $q_1 = 1.0\ \mu C$.	$\left|\vec{F_1}\right| = \left(8.99 \times 10^9\ \mathrm{N \cdot m^2/C^2}\right) \dfrac{\left(1.0 \times 10^{-6}\,\mathrm{C}\right)\left(0.50 \times 10^{-6}\,\mathrm{C}\right)}{\left(0.10\,\mathrm{m}\right)^2}$
Then we resolve the force into x and y components.	$\vec{F_1} = \left[\dfrac{\left(8.99 \times 10^9\ \mathrm{N \cdot m^2/C^2}\right)\left(1.0 \times 10^{-6}\,\mathrm{C}\right)\left(0.50 \times 10^{-6}\,\mathrm{C}\right)}{\left(0.10\,\mathrm{m}\right)^2}\right]$ $\times \left[\cos 60^\circ\, \hat{i} - \sin 60^\circ\, \hat{j}\right]$
The magnitude of the attractive force F_2 on charge $q_3 = 0.50\ \mu C$ due to charge $q_2 = -1.0\ \mu C$ is the same as F_1 since the magnitude of the charges and their separation are the same. The force due to charge B is attractive and therefore points in the negative x direction.	$\vec{F_2} = -\left(8.99 \times 10^9\ \mathrm{N \cdot m^2/C^2}\right) \dfrac{\left(1.0 \times 10^{-6}\,\mathrm{C}\right)\left(0.50 \times 10^{-6}\,\mathrm{C}\right)}{\left(0.10\,\mathrm{m}\right)^2}\ \hat{i}$

The net force on the charge q_3 is the sum of these two forces.	$\vec{F}_{tot} = \vec{F}_1 + \vec{F}_2$ $\vec{F}_{tot} = \dfrac{\left(8.99 \times 10^9 \text{ N} \cdot \text{m}^2/\text{C}^2\right)\left(1.0 \times 10^{-6}\text{C}\right)\left(0.50 \times 10^{-6}\text{C}\right)}{\left(0.10\,\text{m}\right)^2}$ $\times \left[\left(-1 + \cos 60°\right)\hat{i} - \sin 60° \, \hat{j}\right]$ $\vec{F}_{tot} = \boxed{\left(-0.22\,\hat{i} - 0.39\,\hat{j}\right)\text{N}}$

CHECK and THINK

The net force on the charge at corner C points down and to the left as expected. It has magnitude 0.45 N at an angle 60° below the negative x-axis.

77. (A) Two identical spheres of mass m carry identical positive charge q. Static friction between the spheres and the level surface they sit on just barely keeps them at rest. Their center-to-center separation is r. Find an expression for the coefficient of static friction between a sphere and the surface.

INTERPRET and ANTICIPATE

Based on the problem statement, we assume that the spheres don't roll. In this case, the static friction should be at its maximum possible value $\left(F_{s,\max} = \mu_s F_N\right)$ while the sphere is in static equilibrium. We can apply force balance and solve for the coefficient. We might guess that a larger charge or a smaller mass would require a larger coefficient of friction in order to keep the spheres in place.

SOLVE

We first sketch a free-body diagram for the sphere on the right, which includes the electrostatic, static friction, gravitational, and normal forces.	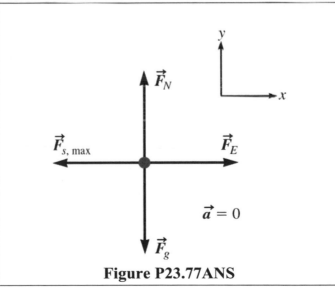

Figure P23.77ANS

Since the sphere is in static equilibrium, the net force on it is zero.	$\sum F_x = 0: \quad F_E - F_{s,\text{max}} = k\dfrac{q^2}{r^2} - \mu_s F_N = 0$ $\sum F_y = 0: \quad F_N - F_g = F_N - mg = 0$
From the y equation, $F_N = mg$. Insert this into the x equation and solve for the coefficient of static friction.	$k\dfrac{q^2}{r^2} - \mu_s mg = 0$ $\boxed{\mu_s = \dfrac{kq^2}{mgr^2}}$

CHECK and THINK

Let's think about the meaning of this equation. This tells us the friction coefficient required to just keep the spheres from moving. As we predicted, the coefficient of friction is larger if q is larger (since more friction is needed to oppose the larger electrostatic force) or if the mass is smaller (since there is a smaller friction force if the normal force is lower). It also looks like the friction force must be larger if the spheres are closer together (smaller r). This makes sense since the electrostatic force will be larger if the spheres are closer together and larger friction is needed.

24

Electric Fields

5. (N) A sphere with a charge of -3.50 nC and a radius of 1.00 cm is located at the origin of a coordinate system.
a. What is the electric field 1.75 cm away from the center of the sphere along the positive y axis?
b. If a particle with a charge of 5.39 nC were placed at that location, what would be the electrostatic force on this charge?

INTERPRET and ANTICIPATE	
These questions can be answered by using Eq. 24.3 and Eq. 24.2, relating the electric field due to a charged particle to the electric force experienced by a nearby second charged particle. Because the sphere at the origin is negatively charged, we expect the electric field to point in the negative y-direction for any location along the positive y axis. The electric force on a positively charged particle at the location of interest should experience an electric force in the same direction.	

SOLVE	
a. Use Eq. 24.3 to find the electric field outside the sphere, where r is the distance from the center of the sphere to the point of interest. The $\hat{r}$ direction will be in the $\hat{j}$ direction, because the $\hat{r}$ direction is always considered to point away from the source.	$\vec{E} = k\dfrac{q}{r^2}\hat{r} = k\dfrac{Q_s}{r^2}\hat{j}$ $\vec{E} = \left(8.99\times10^9 \ \text{N}\cdot\text{m}^2/\text{C}^2\right)\dfrac{-3.50\times10^{-9} \ \text{C}}{\left(0.0175 \ \text{m}\right)^2}\hat{j}$ $\vec{E} = \boxed{-1.03\times10^5 \, \hat{j} \ \text{N/C}}$
b. Use Eq. 24.2 to find the electric force on the 5.39-nC particle.	$\vec{F}_E = q\vec{E} = qk\dfrac{Q_s}{r^2}\hat{j}$ $\vec{F}_E = \left(5.39\times10^{-9}\right)\left(8.99\times10^9 \ \text{N}\cdot\text{m}^2/\text{C}^2\right)\dfrac{-3.50\times10^{-9} \ \text{C}}{\left(0.0175 \ \text{m}\right)^2}\hat{j}$ $\vec{F}_E = \boxed{-5.54\times10^{-4} \, \hat{j} \ \text{N}}$

> **CHECK and THINK**
>
> As expected, the electric field and the electric force on the positively charged object are both directed along the negative y axis. Note that if the 5.39-nC particle was instead negatively charged, the electric force would be in the positive y-direction, though the electric field at that location, due to the source charge, would remain unchanged.

6. (N) Is it possible for a conducting sphere of radius 0.10 m to hold a charge of 4.0 μC in air? The minimum field required to break down air and turn it into a conductor is 3.0×10^6 N/C.

> **INTERPRET and ANTICIPATE**
>
> We need to calculate the magnitude of the electric field produced by the charged sphere on its surface and compare it to the field required to break down air.

> **SOLVE**
> Use Eq. 24.3 to find the magnitude of the electric field on the surface of the sphere.
>
> $$\vec{E}(r) = k\frac{Q_S}{r^2}\hat{r}$$
>
> $$E(r) = k\frac{Q_S}{r^2}$$
>
> $$E(r) = (8.99 \times 10^9 \text{ N}\cdot\text{m}^2/\text{C}^2)\frac{4.0 \times 10^{-6} \text{ C}}{(0.10\,\text{m})^2}$$
>
> $$E(r) = 3.6 \times 10^6 \text{ N/C}$$
>
> Since the electric field on the surface is greater than the break down field of 3.0×10^6 N/C, the sphere will not be able to hold the 4.0 μC charge.

> **CHECK and THINK**
>
> With increase in charge of the conducting sphere the electric field produced by the charge also increases. The charged surface of the conductor experiences an outward electrostatic force due to which the charges on the surface are pulled out. When the electric field becomes greater than the break down strength of air, the air ionizes causing sparking from the metal sphere.

12. (N) A particle with charge $q_1 = +5.0$ μC is located at $x = 0$, and a second particle with charge $q_2 = -3.0$ μC is located at $x = 15$ cm. Determine the location of a third particle with charge $q_3 = +4.0$ μC such that the net electric field at $x = 25$ cm is zero.

<table>
<tr><td colspan="2">

INTERPRET and ANTICIPATE

Since the first two particles are on the x axis, their net field at another point on the x axis will only have an x component, so we must place the third charge somewhere on the x axis as well. For a zero total field at $x = 25$ cm we will make the magnitude and direction of the field due to the third charge equal and opposite to the net field of the first two charges at that point.

</td></tr>
<tr><td>

SOLVE
Find the electric fields due to charges q_1 and q_2 at $x = 25$ cm. We convert distances to meters and note that the distance from charge 2 (at 15 cm) to the point at 25 cm is $(25 - 15)$ cm.

</td><td>

$$E_{1x} = \frac{kq_1}{r_1^2} = \frac{(8.99\times10^9)(5.0\times10^{-6})}{0.25^2} \text{ N/C}$$
$$= 7.19\times10^5 \text{ N/C}$$

$$E_{2x} = \frac{kq_2}{r_2^2} = \frac{(8.99\times10^9)(-3.0\times10^{-6})}{(0.25-0.15)^2} \text{ N/C}$$
$$= -2.70\times10^6 \text{ N/C}$$

</td></tr>
<tr><td>

Sum these to get the net field due to charges 1 and 2.

</td><td>

$$E_{12x} = E_{1x} + E_{2x} = -1.98\times10^6 \text{ N/C}$$

</td></tr>
<tr><td>

Since this field is in the negative x direction, the third charge, which is positive, must be placed at a point $x < 25$ cm to produce an electric field with a positive x component at $x = 25$ cm. Let r_3 be the distance of the charge q_3 from the point $x = 25$ cm. Equate the magnitudes of the fields to find r_3.

</td><td>

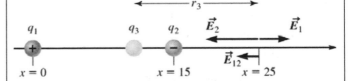

Figure P24.12ANS

$$E_3 = \frac{kq_3}{r_3^2} = \frac{(8.99\times10^9)(4.0\times10^{-6})}{r_3^2} \text{ N/C}$$
$$\frac{(8.99\times10^9)(4.0\times10^{-6})}{r_3^2} \text{ N/C} = +1.98\times10^6 \text{ N/C}$$
$$r_3 = \sqrt{\frac{(8.99\times10^9)(4.0\times10^{-6})}{1.98\times10^6}} \text{ m} = 0.13 \text{ m}$$

</td></tr>
<tr><td>

Finally, since r_3 is the distance from the third charge to the location $x = 25$ cm, use this value to determine the x coordinate of q_3.

</td><td>

$$x_3 = 0.25 - r_3 = 0.25 - 0.13$$
$$x_3 = \boxed{0.12 \text{ m}} = 12 \text{ cm}$$

</td></tr>
</table>

Chapter 24 – Electric Fields

CHECK and THINK

The two positive charges are to the left of the negative charge such that the next electric field at this particular location on the right is zero.

14. (A) A particle with charge q on the negative x axis and a second particle with charge $2q$ on the positive x axis are each a distance d from the origin. Where should a third particle with charge $3q$ be placed so that the magnitude of the electric field at the origin is zero?

INTERPRET and ANTICIPATE

If we use Eq. 24.3 to express the net electric field at the origin due to the two known charged particles, we can then write an equation for the electric field due to the third particle at its unknown location and add this to the already existing net electric field with the knowledge that the total should equal zero. Given that the two existing charged particles are the same distance from the origin, have the same sign of charge, and that the one on the positive x axis has a greater charge, we expect this third similarly charged object must be somewhere along the negative x axis for the net electric field at the origin to be zero.

SOLVE

Begin by finding the net electric field at the origin due to the particles with charge q and $2q$, using Eq. 24.3. The field due to charge q will point in the positive x-direction at the origin, while the field due to $2q$ will point in the negative x-direction at the origin.	$\vec{E}_{net} = \vec{E}_q + \vec{E}_{2q} = k\dfrac{q}{d^2}\hat{i} + k\dfrac{2q}{d^2}\left(-\hat{i}\right) = -k\dfrac{q}{d^2}\hat{i}$
Now, the final net electric field should be 0. Use Eq. 24.3 to express the electric field due to the third particle and solve for the location, x.	$\vec{E}_{3q} = k\dfrac{3q}{x^2}\hat{i}$ $\vec{E}_{net} = 0 = -k\dfrac{q}{d^2}\hat{i} + k\dfrac{3q}{x^2}\hat{i}$ $\dfrac{1}{d^2} = \dfrac{3}{x^2}$ $x = \pm\sqrt{3}d$

Because all of the charged particles have the same type of charge (positive or negative), and that the net electric field at the origin was directed in the negative x-direction before adding the third particle, it must be placed along the negative x axis. Thus, we choose the negative answer.	$x = \boxed{-\sqrt{3}d}$

CHECK and THINK

The third charged particle did reside along the negative x axis, as expected. If the third particle had an opposite charge compared to the other two particles, it would have been located somewhere along the positive x axis in order to make the electric field at the origin equal zero.

17. (N) Figure P24.17 shows a dipole. If the positive particle has a charge of 35.7 mC and the particles are 2.56 mm apart, what is the electric field at point A located 2.00 mm above the dipole's midpoint?

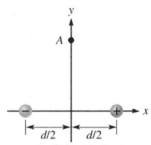

Figure P24.17

INTERPRET and ANTICIPATE
We can use Eq. 24.3 applied to each particle to find the net electric field at A. Note that the y components of the electric field due to each particle will cancel each other because the particles have the opposite sign of charge and are symmetrically located about the point A. The two x components should both point in the negative x-direction so there will be a net electric field in that direction.

SOLVE	
Begin by writing the electric field due to the positively charged particle. If we symbolize the angle between the direction from the particle to the point A and the x axis as θ, we can use the geometry in the	$\vec{E}_+ = k\dfrac{q}{r^2}\hat{r} = k\dfrac{q}{r^2}\left[-\cos\theta\,\hat{i} + \sin\theta\,\hat{j}\right]$ $\vec{E}_+ = k\dfrac{q}{r^2}\left[-\dfrac{d/2}{r}\,\hat{i} + \dfrac{y}{r}\,\hat{j}\right]$

figure to help us express the components of the electric field, where r is the distance between the particle and A, and y is the distance between the origin and A.	
Do the same thing now for the negatively charged particle. Note that the direction of the field is different and that this particle is negatively charged.	$\vec{E}_- = k\dfrac{q}{r^2}\hat{r} = k\dfrac{-q}{r^2}\left[\cos\theta\hat{i} + \sin\theta\hat{j}\right]$ $\vec{E}_- = k\dfrac{q}{r^2}\left[-\dfrac{d/2}{r}\hat{i} - \dfrac{y}{r}\hat{j}\right]$
Find the net electric field at A by performing a vector sum with these two electric fields. Use the Pythagorean theorem and the dimensions in the problem statement to express the distance r. Note that 2.56 mm/2 = 1.28 mm.	$\vec{E}_{net} = \vec{E}_+ + \vec{E}_-$ $\vec{E}_{net} = k\dfrac{q}{r^2}\left[-\dfrac{d/2}{r}\hat{i} + \dfrac{y}{r}\hat{j}\right] + k\dfrac{q}{r^2}\left[-\dfrac{d/2}{r}\hat{i} - \dfrac{y}{r}\hat{j}\right]$ $\vec{E}_{net} = -k\dfrac{q}{r^2}\dfrac{d}{r}\hat{i} = -k\dfrac{qd}{r^3}\hat{i}$ $\vec{E}_{net} = -k\dfrac{\left(35.7\times10^{-3}\text{ C}\right)\left(0.00256\text{ m}\right)}{\left(\sqrt{\left(0.00128\text{ m}\right)^2 + \left(0.00200\text{ m}\right)^2}\right)^3}\hat{i}$ $\vec{E}_{net} = \boxed{-6.14\times10^{13}\hat{i}\text{ N/C}}$

CHECK and THINK
The electric field did point in the negative x-direction. This will be true anywhere along the line that perpendicularly bisects the dipole.

23. **(A)** A positively charged rod with linear charge density λ lies along the x axis (Fig. P24.23). Find an expression for the magnitude of the electric field at the position P a distance x away from the origin, where $x > L$.

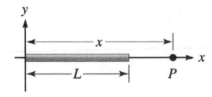

Figure P24.23

INTERPRET and ANTICIPATE We need to use Eq. 24.14 to derive the electric field at the position x to the right of the rod. Here, $dq = \lambda d\ell$, And we must integrate over the length of the rod from $x = 0$ to $x = L$. Here, we only find the magnitude of the electric field because the direction would depend on whether or not the rod is positively or negatively charged.	
SOLVE First, express the differential and choose the form of dq that involves a linear charge density.	$dE = k\dfrac{dq}{r^2} = k\dfrac{\lambda d\ell}{r^2} = k\dfrac{\lambda dr}{r^2}$
The rod is located along the x axis from $x = 0$ to $x = L$, but the value of r represents the distance from the point P to a small element of charge on the rod. We must integrate over the range from one end, where the distance from the right side of the rod to P is x-L, to the other end, where the distance from the left side of the rod to P is x. Remember that x represents the location P relative to the origin of the coordinate system.	$\displaystyle\int dE = \int_{x-L}^{x} k\dfrac{\lambda dr}{r^2} = k\lambda \int_{x-L}^{x} \dfrac{1}{r^2}\,dr$
Now, integrate to derive the expression.	$E = k\lambda\left[-r^{-1}\right]_{x-L}^{x} = k\lambda\left[\dfrac{1}{x-L} - \dfrac{1}{x}\right]$ $E = k\lambda\left[\dfrac{x}{x^2 - xL} - \dfrac{x-L}{x^2 - xL}\right]$ $E = \boxed{k\lambda L/\left(x^2 - xL\right)}$
CHECK and THINK Dimensional analysis of our answer verifies that the units of our expression will be N/C, as expected. Note that this expression is valid anywhere along the positive x axis with $x > L$, but nowhere else.	

27. (A) Find an expression for the position y (along the positive axis perpendicular to the ring and passing through its center) where the electric field due to a charged ring is a maximum. Also find an expression for the electric field at that point.

Chapter 24 – Electric Fields

INTERPRET and ANTICIPATE To find where the function is a maximum, we need to take the derivative of the field due to a ring (Eq. 24.16) with respect to y and set this equal to zero. This will allow us to find the position of the maximum electric field. We can then use this position in Eq. 24.16 to express the electric field. We will work with the magnitudes only in this problem as the direction depends on the sign of the electric charge on the ring.	

SOLVE Begin with Eq. 24.16 for the magnitude of the electric field along the axis of a charged ring.	$\dfrac{kQy}{\left(R^2+y^2\right)^{3/2}}$
Take its derivative and set it equal to zero. Then solve for the value of y where the maximum electric field exists.	$\dfrac{d}{dy}\dfrac{kQy}{\left(R^2+y^2\right)^{3/2}} = kQ\left(R^2+y^2\right)^{-3/2} - kQy\left(R^2+y^2\right)^{-5/2}(3y)$ $kQ\left(R^2+y^2\right)^{-3/2} - kQy\left(R^2+y^2\right)^{-5/2}(3y)=0$ $R^2+y^2=3y^2$ $\boxed{y=\dfrac{R}{\sqrt{2}}}$
Now, express the magnitude of the electric field at this location using Eq. 24.16.	$\vec{E} = \dfrac{kQy_{max}}{\left(R^2+y_{max}^2\right)^{3/2}}\hat{j} = \dfrac{kQ\dfrac{R}{\sqrt{2}}}{\left(R^2+\dfrac{R^2}{2}\right)^{3/2}}\hat{j} = \boxed{\dfrac{2kQ}{3\sqrt{3}R^2}}\hat{j}$

CHECK and THINK

If we ever want to find the maximum value of a function, take the derivative with respect to the independent variable and set the result equal to zero. Then solve for the independent variable. The resulting values are those that maximize the function. Note that if we are very far away from the charged ring, the electric field becomes zero, as expected.

33. (N) If the curved rod in Figure P24.32 has a uniformly distributed charge $Q = 35.5$ nC, radius $R = 0.785$ m, and $\varphi = 60.0°$, what is the electric field at point A?

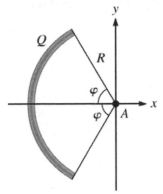

Figure P24.32

INTERPRET and ANTICIPATE
This is not one of our special cases, so we need to integrate the field due to all of the little charge elements (which is what Problem 32 asks us to do) and then substitute values. The field is symmetric about the x axis, so the y components will cancel and it should point in the $+x$ direction.

SOLVE	
First, let's draw a sketch. We represent a small charge element dq in two places, on the top and bottom sides of the arc, and sketch the electric fields due to each. By symmetry, we can see that the y components must cancel since, for every charge element above the center, there is one equally far below the center with a y component pointing in the opposite direction. Therefore, we need only keep track of the x components. θ is the integration variable indicating the location of dq while the end points of the arc span between the fixed angles $\pm\varphi$.	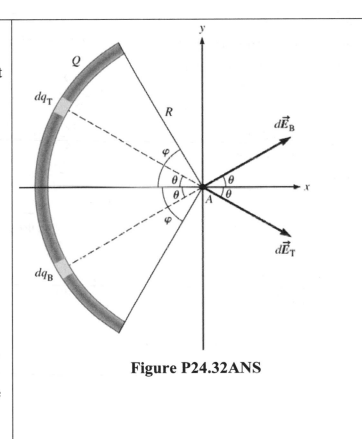 **Figure P24.32ANS**

Using Equation 24.14, we find the x component of the field for each charge element. From the figure, we can use trigonometry to see that $dE_x = E \cos\theta$.	$$dE_x = \frac{k}{r^2}\cos\theta\, dq \qquad (1)$$
We can represent the charge element in terms of the linear charge density, defined as the total charge Q divided by the arc length, $R(2\varphi)$.	$$\lambda = \frac{Q}{2R\varphi}$$
Using Eq. 24.13, we can express the charge element. We also note that the distance dL is a small arc length spanned by $d\theta$, or $Rd\theta$.	$$dq = \lambda\, dL = \lambda R\, d\theta = \frac{Q}{2R\varphi}R\, d\theta$$ $$dq = \frac{Q}{2\varphi}\, d\theta \qquad (2)$$
Every charge element is the same distance R from point A, so $r = R$ for all dq. Using equations 1 and 2, we can integrate θ from $-\varphi$ to $+\varphi$. Since the problem is symmetric, it's also possible to integrate from 0 to $+\varphi$ and multiply by 2.	$$E_{tot,x} = \int_{-\varphi}^{+\varphi} E_x = \frac{kQ}{2\varphi R^2}\int_{-\varphi}^{+\varphi}\cos\theta\, d\theta$$ $$= \frac{kQ}{2\varphi R^2}\sin\theta\Big|_{\theta=-\varphi}^{+\varphi} = \frac{kQ}{2\varphi R^2}\left[\sin\varphi - \sin(-\varphi)\right]$$ $$= \frac{kQ}{2\varphi R^2}\left[2\sin\varphi\right] = \frac{kQ}{\varphi R^2}\sin\varphi$$ or $$E_{tot,x} = 2\int_{0}^{+\varphi} E_x = \frac{kQ}{\varphi R^2}\int_{0}^{+\varphi}\cos\theta\, d\theta = \frac{kQ}{\varphi R^2}\sin\varphi$$
Since this is the x component, we can write the final field as a vector.	$$\vec{E} = \frac{kQ}{\varphi R^2}\sin\varphi\,\hat{i}$$

We have completed Problem 32 at this point and can now substitute values given in Problem 33. We have to be careful when substituting the angle though—for the one in the denominator (where we are not calculating a sine or cosine), we must use the natural units of angle, radians. Specifically, $\varphi = 60° = \dfrac{\pi}{3}$ rad . We can then find the magnitude of the vector.	$\vec{E} = \dfrac{\left(8.99\times10^9 \ \text{N}\cdot\text{m}^2 / \text{C}^2\right)\left(35.5\times10^{-9} \ \text{C}\right)}{\dfrac{\pi}{3}\left(0.785 \ \text{m}\right)^2}\sin\left(\dfrac{\pi}{3}\right)\hat{i}$ $\vec{E} = 428 \ \hat{i} \ \text{N/C}$ $E = \boxed{428 \ \text{N/C}}$

CHECK and THINK

Since this is not a special case, we are able to integrate the field due to all charge elements to determine the total field. The field points to the right as expected. The fact that all charge elements are the same distance R from the point in question makes life easier by removing the r dependence. The resulting integral only depends on the angle.

36. (A) A positively charged disk of radius R and total charge Q_{disk} lies in the xz plane, centered on the y axis (Fig. P24.35). Also centered on the y axis is a charged ring with the same radius as the disk and total charge Q_{ring}. The ring is a distance d above the disk. Determine the electric field at the point P on the y axis, where P is above the ring a distance y from the origin.

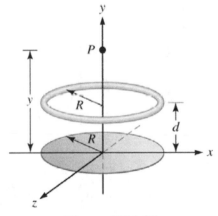

Figure P24.35

INTERPRET and ANTICIPATE Here, we must express the electric field due to the ring and the disk at P separately, and then use the principle of superposition to find the total electric field. The distance between the disk and the point P is the coordinate y, while the distance between the ring and P will be y-d.	
SOLVE Begin by using Eq. 24.17, the electric field due to the charged disk at the point P, a distance y away. The direction will be along the positive y axis, away from the positive charge.	$$\vec{E}_{disk} = \frac{2kQ_{disk}}{R^2}\left[1 - \frac{y}{\sqrt{R^2+y^2}}\right]\hat{j}$$
The ring is negatively charged, and its electric field would be given by Eq. 24.16, where the distance between the center of the ring and P is y-d. The direction will be along the negative y axis, towards the negative charge. This can be seen algebraically when we consider that Q_{ring} is a negative amount of charge.	$$\vec{E}_{ring} = \frac{kQ_{ring}(y-d)}{\left(R^2+(y-d)^2\right)^{3/2}}\hat{j}$$
Now, we use the principle of superposition to add these two electric fields and find the total, or net, electric field.	$$\vec{E}_{total} = \vec{E}_{disk} + \vec{E}_{ring}$$ $$\vec{E}_{total} = \frac{2kQ_{disk}}{R^2}\left[1 - \frac{y}{\sqrt{R^2+y^2}}\right]\hat{j} + \frac{kQ_{ring}(y-d)}{\left(R^2+(y-d)^2\right)^{3/2}}\hat{j}$$ $$\boxed{\vec{E}_{total} = k\left\{\frac{2Q_{disk}}{R^2}\left[1 - \frac{y}{\sqrt{R^2+y^2}}\right] + \frac{Q_{ring}(y-d)}{\left(R^2+(y-d)^2\right)^{3/2}}\right\}\hat{j}}$$

CHECK and THINK
It is difficult to see whether or not the net electric field will be directed along the negative y axis or positive y axis, as the result depends on two primary factors: (1) the amount of excess charge on the disk and ring, and (2) the distance d between the ring and disk.

37. (N) A uniformly charged conducting rod of length $\ell = 30.0$ cm and charge per unit length $\lambda = -3.00 \times 10^{-5}$ C/m is placed horizontally at the origin (Fig. P24.37). What is the electric field at point A with coordinates (0, 0.400 m)?

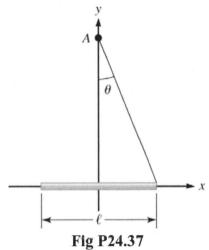

Fig P24.37

INTERPRET and ANTICIPATE
Since this is an extended object, we have two options: (1) use a formula for a special case (if we know of one) which applies for this shape or (2) integrate over all charge elements dq to find the total electric field at point A. We can do both in this case, so let's go ahead and do that. Since this is a negatively charged rod, the electric field should point towards it, or downward.

SOLVE	
(method 1) In the first approach, we apply Eq. 24.15 in Example 24.4 for a charged rod of length 2ℓ and charge Q. The work of integrating over the extended object was done in this problem, so we can take advantage of the result.	$$\vec{E} = \frac{kQ}{y} \frac{1}{\sqrt{\ell^2 + y^2}} \hat{j}$$

To apply this formula, note that the rod is of length 2ℓ, so $\ell = 0.150$ m and the charge is $Q = \lambda L = (-3.00 \times 10^{-5} \text{ C/m})(0.300 \text{ m}) = -9.00 \times 10^{-6}$ C.

$$\vec{E} = \frac{\left(8.99 \times 10^9 \, \text{N} \cdot \text{m}^2 / \text{C}^2\right)\left(-9.00 \times 10^{-6} \, \text{C}\right)}{\left(0.400 \, \text{m}\right)} \frac{1}{\sqrt{\left(0.150 \, \text{m}\right)^2 + \left(0.400 \, \text{m}\right)^2}} \hat{j}$$

$$\vec{E} = -4.73 \times 10^5 \, \text{N/C} \, \hat{j}$$

(**method 2**) First, let's draw a sketch.	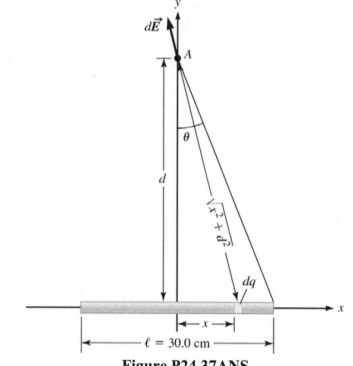 **Figure P24.37ANS**
The electric field at point A due to each element dq is $dE = k\dfrac{dq}{r^2}$, where $r = \sqrt{x^2 + d^2}$.	$dE = k\dfrac{dq}{x^2 + d^2}$
By symmetry, $E_x = \int dE_x = 0$ and using $dq = \lambda dx$, where λ is the charge per unit length, we can calculate the y component.	$E = E_y = \int dE_y = \int dE \cos\theta$
We can express the angle in terms of x and y using trigonometry.	$\cos\theta = \dfrac{d}{\sqrt{x^2 + d^2}}$

Now, plug everything in and integrate. We are integrating over dx from $-\ell/2$ to $+\ell/2$. Since the y component is downward for the entire rod, this is equivalent to integrating from 0 to $+\ell/2$ and doubling it.	$E = k\lambda d \int\limits_{-\ell/2}^{\ell/2} \dfrac{dx}{\left(x^2 + d^2\right)^{3/2}} = 2k\lambda d \int\limits_{0}^{\ell/2} \dfrac{dx}{\left(x^2 + d^2\right)^{3/2}}$ $E = \dfrac{k\lambda\ell}{d\sqrt{\left(\ell/2\right)^2 + d^2}}$
Now, insert numbers to determine the electric field.	$E = \dfrac{\left(8.99 \times 10^9 \ \text{N} \cdot \text{m}^2 / \text{C}^2\right)\left(-3.00 \times 10^{-5} \ \text{C/m}\right)\left(0.300 \ \text{m}\right)}{\left(0.400 \ \text{m}\right)\sqrt{\left(0.150 \ \text{m}\right)^2 + \left(0.400 \ \text{m}\right)^2}}$ $E = -4.73 \times 10^5 \ \text{N/C (downward)}$ $\boxed{\vec{E} = -4.73 \times 10^5 \ \hat{j} \ \text{N/C}}$

CHECK and THINK

We get the same answer in both cases, as we should. After all, Example 24.4 performed the integral that we set up in approach 2. The electric field is also downward, toward the negatively charged rod, as we expected. Because of the symmetry of the problem, there is no x component.

46. (N) A uniform electric field of 725 N/C is produced within a particle accelerator. Starting from rest, what are
a. the speed of an electron placed in this field after 23.0 ns have elapsed, and
b. the speed of a proton placed in this field after 23.0 ns have elapsed?

INTERPRET and ANTICIPATE
A charged particle in an electric field experiences a force, $F = qE$. With the force, we can determine the acceleration and use kinematics to find the final speed. Since the charge and field are the same in both cases, the force will be the same. However, since the electron has a much smaller mass, its acceleration will be much larger, so it should be moving much faster after 23 ns.

SOLVE	
The force depends on the charge and electric field that the charge is in, according to Eq. 24.18.	$F_E = qE$

Newton's 2nd law, $F = ma$, allows us to determine the acceleration.	$ma = qE \quad \rightarrow \quad a = \dfrac{qE}{m}$
Then, using the kinematic equations, we can determine the final speed of the charge. The initial speed is zero.	$v_f = v_0 + at = \dfrac{qE}{m}t$
a. In both cases $E = 725$ N/C, $q = e$ $=1.60 \times 10^{-19}$ C, and $t = 23.0$ ns. We can calculate the force, which is also the same.	$F_E = eE$ $F_E = \left(1.60 \times 10^{-19}\text{ C}\right)\left(7.25 \times 10^2\text{ N/C}\right)$ $F_E = 1.16 \times 10^{-16}$ N
The electron mass is $m = 9.11 \times 10^{-31}$ kg.	$v_f = \dfrac{eE}{m}t$ $v_f = \left(\dfrac{1.16 \times 10^{-16}\text{ N}}{9.11 \times 10^{-31}\text{ kg}}\right)\left(23.0 \times 10^{-9}\text{ s}\right)$ $v_f = \boxed{2.92 \times 10^6\text{ m/s}}$
b. The only difference for the proton is the mass, $m = 1.67 \times 10^{-27}$ kg.	$v_f = \dfrac{eE}{m}t$ $v_f = \left(\dfrac{1.16 \times 10^{-16}\text{ N}}{1.67 \times 10^{-27}\text{ kg}}\right)\left(23.0 \times 10^{-9}\text{ s}\right)$ $v_f = \boxed{1.60 \times 10^3\text{ m/s}}$

CHECK and THINK

Since the proton is nearly 2000 times more massive than the electron, its acceleration and final speed are nearly 2000 times smaller.

49. (N) In Figure P24.49, a charged particle of mass $m = 4.00$ g and charge $q = 0.250\ \mu C$ is suspended in static equilibrium at the end of an insulating thread that hangs from a very long, charged, thin rod. The thread is 12.0 cm long and makes an angle of 35.0° with the vertical. Determine the linear charge density of the rod.

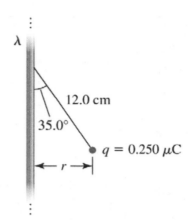

Figure P24.49

INTERPRET and ANTICIPATE	
Since the rod is very long, we can use the reasoning of problem 24.4 to determine the electric field that it produces at the site of the charged particle. Then, the problem is simply to apply the conditions of static equilibrium.	

SOLVE	
Draw a free body diagram of the charged particle. The electric force due to the charged rod will be radially outward from it, which is in the $+x$ direction in the diagram. The other two forces are the tension in the thread and the weight of the particle.	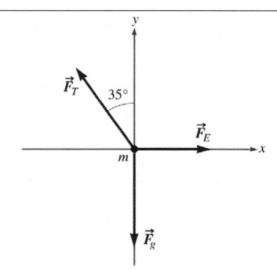 **Figure P24.49ANS**
Since the particle is in static equilibrium, the total x and y components of the forces acting on the particle are zero.	$\sum F_x = F_E - F_T \sin 35° = 0 \qquad (1)$ $\sum F_y = F_T \cos 35° - F_g = 0 \qquad (2)$
Solve for the tension in the thread using Eq. 2 and then insert the result into Eq. 1 to solve for the electric field.	$F_T = \dfrac{F_g}{\cos 35°} = \dfrac{mg}{\cos 35°}$ $F_E = \dfrac{mg}{\cos 35°} \sin 35° = 0.004(9.81)\tan 35°$ $F_E = 0.0275 \text{ N} \qquad (3)$

Write an approximate expression for the electric field of a very long charged rod to find the linear charge density. This approximation can be found from Example 24.4, which derives the expression for the electric field need a charged rod.	$$E = \frac{2k\lambda}{y}\frac{\ell}{\sqrt{\ell^2 + y^2}}$$
In Example 24.2, y is the distance from the charged rod to the point in question (the location of the charged particle), which corresponds to the distance r in this problem. We then assume the length is very large ($\ell \to \infty$) so that in the last step we can use the fact that $r/\ell \approx 0$.	$$E = \frac{2k\lambda}{r}\frac{\ell}{\ell\sqrt{1+\left(\frac{r}{\ell}\right)^2}}$$ $$E = \frac{2k\lambda}{r}\frac{1}{\sqrt{1+\left(\frac{r}{\ell}\right)^2}} \approx \frac{2k\lambda}{r}$$
Use this to write the magnitude of the force on the particle.	$$F_E = q\frac{2k\lambda}{r}$$
Finally, solve this for λ and insert the numerical values, including the strength of the field calculated above (Eq. 3). The distance r of the particle from the rod is found from the triangle in the diagram, using trigonometry.	$r = 0.12\sin 35° = 0.0688$ m $$\lambda = \frac{F_E r}{2kq} = \frac{\left(0.0275\dfrac{N}{C}\right)(0.0688\ m)}{2\left(8.99\times10^9\dfrac{N\cdot m^2}{C^2}\right)(0.250\times10^{-6}C)}$$ $\lambda = \boxed{4.21\times10^{-7}\ \text{C/m}}$

CHECK and THINK

The linear charge density is also positive to produce an electric field and force in the positive x direction. This approximation will be very good as long as the charged rod is much longer than 0.0688 m and the particle is not close to either end of the rod.

51. (N) Figure P24.51 shows four small charged spheres arranged at the corners of a square with side $d = 25.0$ cm.
a. What is the electric field at the location of the sphere with charge +2.00 nC?
b. What is the total electric force exerted on the sphere with charge +2.00 nC by the other three spheres?

Chapter 24 – Electric Fields

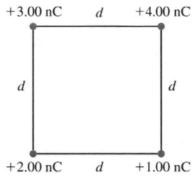

+3.00 nC   *d*   +4.00 nC

d   *d*

+2.00 nC   *d*   +1.00 nC

Figure P24.51

INTERPRET and ANTICIPATE The electric field at this location is the vector sum of the fields due to each of the other charges. So, we need to determine the field due to each of the other charges at this spot and then add them to get the total field. Since all charges are positive and the electric field points away from positive charges, we would expect the total electric field to point down and to the left. The force on a positive charge is along the same direction as the electric field, so the force is down and to the left as well, as we would expect since it's repelled from the other positive charges.	

SOLVE **a.** The field due to an individual charged particle is given by Equation 24.3.	$\vec{E} = \dfrac{kQ}{r^2}\hat{r}$
First, we sketch the radial vectors for each of the three charges and the distances from each to the location of the 2 nC charge.	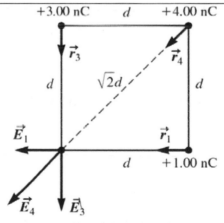 **Figure P24.51ANS**
For the three charges, with charges 1, 3, and 4 nC, we can list the charge and the distance.	$Q_1 = 1.00 \times 10^{-9}\,\text{C},\quad r_1 = d,\quad \hat{r}_1 = -\hat{i}$ $Q_3 = 3.00 \times 10^{-9}\,\text{C},\quad r_3 = d,\quad \hat{r}_3 = -\hat{j}$ $Q_4 = 4.00 \times 10^{-9}\,\text{C},\quad r_4 = \sqrt{2}d,\quad \hat{r}_4 = -\cos 45°\hat{i} - \sin 45°\hat{j}$

45

Now, we add the three fields as vectors.	$\vec{E} = \dfrac{kq_1}{r_1^2}\hat{r}_1 + \dfrac{kq_3}{r_3^2}\hat{r}_3 + \dfrac{kq_4}{r_4^2}\hat{r}_4$
Insert values and calculate: $\vec{E} = -\dfrac{k\left(1.00\times10^{-9}\,\text{C}\right)}{d^2}\hat{i} - \dfrac{k\left(4.00\times10^{-9}\,\text{C}\right)}{2d^2}\left(\cos 45.0°\hat{i} + \sin 45.0°\hat{j}\right) - \dfrac{k\left(3.00\times10^{-9}\,\text{C}\right)}{d^2}\hat{j}$ $\vec{E} = \dfrac{8.99\times10^9}{\left(0.250\text{ m}\right)^2}\left(10^{-9}\text{ C}\right)\left[\left(-1-2\cos 45.0°\right)\hat{i} + \left(-2\sin 45.0° - 3\right)\hat{j}\right]$ $\vec{E} = \boxed{\left(-347\hat{i} - 635\hat{j}\right)\text{ N/C}}$	
b. According to Eq. 24.18, the force on a positive charge points along the direction of the electric field. The charge q is the 2 nC charge and the field E is the field due to the other charges at the location of the charge q.	$\vec{F} = q\vec{E}$
Insert values and calculate the results.	$\vec{F} = \left(2.00\times10^{-9}\,\text{C}\right)\left(\left(-347\hat{i} - 635\hat{j}\right)\text{ N/C}\right)$ $\vec{F} = \boxed{\left(-6.94\times10^{-7}\hat{i} - 1.27\times10^{-6}\hat{j}\right)\text{N}}$

CHECK and THINK

As expected, both the electric field and the force point down and to the left (negative x and y components).

53. (N) A uniform electric field given by $\vec{E} = \left(2.65\,\hat{i} - 5.35\,\hat{j}\right)\times10^5$ N/C permeates a region of space in which a small negatively charged sphere of mass 1.30 g is suspended by a light cord (Fig. P24.53). The sphere is found to be in equilibrium when the string makes an angle $\theta = 23.0°$.

a. What is the charge on the sphere?

b. What is the magnitude of the tension in the cord?

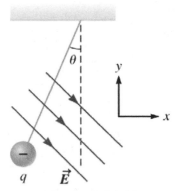

Figure P24.53

INTERPRET and ANTICIPATE Since the sphere is in static equilibrium, the net force on it is zero. We can write the equations for force balance to determine the charge and tension.	

SOLVE **a.** We first sketch a free body diagram.	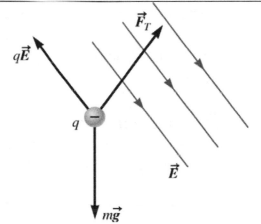 **Figure P24.53ANS**
Use the fact that the net x and y forces are zero. The magnitude of the electrostatic force is given by 24.18, $F = qE$. In this case q is the magnitude of the negative charge.	$\sum F_x = -qE_x + F_T \sin\theta = 0$ $\sum F_y = qE_y + F_T \cos\theta - mg = 0$
Combine these equations to determine the charge. Solve both equations for the tension, set them equal to each other, and solve for q.	$\dfrac{qE_x}{\sin\theta} = \dfrac{mg - qE_y}{\cos\theta}$ $qE_y + \dfrac{qE_x}{\tan\theta} = mg$ $q = \dfrac{mg}{E_y + \dfrac{E_x}{\tan\theta}} = \dfrac{mg}{E_y + E_x \cot\theta}$

Substitute values and solve for the magnitude of the charge. The charge is negative.	$q = \dfrac{\left(1.30\times10^{-3}\ \text{kg}\right)\left(9.80\ \text{m/s}^2\right)}{\left(5.35 + 2.65\cot 23.0°\right)\times10^5}$ $q = \boxed{-1.10\times10^{-8}\ \text{C}}$
b. Solve the x force equation above for tension.	$F_T = \dfrac{qE_x}{\sin 23.0°} = \boxed{7.46\times10^{-3}\ \text{N}}$

CHECK and THINK

Other than using the equation $F = qE$, this is really just a static equilibrium problem.

---

57. (N) A potassium chloride molecule (KCl) has a dipole moment of 8.9×10^{-30} C·m. Assume the KCl molecule is in a uniform electric field of 325 N/C. What is the change in the system's potential energy when the molecule rotates

a. from $\phi = 170°$ to $180°$,

b. from $\phi = 90°$ to $100°$, and

c. from $\phi = 10°$ to $0°$?

INTERPRET and ANTICIPATE

The potential energy of a dipole in a field depends on the dipole moment, the field, and the angle between them. We can use Example 24.11 as a guide.

SOLVE The potential energy of a dipole in an electric field is given by Eq. 24.26. φ is the angle between the dipole moment and the electric field. The energy is lowest when the dipole points along the field direction (in the direction that it "wants" to go), towards $\varphi = 0$. This is similar to the potential energy of a mass decreasing as it falls downward, towards where it "wants" to go.	$U = -\vec{p}\cdot\vec{E} = -pE\cos\varphi$ $\Delta U = -pE\left(\cos\varphi_f - \cos\varphi_i\right)$

a. Apply the equation above for each of the three cases.	$\Delta U = -\left(8.0\times10^{-30}\ \text{C}\cdot\text{m}\right)\left(325\ \text{N/C}\right)\left(\cos180° - \cos170°\right)$ $\Delta U = \boxed{4.4\times10^{-29}\ \text{J}}$
b. In this case, the magnitude is larger because the torque is largest when the dipole is perpendicular to the electric field.	$\Delta U = -\left(8.0\times10^{-30}\ \text{C}\cdot\text{m}\right)\left(325\ \text{N/C}\right)\left(\cos100° - \cos90°\right)$ $\Delta U = \boxed{5.0\times10^{-28}\ \text{J}}$
c. In this case, the change in energy is negative. Since the dipole is rotating from 10 to 0 degrees, it's rotating toward it's preferred orientation and the potential energy is decreasing.	$\Delta U = -\left(8.0\times10^{-30}\ \text{C}\cdot\text{m}\right)\left(325\ \text{N/C}\right)\left(\cos0° - \cos10°\right)$ $\Delta U = \boxed{-4.4\times10^{-29}\ \text{J}}$

CHECK and THINK

We've applied the formula for potential energy of a dipole in an electric field. We see that the energy decreases as it approaches 0 degrees (dipole points in the direction of the electric field) and changes most dramatically when the torque is largest (when the dipole is perpendicular to the field at an angle of 90 degrees).

66. (N) Three identical cylinders made of solid plastic are 15.0 cm in length and 4.50 cm in radius. What is the total charge on

a. the first cylinder, which has a uniform charge density of 235 μC/m^3 throughout the cylinder;

b. the second cylinder, which has a uniform charge density of 12.5 μC/m^2 on its entire surface including its two ends; and

c. the third cylinder, which has a uniform charge density of 12.5 μC/m^2 only on its curved lateral surface, not including its two ends?

INTERPRET and ANTICIPATE

In each case, we must write an expression for the total volume or appropriate amount of area and multiply by the volume or surface charge density. Note that the area of the end of a cylinder is the area of a circle, while the area of the curved lateral surface of the cylinder would be given by $2\pi rh$, where r is the radius of the cylinder and h is its length.

SOLVE	
a. The total charge would be equal to the volume charge density times the volume of the cylinder.	$Q = \rho V = \rho \pi r^2 h$ $$Q = \left(235 \times 10^{-6} \text{ C/m}^3\right)(\pi)(0.0450 \text{ m})^2 (0.150 \text{ m})$$ $Q = \boxed{2.24 \times 10^{-7} \text{ C}}$
b. The total charge would be equal to the surface charge density times the total area of the cylinder. The total area consists of the area of the two ends (circles) and the curved lateral area of the length of the cylinder.	$Q = \sigma A = \sigma \left[2\left(\pi r^2\right) + 2\pi r h \right]$ $$Q = \left(12.5 \times 10^{-6} \text{ C/m}^2\right) \left[\begin{array}{l} (2\pi)(0.0450 \text{ m})^2 \\ + (2\pi)(0.0450 \text{ m})(0.150 \text{ m}) \end{array} \right]$$ $Q = \boxed{6.89 \times 10^{-7} \text{ C}}$
c. The total charge would be equal to the surface charge density times the area of the cylinder, not including the ends.	$Q = \sigma A = \sigma 2\pi r h$ $$Q = \left(12.5 \times 10^{-6} \text{ C/m}^2\right)(2\pi)(0.0450 \text{ m})(0.150 \text{ m})$$ $Q = \boxed{5.30 \times 10^{-7} \text{ C}}$

CHECK and THINK

It is possible that the excess charge on an object could reside on the surface or throughout the volume of an object. Be sure to pay attention to the units of a charge density, as they indicate the dimension of distribution of the charge.

67. (N) The total charge on a uniformly charged ring with diameter 25.0 cm is –54.0 μC. What is the magnitude of the electric field along the ring's axis at a distance of **a.** 2.50 cm, **b.** 12.5 cm, **c.** 25.0 cm, and **d.** 2.00 m from its center?

INTERPRET and ANTICIPATE
To find the electric field near an extended object, we would generally have to integrate to add up the contributions due to each small charge element in the object. The field along the axis of a uniformly charged ring is one of our special cases, Example 24.5, so we can apply that formula here.

SOLVE	
Use Eq. 24.16 from Example 24.5, which is the special case of a field along the axis of a charged ring. In	$$E = \frac{kQy}{\left(R^2 + y^2\right)^{3/2}}$$

this case, Q is the total charge of the ring, R is its radius, and y is the distance from the center of the ring.	
Insert numerical values (other than the distance y). All values are metric, so the electric field will be in units of N/C.	$E = \dfrac{(8.99 \times 10^9)(54.0 \times 10^{-6}) y}{(0.125^2 + y^2)^{3/2}}$
Now, insert each y value: **a.** $y = 0.0250$ m	$E = \boxed{5.86 \times 10^6 \text{ N/C}}$
b. $y = 0.125$ m	$E = \boxed{1.10 \times 10^7 \text{ N/C}}$
c. $y = 0.250$ m	$E = \boxed{5.56 \times 10^6 \text{ N/C}}$
d. $y = 2.00$ m	$E = \boxed{1.21 \times 10^5 \text{ N/C}}$

CHECK and THINK
As a special case (uniformly charged ring), we are able to use the formula derived in the chapter and insert the appropriate numerical values. Notice that the field increases and then decreases as we look at points further from the center of the ring. This is because at the center, the electric fields due to all parts of the ring cancel each other out. As we move further away from the center, there are components along this axis that add together and produce a non-zero field. But of course as we move away towards infinity, far from the charged object, the field eventually has to drop off to zero.

70. (A) Two positively charged spheres are shown in Figure P24.70. Sphere 1 has twice as much charge as sphere 2. Find an expression for the electrostatic field at point A in terms of the parameters shown in the figure.

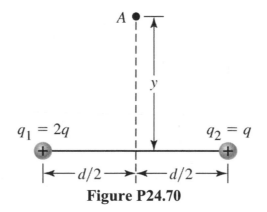

Figure P24.70

Chapter 24 – Electric Fields

<table>
<tr>
<td colspan="2">
INTERPRET and ANTICIPATE
We can use the expression for the electric field due to a point charge, apply it to each of the charges, and add them as vectors to find the result. Since the electric field due to a charge points away from positive charges and the charge on the left is larger, we would expect the net field to be up and a little to the right.
</td>
</tr>
<tr>
<td>
SOLVE
First sketch the electric fields due to each of the charges.
</td>
<td>

Figure P24.70ANS
</td>
</tr>
<tr>
<td>
The electric field due to each charge can be calculated using Equation 24.3.
</td>
<td>

$$\vec{E}(r) = k\frac{Q_S}{r^2}\hat{r}$$
</td>
</tr>
<tr>
<td>
For each charge, the distance to Point A can be calculated using the Pythagorean theorem. We can also calculate the angle θ using trigonometry.
</td>
<td>

$$r^2 = \left(\frac{d}{2}\right)^2 + y^2 \qquad (1)$$

$$\cos\theta = \frac{y}{r} \quad \text{and} \quad \sin\theta = \frac{d}{2r}$$
</td>
</tr>
<tr>
<td>
Calculate for the charge on the left with magnitude $2q$. The field points up and to the right. Substitute the expressions from above.
</td>
<td>

$$\vec{E}_1 = k\frac{2q}{\left(\dfrac{d}{2}\right)^2 + y^2}\left(\sin\theta\,\hat{i} + \cos\theta\,\hat{j}\right)$$

$$\vec{E}_1 = k\frac{8q}{d^2 + 4y^2}\left(\frac{d}{2r}\,\hat{i} + \frac{y}{r}\,\hat{j}\right)$$

$$\vec{E}_1 = k\frac{4q}{d^2 + 4y^2}\left(\frac{d}{r}\,\hat{i} + \frac{2y}{r}\,\hat{j}\right)$$
</td>
</tr>
</table>

52

Now for the charge on the right with magnitude q. This field points up and to the left.	$$\vec{E}_2 = k\frac{q}{\left(\dfrac{d}{2}\right)^2 + y^2}\left(-\sin\theta\,\hat{i} + \cos\theta\,\hat{j}\right)$$ $$\vec{E}_2 = k\frac{4q}{d^2 + 4y^2}\left(-\frac{d}{2r}\,\hat{i} + \frac{y}{r}\,\hat{j}\right)$$
The total field is the sum of these two.	$$\vec{E}_{tot} = \vec{E}_1 + \vec{E}_2$$ $$\vec{E}_{tot} = \frac{4kq}{d^2 + 4y^2}\left(\left(\frac{d}{r} - \frac{d}{2r}\right)\hat{i} + \left(\frac{2y}{r} + \frac{y}{r}\right)\hat{j}\right)$$ $$\vec{E}_{tot} = \frac{4kq}{d^2 + 4y^2}\left(\frac{d}{2r}\,\hat{i} + \frac{3y}{r}\,\hat{j}\right)$$
Finally, from equation (1) above, we know that $$r = \sqrt{\left(\frac{d}{2}\right)^2 + y^2} = \frac{1}{2}\sqrt{d^2 + 4y^2}$$	$$\boxed{\vec{E}_{tot} = \frac{4kq}{\left(d^2 + 4y^2\right)^{3/2}}\left(d\,\hat{i} + 6y\,\hat{j}\right)}$$

CHECK and THINK

This was a fair amount of algebra, but in the end, the field points up and to the right as we would expect.

---

72. (N) Two positively charged spheres are shown in Figure P24.70. Sphere 1 has twice as much charge as sphere 2; $q = 6.55$ nC, $d = 0.250$ m, and $y = 1.25$ m. Suppose an electron is placed at point A and released.

a. What is the electron's acceleration?

b. If the electron were replaced by a proton, what is the proton's acceleration?

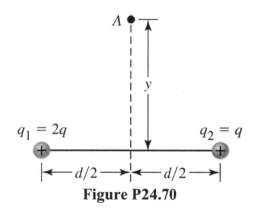

Figure P24.70

Chapter 24 – Electric Fields

INTERPRET and ANTICIPATE
We first calculate the field using the result of Problem 70. (This is actually what Problem 71 asks of us.) Then, with the electric field, it is straightforward to calculate the force on a charge and then calculate its acceleration using Newton's law.

SOLVE
We first apply the result of Problem 70 above with the values given.

$$\vec{E}_{tot} = \frac{4\left(8.99\times10^{9}\,\frac{\text{N}\cdot\text{m}^{2}}{\text{C}^{2}}\right)\left(6.55\times10^{-9}\text{C}\right)}{\left(\left(0.250\text{ m}\right)^{2}+4\left(1.25\text{ m}\right)^{2}\right)^{3/2}}\left(0.250\text{ m }\hat{i}+6\left(1.25\text{ m}\right)\hat{j}\right)$$

$$\vec{E}_{tot} = 14.9\frac{\text{N}}{\text{C}\cdot\text{m}}\left(0.250\text{ m }\hat{i}+7.25\text{ m}\hat{j}\right)$$

$$\vec{E}_{tot} = \left(3.71\,\hat{i}+111\,\hat{j}\right)\frac{\text{N}}{\text{C}}$$

We can now use Eq. 24.18 to determine the force and then find the acceleration using Newton's law.	$\vec{F} = q\vec{E} = m\vec{a}$ $$\vec{a} = \frac{q\vec{E}}{m}$$

Calculate this quantity for both the proton and the electron placed at point A.

$$\vec{a}_{p} = \frac{\left(1.6\times10^{-19}\text{C}\right)\left(\left(3.71\,\hat{i}+111\,\hat{j}\right)\frac{\text{N}}{\text{C}}\right)}{1.67\times10^{-27}\text{kg}} = \boxed{\left(3.6\times10^{8}\hat{i}+1.1\times10^{10}\,\hat{j}\right)\text{ m/s}^{2}}$$

$$\vec{a}_{e} = \frac{\left(-1.6\times10^{-19}\text{C}\right)\left(\left(3.71\,\hat{i}+111\,\hat{j}\right)\frac{\text{N}}{\text{C}}\right)}{9.11\times10^{-31}\text{kg}} = \boxed{\left(-6.5\times10^{11}\hat{i}-2.0\times10^{13}\,\hat{j}\right)\text{ m/s}^{2}}$$

CHECK and THINK
Note that the acceleration is in the opposite direction of the proton and much larger, since the force on each is the same but the mass of the electron is much smaller.

74. (N) Two parallel plates are placed 5.60 cm apart, creating a uniform electric field of magnitude 525 N/C between the plates. At the same instant in time, a proton is released from rest from the positive plate and an electron is released from rest from the negative plate. If we ignore the electrical attraction between the two particles, at what distance from the positive plate do the particles pass each other?

54

© 2016 Cengage Learning. All Rights Reserved. May not be scanned, copied or duplicated, or posted to a publicly accessible website, in whole or in part.

INTERPRET and ANTICIPATE

In the presence of an electric field, both charges experience a force $F = qE$. Since they have the same charge, they experience the same force. However, the more massive proton will have a smaller acceleration ($a = F/m$) and make it a smaller distance than the electron during the same amount of time. We can use kinematics to find the precise position, but we expect it to be closer to the plate where the proton started (i.e. closer to the positive plate).

SOLVE First, draw a sketch. We call the distance from the positive plate to the location of the collision d.	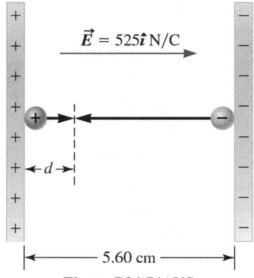 **Figure P24.74ANS**
The force on each charge can be found using Eq. 24.18. The positive charge experiences a force to the right and the negative to the left.	$F_E = qE$
This force causes an acceleration according to Newton's second law, $F = ma$.	$ma = qE \quad \rightarrow \quad a = \dfrac{qE}{m}$
Calculate the proton acceleration to the right using $E = 525$ N/C, $q = e = 1.60 \times 10^{-19}$ C, and $m = 1.67 \times 10^{-27}$ kg.	$\left\|a_p\right\| = \dfrac{qE}{m_p}$ $\left\|a_p\right\| = \dfrac{\left(1.60 \times 10^{-19}\ \text{C}\right)\left(525\ \text{N/C}\right)}{1.673 \times 10^{-27}\ \text{kg}}$ $\left\|a_p\right\| = 5.02 \times 10^{10}\ \text{m/s}^2$

Now calculate the electron acceleration, using its mass of $m = 9.11 \times 10^{-31}$ kg. Although they experience the same force, the electron is 1836 times less massive and therefore has an acceleration 1836 times larger.	$$\left\lvert a_e \right\rvert = \frac{qE}{m_e}$$ $$\left\lvert a_e \right\rvert = \frac{\left(1.60 \times 10^{-19} \text{ C}\right)\left(525 \text{ N/C}\right)}{9.110 \times 10^{-31} \text{ kg}}$$ $$\left\lvert a_e \right\rvert = 9.22 \times 10^{13} \text{ m/s}^2$$ $$\left\lvert \frac{a_e}{a_p} \right\rvert = 1836$$
We also know that each charge travels for the same amount of time before colliding. Using kinematic equations, the distance traveled by each is $\Delta x = \frac{1}{2}at^2$. The total distance is 5.60 cm. Since the electron's acceleration is much greater, we can make the approximation shown. (Of course, this is not necessary. You can keep the other term as well and should get almost exactly the same answer!)	$$5.60 \text{ cm} = \frac{1}{2}a_p t^2 + \frac{1}{2}a_e t^2$$ $$5.60 \text{ cm} \approx \frac{1}{2}a_e t^2$$ $$t^2 \approx \frac{2\left(5.60 \text{ cm}\right)}{a_e}$$
We can now calculate the distance traveled by the proton.	$$d = \frac{1}{2}a_p t^2$$ $$d = \frac{1}{2}a_p \left(\frac{2\left(5.60 \text{ cm}\right)}{a_e}\right) = \left(5.60 \text{ cm}\right)\frac{a_p}{a_e}$$ $$d = \frac{5.60 \text{ cm}}{1836} = \boxed{3.05 \times 10^{-3} \text{ cm}} = 3.05 \times 10^{-5} \text{ m}$$

CHECK and THINK

As expected, the particles travel a small distance from the positive plate before colliding—a *very* small distance in fact! Since the electron is 1836 times less massive, its acceleration is 1836 times larger, and it travels almost the entire distance before it meets the proton.

25

Gauss's Law

7. (C) A positively charged sphere and a negatively charged sphere are in a sealed container. The only way the charged spheres can be examined is by observing the electric field outside the container.

a. Given the depiction of the electric fields in Figure P25.7A, is the net electric flux through the container zero, positive, or negative? Explain your answer.

b. Two different spheres are placed inside a container. Given the depiction of the electric fields in Figure P25.7B, is the net electric flux through the container zero, positive, or negative? Explain your answer.

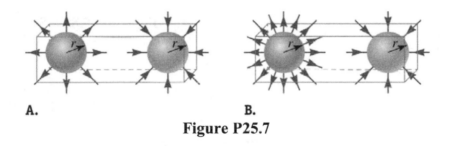

A. **B.**

Figure P25.7

---

a. Note that the number of electric field lines that are shown entering the container is equal to the number of lines that are shown leaving the container. The net electric flux through the walls of the container must then be zero.

b. In the second figure, there are more electric field lines shown exiting the container than there are shown entering the container. This means that the net electric charge enclosed by the container is positive and that the net electric flux through the container is positive.

---

8. (N) A circular hoop of radius 0.50 m is immersed in a uniform electric field of 12.0 N/C. The electric field is at an angle of 30.0° to the plane of the hoop. Determine the electric flux through the hoop.

INTERPRET and ANTICIPATE
The flux depends on the electric field and the area and orientation of the surface. We have all the pieces needed to calculate the flux.

SOLVE For a flat surface in a uniform electric field, we can use Equation 25.3 or 25.4.	$\Phi_E = \vec{E} \cdot \vec{A}$ $\Phi_E = EA\cos\theta$
The electric field is given and the area of the circular hoop can be calculated using the radius ($A = \pi r^2$). For the angle, we need to be careful to use the angle between the electric field and the area vector, which is a vector perpendicular to the surface. In this case, it is 60°.	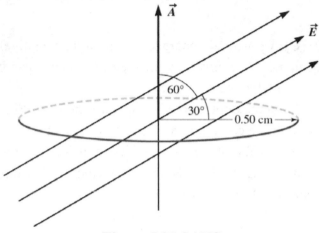 **Figure P25.8ANS**
Insert values.	$\Phi_E = EA\cos\theta$ $\Phi_E = (12.0 \text{ N/C})\left(\pi(0.50\text{ m})^2\right)\cos 60.0°$ $\Phi_E = \boxed{4.7 \text{ N}\cdot\text{m}^2/\text{C}}$
CHECK and THINK Notice that the electric field is constant through the surface, so there is no need to integrate.	

15. (N) For the pyramid in Problem 13, what is the electric flux through one of the four walls?

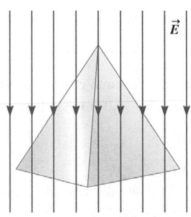

Figure P25.13

INTERPRET and ANTICIPATE	

INTERPRET and ANTICIPATE

In this case, we can determine the area vector and apply Eq. 25.3, but there is an easier way. The total flux through the four triangular walls must be equal to the flux through the square base, which is easy to calculate. By symmetry, the flux through each wall then is a quarter of this result.

SOLVE	
The total flux through the four walls must be equal to the flux through the square base. To see this, imagine looking straight through the bottom of the pyramid. All of the field lines that point through the base must also pass through one of the four walls. Another way to see this is that the net flux through the pyramid must be zero (Gauss's law, Eq. 25.11, using the pyramid as a Gaussian surface and the fact that there is no charge enclosed inside.) So, let's first calculate the flux through the square base using Equation 25.4.	$\Phi_E = EA\cos\varphi$
The area vector for the base (which is perpendicular to the surface) is parallel to the electric field vector ($\varphi = 0$), so the flux through the surface is just EA.	$\Phi_{E,\text{base}} = EA_{\text{base}} = (655 \text{ N/C})(4.00 \text{ m}^2)$ $\Phi_{E,\text{base}} = 2620 \text{ N} \cdot \text{m}^2/\text{C}$
Based on symmetry, the flux through each triangular side must equal one quarter of the flux through the base.	$\Phi_{E,\text{side}} = \dfrac{1}{4}\Phi_{E,\text{base}} = \dfrac{2620 \text{ N} \cdot \text{m}^2/\text{C}}{4} = \boxed{655 \text{ N} \cdot \text{m}^2/\text{C}}$

CHECK and THINK

Using symmetry makes this into a couple line problem. Of course, it is possible to find the angle between the side and the base using trigonometry, which is also the angle between the area vector (perpendicular to the surface) and the electric field, and apply Eq. 25.4, but this involves many more chances for mistakes. To quickly sketch out this approach:

The area of 4 m^2 must have sides of 2 m in length. The distance from the middle of the triangular base to the center of the square is then 1m. This allows us to calculate the angle between the side and the base (using the triangle from the middle of the base to the center of the square and to the top of the pyramid) as $\theta = \tan^{-1}\left(\dfrac{3.5}{1}\right) = 74.1°$. As mentioned above this is also the angle between the area vector and the electric field. The area of the triangular side is $\dfrac{1}{2}bh$ where the base is 2 cm (the side of the square) and the height of the triangle forming the side is the hypotenuse of the triangle just used the find the angle can be found using the Pythagorean theorem.

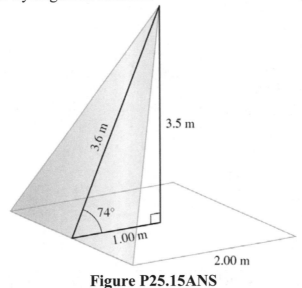

Figure P25.15ANS

18. (N) The net electric flux through a Gaussian surface is –456 N·m^2/C. What is the net charge of the source inside the surface?

INTERPRET and ANTICIPATE
This is an application of Gauss's law, which relates the flux through a Gaussian surface to the charge enclosed inside the surface.

SOLVE	
Apply Gauss's law by rearranging Equation 25.7.	$\Phi_E = \dfrac{q_{in}}{\varepsilon_0} \quad \rightarrow \quad q_{in} = \varepsilon_0 \Phi_E$

| Substitute values. | $q_{in} = \left(8.85 \times 10^{-12} \ C^2 / N \cdot m^2\right)\left(-456 \ N \cdot m^2 / C\right)$ |
| | $q_{in} = \boxed{-4.04 \times 10^{-9} C} = -4.04 \ nC$ |

CHECK and THINK

This is a fairly straightforward application of Gauss's law, relating the flux to the enclosed charge.

20. An isolated system consists of a single particle with charge 56.0 μC placed at the center of a cube of side 1.25 m.

a. (N) What is the flux through each of the faces of the cube?

INTERPRET and ANTICIPATE

The total flux through the surface of the cube can be found based on the enclosed charge using Gauss's law. By symmetry, the flux through each face of the cube is the same.

SOLVE Gauss's law relates the total flux of a closed Gaussian surface to the charge contained inside (Eq. 25.7).	$\Phi_E = \dfrac{q_{in}}{\varepsilon_0}$
	$\Phi_E = \left(\dfrac{56.0 \times 10^{-6} \ C}{8.85 \times 10^{-12} \ C^2 / N \cdot m^2}\right)$
	$\Phi_E = 6.33 \times 10^6 \ N \cdot m^2 / C$
Since the system is symmetric, the flux through each side must be the same. Therefore the flux through each side is one sixth of the total.	$\Phi_{E, face} = \dfrac{1}{6}\Phi_E$
	$\Phi_{E, face} = \dfrac{1}{6}\left(\dfrac{56.0 \times 10^{-6} \ C}{8.85 \times 10^{-12} \ C^2 / N \cdot m^2}\right)$
	$\Phi_{E, face} = \boxed{1.05 \times 10^6 \ N \cdot m^2 / C}$

CHECK and THINK

Gauss's law allows us to easily relate the total flux with the enclosed charge. Taking advantage of symmetry allows us to state that the flux through each face is the same.

b. (C) Would the answer to part (a) change if the particle was moved away from the center of the cube?

Yes. Gauss's law only tells us the *total* flux. We needed to rely on symmetry to say that each side has one sixth of the total passing through it. If the charge is off center, one side might have a larger field and a larger flux than another. That is, the total flux out of the cube is constant, but the amount passing through each side depends on the location of the charge.

23. (A) A particle with charge q is placed at a corner of a cube with side length a. Determine the net electric flux through the cube.

INTERPRET and ANTICIPATE

Integrating the flux through the surfaces of the cube is possible, but not at all simple. Gauss's law cannot be applied directly to the given cube, since the charge is not inside. However, we can construct a new, symmetric Gaussian surface that we can calculate that includes the given cube. By symmetry, we can then determine what fraction of the total flux passes through the cube.

SOLVE

First note that the flux through the three sides that form the corner where the charge is located is zero. This is because the electric field is parallel to each of the surfaces, which are shaded in the figure.

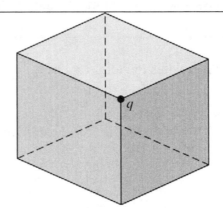

Figure P25.23aANS

Now use this as a building block to form a larger cube with side $2a$ and the charge at the center of it. There are 8 of the smaller cubes needed to form the larger cube.

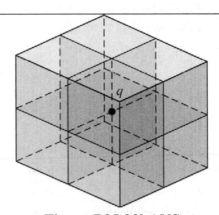

Figure P25.23bANS

The original cube forms one of 8 identical corners of this larger cube, so using this symmetry, 1/8th of the total flux passes through it. Use Gauss's law, Equation 25.11, to relate the total flux to the charge enclosed by the Gaussian surface.	$$\Phi_E = \frac{\Phi_{E,net}}{8} = \boxed{\frac{q}{8\varepsilon_0}}$$

CHECK and THINK

We did not actually have to perform the integral to find the flux since Gauss's law relates the flux to the enclosed charge. By clever construction and symmetry we could find the flux in what started as a problem without a simple solution.

26. (N) Three point charges $q_1 = 2.0$ nC, $q_2 = -4.0$ nC, and $q_3 = -3.0$ nC are placed as shown in Figure P25.26. Find the electric flux through each of the closed Gaussian surfaces C_1, C_2, C_3, and C_4.

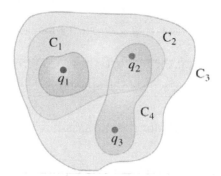

Figure P25.26

INTERPRET and ANTICIPATE	
This is an application of Gauss's law (Equation 25.11). The flux through each closed surface depends on the charge enclosed inside the surface, which we can determine easily.	
SOLVE The flux through each closed surface depends on the charge enclosed inside the surface (Eq. 25.11).	$$\Phi_E = \oint \vec{E} \cdot d\vec{A} = \frac{q_{in}}{\varepsilon_0}$$
Closed surface C_1: This surface only includes the charge $q_1 = 2.0$ nC.	$$\Phi_{E,1} = \frac{q_1}{\varepsilon_0} = \frac{2.0 \times 10^{-9} \text{ C}}{8.85 \times 10^{-12} \dfrac{\text{C}^2}{\text{N} \cdot \text{m}^2}} = \boxed{2.3 \times 10^2 \text{ N} \cdot \text{m}^2/\text{C}}$$

Closed surface C_2: This surface includes two charges $q_1 = 2.0$ nC and $q_2 = -4.0$ nC. So net charge enclosed is $q_{in} = q_1 + q_2 = -2.0$ nC.	$\Phi_{E,2} = \dfrac{q_1 + q_2}{\varepsilon_0} = \dfrac{-2.0 \times 10^{-9}\,\text{C}}{8.85 \times 10^{-12}\,\dfrac{\text{C}^2}{\text{N}\cdot\text{m}^2}}$ $\Phi_{E,2} = \boxed{-2.3 \times 10^2\,\text{N}\cdot\text{m}^2/\text{C}}$
Closed surface C_3: This surface includes all three charges $q_1 = 2.0$ nC, $q_2 = -4.0$ nC and $q_3 = -3.0$ nC. So net charge enclosed is $q_{in} = q_1 + q_2 + q_3 = -5.0$ nC.	$\Phi_{E,3} = \dfrac{q_1 + q_2 + q_3}{\varepsilon_0} = \dfrac{-5.0 \times 10^{-9}\,\text{C}}{8.85 \times 10^{-12}\,\dfrac{\text{C}^2}{\text{N}\cdot\text{m}^2}}$ $\Phi_{E,3} = \boxed{-5.6 \times 10^2\,\text{N}\cdot\text{m}^2/\text{C}}$
Closed surface C_4: This surface includes two charges $q_2 = -4.0$ nC and $q_3 = -3.0$ nC. So net charge enclosed is $q_{in} = q_2 + q_3 = -7.0$ nC.	$\Phi_{E,4} = \dfrac{q_2 + q_3}{\varepsilon_0} = \dfrac{-7.0 \times 10^{-9}\,\text{C}}{8.85 \times 10^{-12}\,\dfrac{\text{C}^2}{\text{N}\cdot\text{m}^2}}$ $\Phi_{E,4} = \boxed{-7.9 \times 10^2\,\text{N}\cdot\text{m}^2/\text{C}}$

CHECK and THINK

The essence of this exercise is that the electric flux through a closed surface is proportional to the net charge enclosed by the surface.

28. (N) A very long, thin wire fixed along the *x* axis has a linear charge density of 3.2 μC/m.

a. Determine the electric field at point *P* a distance of 0.50 m from the wire.

b. If there is a test charge $q_0 = +2.0$ μC at point *P*, what is the magnitude of the net force on this charge? In which direction will the test charge accelerate?

INTERPRET and ANTICIPATE

Equation 25.13 allows us to calculate the electric force due to an infinitely long linear charge at a particular distance from the line of charge. Then, with $F = qE$, we can determine the force on a nearby point charge.

SOLVE **a.** Apply Eq. 25.13 to find the magnitude of the electric field.	$E_P = \dfrac{1}{2\pi\varepsilon_0}\dfrac{\lambda}{r}$ $E_P = \dfrac{1}{2\pi\left(8.85 \times 10^{-12}\,\dfrac{\text{C}^2}{\text{N}\cdot\text{m}^2}\right)}\dfrac{3.2 \times 10^{-6}\,\text{C/m}}{(0.50\,\text{m})}$ $E_P = \boxed{1.2 \times 10^5\,\text{N/C}}$

b. Now, use $F = qE$.	$F = q_0 E_P$ $F = (2.0 \times 10^{-6}\,\text{C})(1.2 \times 10^5\,\text{N/C})$ $F = \boxed{0.24\,\text{N}}$

CHECK and THINK

The electric force and charge are radially outward. Using the formula derived for the specific geometry (line charge) makes calculating the electric field straightforward.

32. (N) Two long, thin rods each have linear charge density $\lambda = 6.0\ \mu\text{C/m}$ and lie parallel to each other, separated by 20.0 cm as shown in Figure P25.32. Determine the magnitude and direction of the net electric field at point P, a distance of 15.0 cm directly above the right rod.

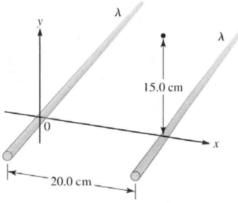

Figure P25.32

INTERPRET and ANTICIPATE

We have found a general expression for the electric field produced by an infinite rod (Eq. 25.13). The net field at any point for two such rods will be the vector sum of the fields of each at that point. Since the rods are positively charged and the electric field due to each points radially away from the rod, the total field should point up and a little to the right (i.e. positive x and y components).

SOLVE Let's first draw a sketch that includes the geometry and electric fields due to each rod at the location of point P as well as an approximate vector for the total electric field at this point.	 **Figure P25.32ANS**
Use Equation 25.13 to find the magnitude of the electric field from the rod on the right at point P, which is 15 cm = 0.15 m above the rod.	$E_R = \dfrac{\lambda}{2\pi\varepsilon_0 r_R} = \dfrac{(6.0\times10^{-6})}{2\pi(8.85\times10^{-12})(0.15)}$ $E_R = 7.2\times10^5$ N/C
This field points straight up along the y axis.	$\vec{E}_R = 7.2\times10^5\,\hat{\boldsymbol{j}}$ N/C
Use the same expression for the field due to the rod on the left. From the Pythagorean theorem, the distance to point P is the hypotenuse of the triangle with sides 0.20 m and 0.15 m.	$E_L = \dfrac{\lambda}{2\pi\varepsilon_0 r_L} = \dfrac{(6.0\times10^{-6})}{2\pi(8.85\times10^{-12})\sqrt{0.15^2+0.20^2}}$ $E_L = 4.3\times10^5$ N/C
This field points at an angle θ above the x axis that we can determine using trigonometry.	$\theta = \tan^{-1}\dfrac{0.15}{0.20} = 36.9°$
This allows us to determine the x and y components of the field due to the rod on the left.	$\vec{E}_L = 4.3\times10^5\left(\cos 36.9°\hat{\boldsymbol{i}} + \sin 36.9°\hat{\boldsymbol{j}}\right)$N/C $\vec{E}_L = \left(3.4\times10^5\hat{\boldsymbol{i}} + 2.6\times10^5\hat{\boldsymbol{j}}\right)$N/C

The total field at point P is the vector sum.	$\vec{E}_{tot} = \vec{E}_R + \vec{E}_L$ $\vec{E}_{tot} = \left(3.4\times10^5\hat{i} + \left[7.2\times10^5 + 2.6\times10^5\right]\hat{j}\right)$ N/C $\vec{E}_{tot} = \left(3.4\times10^5\hat{i} + 9.8\times10^5\hat{j}\right)$ N/C or $E_x = 3.4\times10^5$ N/C and $E_y = 9.8\times10^5$ N/C
Let φ be the angle between the resultant electric field vector and the x axis. Find φ and E from the E_x and E_y.	$E = \sqrt{E_x^2 + E_y^2} = \boxed{1.0\times10^6 \text{ N/C}}$ $\varphi = \tan^{-1}\dfrac{E_y}{E_x} = \boxed{71° \text{ above the } x \text{ axis}}$

CHECK and THINK

Because point P is closer to the right rod, the net field makes an angle greater than 45° with the x axis. It's consistent with our sketch.

37. (N) A particle with a charge of 55.0 μC is at the center of a thin spherical shell of radius $R = 12.0$ cm that has uniform surface charge density σ. Determine the value of σ such that the net electric field outside the shell is zero.

INTERPRET and ANTICIPATE

Gauss's law tells us that the total electric flux through a surface is proportional to the enclosed charge. Imagine a spherical Gaussian surface around the shell. If the electric field is zero, the flux through the surface must be zero, which means that the net charge inside must be zero. So, the total charge on the spherical shell must be exactly equal and opposite the charge at the center of the shell. That suddenly makes the problem much simpler.

SOLVE A spherical Gaussian surface of radius $r > 12$ cm will have a net flux of zero if the electric field is zero since the integral must be zero if the electric field E is zero everywhere on the surface. From Gauss's law (Eq. 25.11), this means the total charge inside the surface is zero.	$\oint \vec{E} \cdot d\vec{A} = 0 = \dfrac{q_{in}}{\varepsilon_0}$

Since the total charge is zero, the charge on the spherical shell must be equal and opposite the charge at the center.	$q_{in} = q_{shell} + q_{particle} = 0$ $q_{shell} = -q_{particle} = -5.50 \times 10^{-5}$ C
The charge on the spherical shell is uniformly distributed on the surface, so we can find the density σ for a shell of radius R.	$\sigma = \dfrac{q_{shell}}{4\pi R^2} = \dfrac{-5.50 \times 10^{-5}}{4\pi (0.120)^2}$ $\sigma = \boxed{-3.04 \times 10^{-4} \text{ C/m}^2}$

CHECK and THINK

Since there is a positive charge inside, it makes sense that the shell is negatively charged in order to have zero field outside.

42. (N) Two uniform spherical charge distributions (Fig. P25.41) each have a total charge of 45.3 mC and radius $R = 15.2$ cm. Their center-to-center distance is 37.50 cm. Find the magnitude of the electric field at point B, 7.50 cm from the center of one sphere and 30.0 cm from the center of the other sphere.

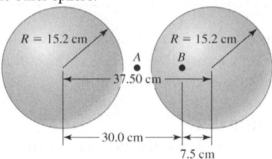

$R = 15.2$ cm $R = 15.2$ cm

A B

37.50 cm

30.0 cm

7.5 cm

Figure P25.41

INTERPRET and ANTICIPATE
Point B is outside of the sphere on the left, so it acts like a point charge concentrated at its center, and within the charge on the right (as in Example 25.7). We can determine the electric field due to each and add them as vectors to find the total electric field.

| **SOLVE**
 Outside of a spherical charge distribution, we can apply Eq. 25.14, which is just the field due to a point charge. Point B is outside of sphere on the left and 30.0 cm from its center. The field due to this positive charge on the left points to the right at B. | $E_L = \dfrac{1}{4\pi\varepsilon_0} \dfrac{q}{R^2}$

 $E_L = \left(8.99 \times 10^9 \dfrac{\text{N} \cdot \text{m}^2}{\text{C}^2} \right) \dfrac{45.3 \times 10^{-3} \text{ C}}{(0.300 \text{ m})^2}$

 $E_L = 4.52 \times 10^9$ N/C to the right |

Point B is inside the sphere on the right, so we can use Equation 25.15 from Example 25.7 with $r = 7.50$ cm $= 0.0750$ m. The radius of the sphere is $R = 0.152$ m. The field at B due to the charge on the right is to the left.	$E_R = \dfrac{1}{4\pi\varepsilon_0}\dfrac{q}{R^3}r$ $E_R = \left(8.99\times10^9\,\dfrac{\text{N}\cdot\text{m}^2}{\text{C}^2}\right)\dfrac{45.3\times10^{-3}\,\text{C}}{(0.152\text{ m})^3}(0.0750\text{ m})$ $E_R = 8.70\times10^9$ N/C to the left
The total electric field is the vector sum of these two. We can indicate "to the right" as positive.	$E_{tot} = E_L + E_R = 4.52\times10^9$ N/C -8.70×10^9 N/C $E_{tot} = \boxed{-4.18\times10^9\text{ N/C} \quad\text{or } 4.18\times10^9\text{ N/C to the left}}$

CHECK and THINK
We need to be careful about whether the point was within or outside of the spherical charge distribution and use the appropriate formula, but the total electric field can be calculated at this point for each sphere and added together as vectors.

47. (N) The infinite sheets in Figure P25.47 are both positively charged. The sheet on the left has a uniform surface charge density of 48.0 μC/m^2, and the one on the right has a uniform surface charge density of 24.0 μC/m^2.
a. What are the magnitude and direction of the net electric field at points A, B, and C?
b. What is the force exerted on an electron placed at points A, B, and C?

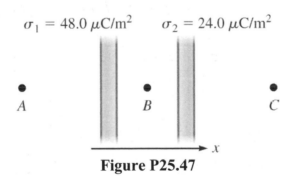

$\sigma_1 = 48.0\ \mu$C/m^2 $\quad\quad$ $\sigma_2 = 24.0\ \mu$C/m^2

$A \quad\quad\quad B \quad\quad\quad C$

Figure P25.47

INTERPRET and ANTICIPATE
This is very similar to Example 25.9, so we can follow this as a model. Each sheet creates a uniform electric field that points away from it (since they are positively charged). We need only determine the field caused by each plate and find the vector sum at each location. Once we have the field, we can calculate the force using $F = qE$.

SOLVE **a.** The field created by an infinite sheet is given by Eq. 25.16. Use this to determine the field created by the plate on the left and the plate on the right.	$E = \dfrac{\sigma}{2\varepsilon_0}$ $E_1 = \dfrac{48.0 \times 10^{-6}\ \text{C/m}^2}{2\left(8.85 \times 10^{-12}\ \dfrac{\text{C}^2}{\text{N} \cdot \text{m}^2}\right)} = 2.71 \times 10^6\ \text{N/C}$ $E_2 = \dfrac{24.0 \times 10^{-6}\ \text{C/m}^2}{2\left(8.85 \times 10^{-12}\ \dfrac{\text{C}^2}{\text{N} \cdot \text{m}^2}\right)} = 1.36 \times 10^6\ \text{N/C}$
The field due to each is constant and points away from the plate everywhere since both are positively charged.	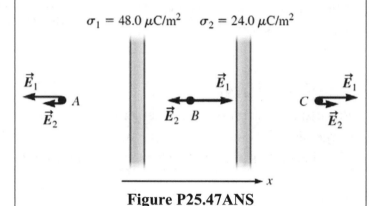 $\sigma_1 = 48.0\ \mu\text{C/m}^2 \qquad \sigma_2 = 24.0\ \mu\text{C/m}^2$ **Figure P25.47ANS**
Now, find the vector sum at each location.	$\vec{E}_A = -E_1\hat{i} - E_2\hat{i} = \boxed{-4.07 \times 10^6\,\hat{i}\ \text{N/C}}$ $\vec{E}_B = +E_1\hat{i} - E_2\hat{i} = \boxed{+1.36 \times 10^6\,\hat{i}\ \text{N/C}}$ $\vec{E}_C = +E_1\hat{i} + E_2\hat{i} = \boxed{+4.07 \times 10^6\,\hat{i}\ \text{N/C}}$

b. The force can now be calculated with $F = qE$, the fields from part (a), and the charge of the electron $(-e)$.

$$\vec{F}_A = -e\vec{E}_A = \left(-1.6 \times 10^{-19}\ \text{C}\right)\left(-4.07 \times 10^6\,\hat{i}\ \text{N/C}\right) = \boxed{6.51 \times 10^{-13}\,\hat{i}\ \text{N}}$$

$$\vec{F}_B = -e\vec{E}_B = \left(-1.6 \times 10^{-19}\ \text{C}\right)\left(1.36 \times 10^6\,\hat{i}\ \text{N/C}\right) = \boxed{-2.17 \times 10^{-13}\,\hat{i}\ \text{N}}$$

$$\vec{F}_C = -e\vec{E}_C = \left(-1.6 \times 10^{-19}\ \text{C}\right)\left(4.07 \times 10^6\,\hat{i}\ \text{N/C}\right) = \boxed{-6.51 \times 10^{-13}\,\hat{i}\ \text{N}}$$

CHECK and THINK

The field due to an infinite sheet is constant, which makes things a lot easier. With the field due to each plate, we can write them as vectors to find the total field and then the force on the electron.

53. (N) The electric field at a point 1.0 m from the center of a charged spherical conductor of radius 0.20 m is –12 N/C. Determine the number of excess electrons in the conductor.

INTERPRET and ANTICIPATE
For a spherical charge distribution, like the charge on this spherical conductor, the electric fields at points outside are the same as if the charge was concentrated at the center of the charge distribution (Eq. 25.14). Using the electric field and distance, we can determine the corresponding charge and then the number of excess electrons required to have that charge.

SOLVE	
The magnitude of the electric field at a distance r from a spherical charge distribution is given by Equation 25.14. It is as if the charge is concentrated at the center.	$E = \dfrac{1}{4\pi\varepsilon_0}\dfrac{Q}{r^2}$ $-12\,\text{N/C} = \dfrac{1}{4\pi\left(8.85\times10^{-12}\,\dfrac{\text{C}^2}{\text{N}\cdot\text{m}^2}\right)}\dfrac{Q}{(1.0\,\text{m})^2}$ $Q = -1.33\times10^{-9}\,\text{C}$
We can now calculate the number of excess electrons N of charge $e = -1.6\times10^{-19}$ C needed to produce this total charge.	$Q = Ne$ $N_e = \dfrac{-1.33\times10^{-9}\,\text{C}}{-1.6\times10^{-19}\,\text{C}} = \boxed{8.3\times10^9}$

CHECK and THINK
Note that for this spherical conductor all the charges reside on the surface of the conductor. The electric field inside the conductor is zero but outside it's as if all the charge is at the center. As a result, the radius of the sphere did not enter into the calculation.

56. (N) A spherical conducting shell with a radius of 0.200 m has a very small charged sphere suspended in its center. The sphere has a charge of 24.6 mC, and the conducting shell has a charge of –24.6 mC. What is the magnitude of the electric field at distances of
a. 0.100 m and
b. 0.300 m from the center of the shell?

Chapter 25 – Gauss's Law

INTERPRET and ANTICIPATE We should apply Gauss's Law within the shell and then outside the shell, substituting the distance from the center for the radius of the Gaussian sphere. Because of the spherical symmetry, we expect that the electric field will appear to be that of a point charge (since the distances from the center are not within a continuous charge distribution), where the charge is equal to that enclosed by the Gaussian sphere. Because the shell has an equal but opposite electric charge compared to the sphere, we expect that the electric field outside the sphere is zero since the net charge enclosed by a Gaussian surface at that point will be zero.	

SOLVE **a.** First, apply Gauss's Law when $r = 0.100$ m, using Eq 25.11, and the spherical symmetry for this scenario. This Gaussian sphere would only enclose the charged sphere at the center and it would have an area of $4\pi(0.100 \text{ m})^2$.	$\oint \vec{E} \cdot d\vec{A} = \dfrac{q_{in}}{\varepsilon_0}$ $E\left(4\pi(0.100 \text{ m})^2\right) = \dfrac{24.6 \times 10^{-3} \text{ C}}{8.85 \times 10^{-12} \text{ N} \cdot \text{m}^2 / \text{C}^2}$
Solving for E, we get the magnitude of the electric field at $r = 0.100$ m.	$E = \dfrac{24.6 \times 10^{-3} \text{ C}}{8.85 \times 10^{-12} \text{ N} \cdot \text{m}^2 / \text{C}^2 \left(4\pi(0.100 \text{ m})^2\right)}$ $E = \boxed{2.21 \times 10^{10} \text{ N/C}}$
b. For $r = 0.300$ m, we can again Gauss's Law and Eq 25.11, where the net charge enclosed by our Gaussian surface outside the shell will include the sphere at the center and the shell.	$\oint \vec{E} \cdot d\vec{A} = \dfrac{q_{in}}{\varepsilon_0}$ $E\left(4\pi(0.300 \text{ m})^2\right) = \dfrac{24.6 \times 10^{-3} \text{ C} + \left(-24.6 \times 10^{-3} \text{ C}\right)}{8.85 \times 10^{-12} \text{ N} \cdot \text{m}^2 / \text{C}^2}$
Since the net charge enclosed is zero, the electric field will be zero.	$E = \boxed{0}$

CHECK and THINK
When we have spherical symmetry, we can imagine these Gaussian surfaces with certain radii, and look to see the charge enclosed by the surface. As long as the surface of our imaginary Gaussian sphere does not fall inside a volume charge distribution, we expect that form of the electric field to look like that due to a point charge.

59. (N) A thick spherical conducting shell with an inner radius of 0.200 m and an outer radius of 0.250 m has a very small charged sphere suspended in its center. The sphere has a charge of 24.6 mC, and the conducting shell has a net charge of –24.6 mC. What is the net amount of charge that resides on

a. the inner and

b. the outer surface of the shell?

What is the charge density on

c. the inner and

d. the outer surface of the shell?

INTERPRET and ANTICIPATE
Because the spherical shell is conducting, the electric field within the shell must be equal to zero. This means that the charge enclosed by a Gaussian surface with its radii falling within the shell must enclose zero net electric charge. If we can find the charge on the inner surface of the shell, we can find the charge on the outer surface by using the total charge of the shell. The surface charge densities can be found by dividing the total charge on either surface by its area.

SOLVE	
a. Considering a Gaussian surface with a radius between 0.200 m and 0.250 m, the charge enclosed would be the sum of the charge on the sphere at the center and the inner surface of the conducting shell. However, the sum must equal zero because this Gaussian surface is inside a conductor. Solve for the charge on the inner surface of the shell.	$q_{sphere} + q_{inner} = 0$ $24.6 \text{ mC} + q_{inner} = 0$ $q_{inner} = \boxed{-24.6 \text{ mC}}$
b. We know the total charge of the conducting shell is –24.6 mC. Since the charge on the inner surface is –24.6 mC, the charge on the outer surface must be $\boxed{0}$.	
c. Divide the total charge on the inner surface by the area of the inner surface.	$\sigma_{inner} = \dfrac{q_{inner}}{A_{inner}} = \dfrac{-24.6 \times 10^{-3} \text{ C}}{4\pi (0.200 \text{ m})^2} = \boxed{-4.89 \times 10^{-2} \text{ C/m}^2}$

d. Since there is no electric charge on the outer surface, its charge density is $\boxed{0}$.	

CHECK and THINK

We can imagine a similar scenario where the net electric charge of the conducting shell is not equal and opposite to that of the central sphere. It would still be true that the charge on the inner surface must be equal and opposite to that of the central sphere, for the reasons stated above, but there would be an excess charge that would exist on the outer surface of the shell. There would then be a surface charge density on the outer surface. This would also mean that an electric field would exist when all the way outside the shell, whereas that is not the case in this problem (consider Gauss's Law outside the shell to prove this to yourself!).

---

61. (N) A rectangular plate with sides 0.60 m and 0.40 m long is lying in the *xy* plane. The plate is placed in a uniform electric field $\vec{E} = \left(-4.0\,\hat{i} + 5.0\,\hat{j} + 3.0\,\hat{k}\right)$ N/C. Calculate the flux through the plate.

INTERPRET and ANTICIPATE

The flux depends on the electric field and the area and orientation of the surface. We have all the pieces needed to calculate the flux.

SOLVE For a flat surface in a uniform electric field, we can use Equation 25.3.	$\Phi_E = \vec{E} \cdot \vec{A}$
The area vector has a magnitude equal to the area of the place and points normal to the plate, along the *z*-axis as shown.	**Figure P25.61ANS**

	$\vec{A} = (0.60 \text{ m})(0.40 \text{ m})\hat{k} = 0.24 \, \hat{k} \text{ m}$
Calculate the flux using Eq. 25.3.	$\Phi_E = \vec{E} \cdot \vec{A}$ $\Phi_E = \left(-4.0\hat{i} + 5.0\hat{j} + 3.0\hat{k}\right)\dfrac{N}{C} \cdot \left(0.24 \, \hat{k} \text{ m}\right)$ $\Phi_E = \boxed{0.72 \text{ N} \cdot \text{m}^2/\text{C}}$

CHECK and THINK

Notice that the x and the y components of the electric field do not contribute to the net flux.

---

66. (N) A uniform electric field $\vec{E} = 1.57 \times 10^4 \, \hat{i}$ N/C passes through a closed surface with a slanted top as shown in Figure P25.66.

a. Given the dimensions and orientation of the closed surface, what is the electric flux through the slanted top of the surface?

b. What is the net electric flux through the entire closed surface?

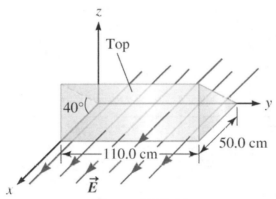

Figure P25.66

INTERPRET and ANTICIPATE

Note that all of the electric field lines that enter the top surface also exit the vertical side of the closed surface (the side parallel the z axis). Also note that the area vector of the vertical side is parallel to the electric field. This means that we can find the electric flux through the vertical side and this will be equal and opposite to the flux through the slanted top. Because there is no indication that an electric charge is contained within the closed surface, we expect that the net electric flux through the entire closed surface to be zero.

SOLVE **a.** First, we find the length of the vertical side (call it z) by using the base of the container and the angle shown.	$\tan(40°) = \dfrac{z}{0.500 \text{ m}}$ $z = (0.500 \text{ m})\tan(40°)$
Now, use that and the 1.100 m side to express the area of the vertical side and find the net electric flux through the vertical wall of the container. Since the area vector points outward from the container, it is parallel to the electric field.	$\Phi_E = \vec{E} \cdot \vec{A} = EA\cos\theta$ $\Phi_E = (1.57 \times 10^4 \text{ N/C})(1.100 \text{ m})(\tan(40°)(0.500 \text{ m}))\cos(0°)$ $\Phi_E = 7.25 \times 10^3 \text{ N} \cdot \text{m}^2/\text{C}$
Then, because the field lines enter the container through the slanted top, and exit through the vertical side, the electric flux through the slanted top must be equal and opposite to that through the vertical side.	$\Phi_E = \boxed{-7.25 \times 10^3 \text{ N} \cdot \text{m}^2/\text{C}}$
b. Since all the field lines that enter the closed surface also exit it, the charge enclosed must be zero and the net electric flux is zero.	$\Phi_E = \boxed{0}$

CHECK and THINK

For part (b), we could also imagine adding together the flux through the slanted top and the flux through the vertical side, which are equal and opposite, to arrive at zero net electric flux. The bottom of the container is parallel to the field and thus has no electric flux through it.

67. (N) A solid plastic sphere of radius $R_1 = 8.00$ cm is concentric with an aluminum spherical shell with inner radius $R_2 = 14.0$ cm and outer radius $R_3 = 17.0$ cm (Fig.

P25.67). Electric field measurements are made at two points: At a radial distance of 34.0 cm from the center, the electric field has magnitude 1.70×10^3 N/C and is directed radially outward, and at a radial distance of 12.0 cm from the center, the electric field has magnitude 9.10×10^4 N/C and is directed radially inward. What are the net charges on

a. the plastic sphere and

b. the aluminum spherical shell?

c. What are the charges on the inner and outer surfaces of the aluminum spherical shell?

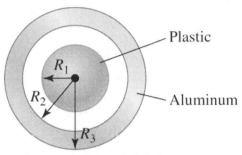

Figure P25.67

INTERPRET and ANTICIPATE
Since this is a spherically symmetric charge distribution, we have a couple tools we can use. For points outside the object, the electric field will be the same as if there was instead a point charge with the same net charge located at the center of the object (Equation 25.14). For any point in space (including inside the object), we can use Gauss's law (Equation 25.11) with a spherical Gaussian surface to relate the electric field and the enclosed charge. These will allow us to determine the charge needed to produce the fields given.

SOLVE	
a. Use Gauss's law (Eq. 25.11) by choosing a Gaussian surface at a radius of 12.0 cm, which is between the plastic sphere and the aluminum shell. This will relate the field at this distance (which is given) to the enclosed charge (that is, the charge on the plastic sphere). The electric field and area vector are both radial at each point on the Gaussian surface, so the integral is equal to the magnitude of the electric field times the area of the sphere.	$$\oint \vec{E} \cdot d\vec{A} = E\left(4\pi r^2\right) = \frac{q_{\text{in}}}{\varepsilon_0}$$

Plug in numbers:

$$\left(-9.10\times10^{4}\ \text{N/C}\right)4\pi\left(0.120\ \text{m}\right)^{2}=\frac{q_{\text{sphere}}}{8.85\times10^{-12}\ \text{C}^{2}/\text{N}\cdot\text{m}^{2}}$$

$$q_{\text{sphere}}=\boxed{-1.46\times10^{-7}\ \text{C}}=-146\ \text{nC}$$

---

b. We can now choose a Gaussian surface with radius 34.0 cm. This will allow us to relate the electric field that is given to the total charge enclosed, $q_{\text{sphere}}+q_{\text{shell}}$.

$$\oint\vec{E}\cdot d\vec{A}=E\left(4\pi r^{2}\right)=\frac{q_{\text{sphere}}+q_{\text{shell}}}{\varepsilon_{0}}$$

$$\left(+1.70\times10^{3}\ \text{N/C}\right)4\pi\left(0.340\ \text{m}\right)^{2}=\frac{q_{\text{sphere}}+q_{\text{shell}}}{8.85\times10^{-12}\ \text{C}^{2}/\text{N}\cdot\text{m}^{2}}$$

$$q_{\text{sphere}}+q_{\text{shell}}=2.19\times10^{-8}\text{C}=21.9\ \text{nC}$$

$$q_{\text{shell}}=21.9\ \text{nC}-\left(-146\ \text{nC}\right)=168\ \text{nC}=\boxed{1.68\times10^{-7}\ \text{C}}$$

---

c. Finally, we choose a Gaussian surface within the aluminum shell, $R_{2}<r<R_{3}$. Within the conductor $E=0$, so the flux is zero, and therefore $q_{\text{sphere}}+Q_{\text{inner shell surface}}=0$

$$\oint\vec{E}\cdot d\vec{A}=0=\frac{q_{\text{sphere}}+Q_{\text{inner shell surface}}}{\varepsilon_{0}}$$

$$Q_{\text{inner shell surface}}=-q_{\text{sphere}}=146\ \text{nC}=\boxed{1.46\times10^{-7}\ \text{C}}$$

Since the shell has a charge of 168 nC and the inner surface has a charge of 146 nC, the rest of the charge must be on the outer surface.

$$Q_{\text{inner shell surface}}+Q_{\text{outer shell surface}}=q_{\text{shell}}$$

$$Q_{\text{outer shell surface}}=168\ \text{nC}-146\ \text{nC}=22\ \text{nC}$$

$$Q_{\text{outer shell surface}}=\boxed{2.2\times10^{-8}\ \text{C}}$$

CHECK and THINK

Using symmetry and Gauss's law, we can determine the charges on each object needed to produce the electric fields measured.

---

72. (A) A coaxial cable is formed by a long, straight wire and a hollow conducting cylinder with axes that coincide. The wire has charge per unit length $\lambda=2\lambda_{0}$, and the hollow cylinder has net charge per unit length $\lambda=3\lambda_{0}$. Use Gauss's law to answer these questions: What are the charges per unit length on
a. the inner surface and
b. the outer surface of the hollow cylinder?

c. What is the electric field a radial distance d from the axis of the coaxial cable?

INTERPRET and ANTICIPATE We can apply Gauss's law using the fact that the electric field within the conductors is zero. One consequence is that all charges reside on the surfaces of the conductors.	
SOLVE First, draw a cross-section of the wire and cylindrical shell. A Gaussian surface is drawn which is also a cylinder with a radius between the inner and outer radii of the cylindrical conductor and an arbitrary length ℓ.	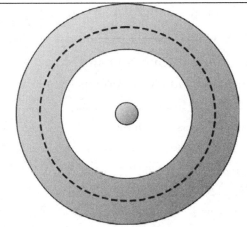 **Figure P25.72aANS**
a. According to Gauss's law (Equation 25.11), since E inside the conducting shell is zero, the flux through the Gaussian surface is zero, and therefore the total charge contained inside the Gaussian surface must also be zero. Since the wire has a charge density of $2\lambda_0$, the inside surface of the cylindrical shell must have a charge density of $\boxed{-2\lambda_0}$ to have a total charge of zero.	$$\oint \vec{E} \cdot d\vec{A} = \frac{q_{in}}{\varepsilon_0}$$
b. Since the cylinder has a total charge density of $3\lambda_0$ and a charge on the interior surface of $-2\lambda_0$, the exterior surface must have a charge density of $\boxed{5\lambda_0}$ (so that the interior and exterior surface charges add up to the total).	

c. Apply Gauss's law (Eq. 25.11). For d larger than the outside radius of the cylinder, the total charge per unit length contained is $5\lambda_0$, the $2\lambda_0$ net charge on the wire and the $3\lambda_0$ net charge on the cylinder. The total charge contained in our Gaussian surface of length ℓ is then $5\lambda_0 \cdot \ell$. The electric field points radially outward in the same direction as the area vector and the area of the cylindrical Gaussian surface is $2\pi r\ell$. (The two end caps on the Gaussian surface have zero flux since the electric field is perpendicular to the area vector.)

$$\oint \vec{E} \cdot d\vec{A} = \frac{q_{in}}{\varepsilon_0}$$

$$E2\pi r\ell = \frac{5\lambda_0 \ell}{\varepsilon_0}$$

$$E = 2\frac{5\lambda_0}{4\pi\varepsilon_0 r}$$

For d larger than the outside radius of the cylinder:

$$E = \boxed{\frac{10\lambda_0 k}{r}}, \text{ radially outward}$$

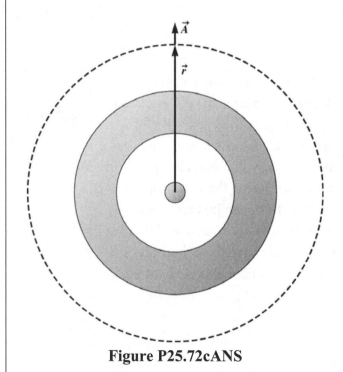

Figure P25.72cANS

For d within the conducting cylinder, the electric field is zero.

For d within the cylinder:

$$E = \boxed{0}$$

For d smaller than the inside radius of the cylinder, the enclosed charge is only the wire with charge density $2\lambda_0$. Repeat the Gauss's law calculation.	$E2\pi r\ell = \dfrac{2\lambda_0\ell}{\varepsilon_0}$ For d smaller than the inside radius of the cylinder: $E = 2\dfrac{2\lambda_0}{4\pi\varepsilon_0 r} = \boxed{\dfrac{4\lambda_0 k}{r}}$, radially outward

CHECK and THINK

Gauss's law along with the fact that the electric field within the conductor is zero allows us to calculate the charges on the surfaces of the conductor.

75. (A) A solid sphere of radius R has a spherically symmetrical, nonuniform volume charge density given by $\rho(r) = A/r$, where r is the radial distance from the center of the sphere in meters, and A is a constant such that the density has dimensions M/L^3.
a. Calculate the total charge in the sphere.

INTERPRET and ANTICIPATE

The total charge on the sphere can be found by integrating the charge element $dq = \rho dV$ over the whole volume.

SOLVE	
Because the charge density ρ is a function of only r, it is spherically symmetric and a suitable volume element dV is a spherical shell of radius r and thickness dr. The volume of this infinitesimally thin shell can be found by the product of the surface area and the thickness.	$dV = 4\pi r^2 dr$
Integrate the charge density per volume over the volume of the sphere to determine the total charge.	$Q = \int \rho\, dV$ $Q = \int_0^R \dfrac{A}{r} 4\pi r^2 dr$ $Q = 4\pi A \int_0^R r\, dr = 2\pi A r^2 \Big|_0^R = \boxed{2\pi A R^2}$

Chapter 25 – Gauss's Law

CHECK and THINK
We can check that this solution has the proper units. From the definition of ρ we see that A must have the units of C/m^2. Multiplying by R^2 with units of m^2 then gives Q the proper units of C.

b. Using the answer to part (a), write an expression for the magnitude of the electric field outside the sphere—that is, for some distance $r > R$.

INTERPRET and ANTICIPATE
Outside the spherically symmetric distribution, the field will be like that of a particle with total charge Q.

SOLVE	
We use the results of part (a) with Equation 25.14 to find the electric field outside the sphere.	$\vec{E} = \dfrac{q}{4\pi\varepsilon_0 r^2}\hat{r} = \dfrac{2\pi A R^2}{4\pi\varepsilon_0 r^2}\hat{r} = \dfrac{A R^2}{2\varepsilon_0 r^2}\hat{r}$ $E = \boxed{\dfrac{A R^2}{2\varepsilon_0 r^2}}$

CHECK and THINK
As in all cases of spherically symmetric charge distributions, the field is radially outward and falls off as $1/r^2$ outside the charge.

c. Find an expression for the magnitude of the electric field inside the sphere at position $r < R$.

INTERPRET and ANTICIPATE
We apply Gauss's law for spherical symmetry and carefully calculate the enclosed charge using the same method we used in part (a).

SOLVE	
Apply Gauss's law (Equation 25.11). Choose as the Gaussian surface a sphere of radius $r < R$.	$\Phi = \oint \vec{E}\cdot d\vec{A} = \dfrac{q_{in}}{\varepsilon_0}$
Since the charge is spherically symmetric, we again expect the electric field points radially and depends only on r. The direction of the area vector at each location of	$\Phi = \oint \vec{E}\cdot d\vec{A} = E4\pi r^2$

the surface is also radial. The flux is then just the magnitude of the electric field at that r times the area of the spherical surface.	
The charge enclosed inside the Gaussian surface is found as in part (a), but the upper limit of the integral is now r, the radius of the Gaussian surface.	$q_{in} = 4\pi A \int_0^r r' \, dr' = 2\pi A r'^2 \Big|_0^r = 2\pi A r^2$
Apply Gauss's law to find the magnitude of the field.	$\Phi = \dfrac{q_{in}}{\varepsilon_0}$ $E 4\pi r^2 = \dfrac{2\pi A r^2}{\varepsilon_0}$ $E = \boxed{\dfrac{A}{2\varepsilon_0}}$
For positive values of A, the electric field points radially outward. Though we're only asked for the magnitude, we could also write this in vector form.	$\vec{E} = \dfrac{A}{2\varepsilon_0} \hat{r}$

CHECK and THINK

Interestingly, this non-uniform spherical charge distribution with a $1/r$ dependence produces an electric field that points radially outward, but has the same magnitude everywhere inside the sphere!

26

Electric Potential

7. (N) Consider the final arrangement of charged particles shown in Figure P26.7. What is the work necessary to build such an arrangement of particles, assuming they were originally very far from one another?

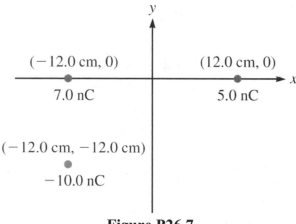

Figure P26.7

INTERPRET and ANTICIPATE

The work required to build the arrangement of charges is equal to the total electric potential energy in the arrangement of charges. We can use Eq. 26.3 to calculate the electric potential energy between each pair of charges and then sum the results. It is possible that the work required could be positive or negative. Here, given the similarity in the magnitude of the charges and the similar distances between the particles, we expect the work required will be negative because of the attractive force between each of the positively charged particles and the negatively charged particle. The only repulsive force is between the two positive particles.

SOLVE	
First, calculate the electric potential energy between the 7.0 nC particle and the 5.0 nC particle, using Eq. 26.3. The distance between the two particles is 0.240 m.	$U_{E1} = \dfrac{kQq}{r}$ $U_{E1} = \dfrac{\left(8.99\times10^9 \text{ N}\cdot\text{m}^2/\text{C}^2\right)\left(7.0\times10^{-9} \text{ C}\right)\left(5.0\times10^{-9} \text{ C}\right)}{0.240 \text{ m}}$ $U_{E1} = 1.3\times10^{-6} \text{ J}$

84

Second, calculate the electric potential energy between the 7.0 nC particle and the −10.0 nC particle. The distance between the two particles is 0.120 m.	$U_{E2} = \dfrac{kQq}{r}$ $U_{E2} = \dfrac{(8.99 \times 10^9 \text{ N} \cdot \text{m}^2/\text{C}^2)(7.0 \times 10^{-9} \text{ C})(-10.0 \times 10^{-9} \text{ C})}{0.120 \text{ m}}$ $U_{E2} = -5.2 \times 10^{-6} \text{ J}$
Third, calculate the electric potential energy between the −10.0 nC particle and the 5.0 nC particle. The distance between the two particles can be found using the formula for the distance between two points.	$d = \sqrt{(0.240 \text{ m})^2 + (0.120 \text{ m})^2}$ $U_{E3} = \dfrac{kQq}{r}$ $U_{E3} = \dfrac{(8.99 \times 10^9 \text{ N} \cdot \text{m}^2/\text{C}^2)(5.0 \times 10^{-9} \text{ C})(-10.0 \times 10^{-9} \text{ C})}{\sqrt{(0.240 \text{ m})^2 + (0.120 \text{ m})^2}}$ $U_{E3} = -1.7 \times 10^{-6} \text{ J}$
Then, sum the results to get the total electric potential energy for the system of charged particles. This is the work required to build the arrangement.	$W = U_{E1} + U_{E2} + U_{E3}$ $W = 1.3 \times 10^{-6} - 5.2 \times 10^{-6} - 1.7 \times 10^{-6} = \boxed{-5.6 \times 10^{-6} \text{ J}}$

CHECK and THINK

The fact that the work required turned out to be negative means that the work will really be done by the electric force between the charges. For example, to move a positive particle nearer to a negative particle, no external force is required; the two particles are attracted to one another.

9. **(A)** Find an expression for the electric potential energy associated with each system in Figure P26.8 in terms of the quantities provided on the figure.

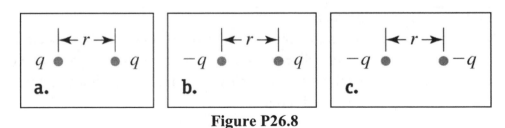

Figure P26.8

INTERPRET and ANTICIPATE

The electric potential energy depends on the two charges and their separations. Starting at zero energy for charges infinitely far apart, two like charges require work to push together and have a potential energy larger than zero. Opposite charges attract each other (they would tend to attract and "fall together" when they are apart) so they have a negative potential energy when close together.

SOLVE Use Equation 26.3 to determine the potential energy for two charges.	$U_E(r) = k\dfrac{Qq}{r}$
a. Two charges of $+q$ are separated by a distance r.	$U_E = \boxed{k\dfrac{q^2}{r}}$
b. Charges of $+q$ and $-q$ are separated by a distance r.	$U_E = k\dfrac{(q)(-q)}{r} = \boxed{-k\dfrac{q^2}{r}}$
c. Charges of $-q$ and $-q$ are separated by a distance r.	$U_E = k\dfrac{(-q)(-q)}{r} = \boxed{k\dfrac{q^2}{r}}$

CHECK and THINK

When both charges are positive or both negative, there is a repulsive force between them and the potential energy is greater than zero while the potential energy for two opposite charges is negative. Since they are the same magnitude charges and separation, the magnitude of the energy is the same in each case.

13. (N) A proton is fired from very far away directly at a fixed particle with charge $q = 1.28 \times 10^{-18}$ C. If the initial speed of the proton is 2.4×10^5 m/s, what is its distance of closest approach to the fixed particle? The mass of a proton is 1.67×10^{-27} kg.

INTERPRET and ANTICIPATE

The proton slows as it approaches the charge until it stops momentarily at its closest distance and is then repelled away. Since the electric force is conservative, we can use conservation of energy to find the distance of closest approach—the initial kinetic energy must equal the potential energy at this point.

SOLVE We start with conservation of energy	$U_i + K_i = U_f + K_f$

for the two charges. Since one is fixed, only the proton has kinetic energy. The proton starts from a great distance away, so the initial potential energy of the two charges is essentially zero. At closest approach, the kinetic energy of the proton is instantaneously zero. Equation 26.3 then can be used to determine the potential energy of the system.	$U_i = 0$ and $K_f = 0$ $K_i = \dfrac{1}{2}mv_i^2 = U_f = U_E = \dfrac{kqe}{r}$
Solve for the distance of closest approach and insert the numerical values.	$r = \dfrac{2kqe}{mv_i^2}$ $r = \dfrac{2(8.99 \times 10^9)(1.60 \times 10^{-19})(1.28 \times 10^{-18})}{(1.67 \times 10^{-27})(2.4 \times 10^5)^2}$ $r = \boxed{3.8 \times 10^{-11} \text{ m}}$

CHECK and THINK

Although the initial speed is quite large, the distance of closest approach is much greater than the size of an atom. This is one indication of why particle accelerators need to produce such high energy particles to be able to penetrate the atom.

14. Four charged particles are at rest at the corners of a square (Fig. P26.14). The net charges are $q_1 = q_2 = 2.65 \ \mu C$ and $q_3 - q_4 - 5.15 \ \mu C$. The distance between particle 1 and particle 3 is $r_{13} = 1.75$ cm.

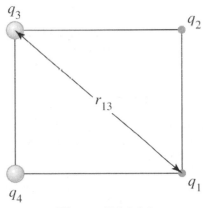

Figure P26.14

a. (N) What is the electric potential energy of the four-particle system?

INTERPRET and ANTICIPATE

The potential energy of the four-particle system is just the total potential energy due to each pair of charges, where the potential energy of two charges is kQq/r. Note that energies are scalars rather than vectors, which means we only need to know the charges and the distances between them (as opposed to a force calculation, where we need to consider their orientation). This is similar to Example 26.2.

SOLVE

First, we note that charges 1 and 2 are the same as are 3 and 4. In addition, since it's a square, the diagonal distances r_{13} (distance between charges 1 and 3) and r_{14} are the same. Finally, using geometry, the diagonal is $\sqrt{2}$ times larger than the length of one of the edges. To see this, we could use the Pythagorean theorem.

$$\left(r_{34}\right)^2 + \left(r_{41}\right)^2 = \left(r_{13}\right)^2$$

and since $r_{34} = r_{43}$,

$$\left(r_{13}\right)^2 = 2\left(r_{34}\right)^2 \text{ and } r_{13} = \sqrt{2}\ r_{34}.$$

$$q_1 = q_2 \qquad q_3 = q_4$$

$$r_{13} = r_{24} \qquad r_{12} = r_{23} = r_{34} = r_{14} = 1.75\ \text{cm}/\sqrt{2}$$

Now, apply Eq. 26.3 for each pair of charges.

$$U_E = k\frac{q_1 q_2}{r_{12}} + k\frac{q_1 q_3}{r_{13}} + k\frac{q_1 q_4}{r_{14}} + k\frac{q_2 q_3}{r_{23}} + k\frac{q_2 q_4}{r_{24}} + k\frac{q_3 q_4}{r_{34}}$$

Use the relationships above to combine terms:

$$U_E = k\left(\frac{q_1^2}{r_{12}} + 2\frac{q_1 q_3}{r_{13}} + 2\frac{q_1 q_3}{r_{12}} + \frac{q_3^2}{r_{12}} \right)$$

Substitute values to solve for a numerical answer:

$$U_E = \left(8.99\times10^9\ \text{N}\cdot\text{m}^2/\text{C}\right)\left(\begin{array}{l} \dfrac{\left(2.65\times10^{-6}\ \text{C}\right)^2}{\left(0.0175\ \text{m}/\sqrt{2}\right)} + 2\dfrac{\left(2.65\times10^{-6}\ \text{C}\right)\left(5.15\times10^{-6}\ \text{C}\right)}{0.0175\ \text{m}} \\[4mm] + 2\dfrac{\left(2.65\times10^{-6}\ \text{C}\right)\left(5.15\times10^{-6}\ \text{C}\right)}{\left(0.0175\ \text{m}/\sqrt{2}\right)} + \dfrac{\left(5.15\times10^{-6}\ \text{C}\right)^2}{\left(0.0175\ \text{m}/\sqrt{2}\right)} \end{array} \right)$$

$$U_E = \boxed{58.2\ \text{J}}$$

CHECK and THINK

The charges are all positive (and therefore the same sign), so it makes sense that the energy is greater than zero. It would take work to push all of these repelling particles together from infinitely far away!

b. (C) If the particles are released from rest, what will happen to the system? In particular, what will happen to the system's kinetic energy as their separations become infinite?

Since the potential energy is greater than zero, they will fly apart leading the potential energy to decrease and the kinetic energy to increase. This is an energy approach to what we also know based on the forces—the charges are repulsive, so they will push each other apart and fly away with some kinetic energy. In the end, if they are infinitely far apart, the potential energy will be zero and the total kinetic energy must be 58.2 J.

19. (N) The speed of an electron moving along the y axis increases from 4.40×10^6 m/s at $y = 10.0$ cm to 7.00×10^6 m/s at $y = 2.00$ cm.
a. What is the electric potential difference between these two points?
b. Which of the two points is at a higher electric potential?

INTERPRET and ANTICIPATE

A charge that moves between two different potentials will gain or lose potential energy. Using energy conservation, we can relate the change in kinetic energy to the change in potential energy and determine the potential difference. A positive charge will accelerate as it falls from high to low potential and a negative charge behaves in the opposite way.

SOLVE	
a. The electron-field system is isolated and the total mechanical energy is conserved. The change in potential energy is related to the change in electric potential according to Eq. 26.7, $\Delta U_E = q\Delta V_E$.	$\Delta K + \Delta U = 0$ $$\frac{1}{2}m\left(v_f^2 - v_i^2\right) + \left(-e\right)\Delta V = 0$$
Now solve for the potential difference.	$$\Delta V = \frac{m\left(v_f^2 - v_i^2\right)}{2e}$$

Chapter 26 – Electric Potential

Now insert numerical values:

$$\Delta V = \frac{\left(9.11\times10^{-31}\ \text{kg}\right)\left[\left(7.00\times10^{6}\ \text{m/s}\right)^{2}-\left(4.40\times10^{6}\ \text{m/s}\right)^{2}\right]}{2\left(1.60\times10^{-19}\ \text{C}\right)}$$

$$\Delta V = \boxed{84.4\ \text{V}}$$

b. A positive charge would "fall" from high to low potential. A negative charge behaves in the opposite way and would accelerate from a low to a high potential. Therefore, $\boxed{y = 2.00\ \text{cm is at the higher potential}}$.

CHECK and THINK

Using energy conservation, we can relate the change in kinetic energy to the electric potential energy (and the potential difference) that a charge experiences.

---

24. Two point charges, $q_1 = -2.0\ \mu\text{C}$ and $q_2 = 2.0\ \mu\text{C}$, are placed on the x axis at $x = 1.0$ m and $x = -1.0$ m, respectively (Fig. P26.24).

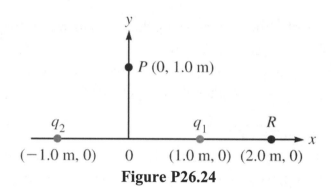

Figure P26.24

a. (N) What are the electric potentials at the points P (0, 1.0 m) and R (2.0 m, 0)?

INTERPRET and ANTICIPATE

The electric potential at a point is the scalar sum of the potentials due to each charge. A positive charge produces a positive potential and a negative charge produces a negative potential. The potential decreases in magnitude for points further from the charge.

SOLVE

At points P and R the potential is the sum of the potentials due to the charges q_1 and q_2, as given by Eq. 26.9. The charges are equal in magnitude but have opposite signs, $q_1 = -q_2$.

$$V = k\frac{q_1}{r_1} + k\frac{q_2}{r_2}$$

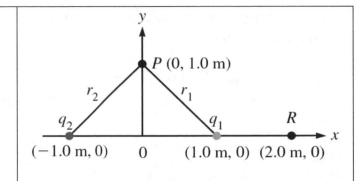

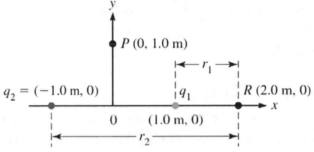

Figure P26.24ANS

Point P is equidistant from the equal and opposite charges, so the total potential is zero.	$V_P = \dfrac{kq}{r} + \dfrac{k(-q)}{r} = 0$ or, performing the calculation with the numbers given, $r_2 = r_1 = \sqrt{(1.0\text{ m})^2 + (1.0\text{ m})^2} = 1.4\text{ m}$ $V_P = \left(8.99 \times 10^9\right)\left(\dfrac{-2.0 \times 10^{-6}}{1.4} + \dfrac{2.0 \times 10^{-6}}{1.4}\right) = \boxed{0}$

For point R, $r_1 = 1.0$ m and $r_2 = 3.0$ m.

$$V_R = \frac{kq_1}{r_1} + \frac{kq_2}{r_2} = \left(8.99 \times 10^9 \text{ N} \cdot \text{m}^2/\text{C}^2\right)\left(\frac{-2.0 \times 10^{-6} \text{ C}}{1.0 \text{ m}} + \frac{2.0 \times 10^{-6} \text{ C}}{3.0 \text{ m}}\right)$$

$$V_R = \boxed{-1.2 \times 10^4 \text{ V}}$$

CHECK and THINK

Determining the potential due to a collection of charges requires knowing the charges and distance from each charge to the point in question. In an example like point P, a point that is equidistant from equal and opposite charges produces a total potential of zero. Point R is closer to the negative charge and therefore the total potential is negative.

b. (N) Find the work done in moving a 1.0-μC charge from P to R along a straight line joining the two points.

INTERPRET and ANTICIPATE	
The work done is equal to the charge times the potential difference, which we determined in part (a).	
SOLVE The work done in carrying a 1.0 μC charge from point P to R is equal to the change in the potential energy (Eq. 26.7).	$W_{P \to R} = qV_{PR} = q\left(V_R - V_P\right)$ $W_{P \to R} = \left(1.0 \times 10^{-6} \text{ C}\right)\left(-1.2 \times 10^4 \text{ V}\right)$ $W_{P \to R} = \boxed{-1.2 \times 10^{-2} \text{ V}}$

CHECK and THINK

Since a positive charge is being moved from a higher to a lower potential (i.e. the direction it "wants" to go), the work is negative. It would pull in that direction and would not require work to move it from P to R.

c. (C) Is there any path along which the work done in moving the charge from P to R is less than the value from part (b)? Explain.

$\boxed{\text{No}}$. The work done in moving a charge between two points (here P and R) depends only on the change in potential between the two points, so it is independent of the path chosen. This is because the electrostatic force is a conservative force.

28. (N) Find the electric potential at the origin given the arrangement of charged particles shown in Figure P26.7.

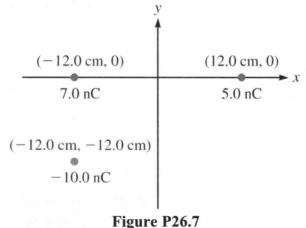

Figure P26.7

INTERPRET and ANTICIPATE

To find the electric potential at the origin, we need to find the electric potential due to each individual charged particle and sum the effects as in Eq. 26.9. The electric potential due to a single charged particle can be positive or negative, depending on the sign of the charge. Here, given the similarity in the magnitude of the charges, the similar distances between the particles, and the fact that two of the three charged particles are positive, we expect that the electric potential at the origin will be positive.

SOLVE	
First, we must find the distance between the −10.0 nC particle and the origin using the formula for the distance between two points.	$r = \sqrt{(0.120 \text{ m})^2 + (0.120 \text{ m})^2}$
Now, we can build Eq. 26.9 using all three charged particles and solve for the electric potential at the origin.	$V = k \sum_{i=1}^{3} \frac{q_i}{r_i}$ $V = (8.99 \times 10^9 \text{ N} \cdot \text{m}^2/\text{C}^2) \left[\frac{-10.0 \times 10^{-9} \text{ C}}{\sqrt{(0.120 \text{ m})^2 + (0.120 \text{ m})^2}} + \frac{5.0 \times 10^{-9} \text{ C}}{0.120 \text{ m}} + \frac{7.0 \times 10^{-9} \text{ C}}{0.120 \text{ m}} \right]$ $V = \boxed{3.7 \times 10^2 \text{ V}}$

CHECK and THINK

The electric potential at the origin is positive, as expected. If we wanted to find the electric potential at a second location, we would need to apply Eq. 26.9 again, using the appropriate distances from each particle to the point of interest. If we were to examine the electric potential very near the negatively charged particle, we would expect to find that the electric potential is negative, since the term corresponding to that particle in Eq. 26.9 will approach negative infinity, as the distance from the negative particle approaches 0.

33. (N) A source consists of three charged particles located at the vertices of a square (Fig. P26.32), where the square has sides of length 0.243 m. The charges are $q_1 = 35.0$ nC, $q_2 = -65.0$ nC, and $q_3 = 56.5$ nC. Find the electric potential at point A located at the fourth vertex.

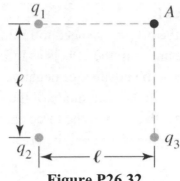

Figure P26.32

INTERPRET and ANTICIPATE	
The electric potential at point A is the sum of the potentials due to each of the three charges. The positive charges will contribute a positive potential while the negative charge will contribute a negative potential. We will essentially solve problem 32 and then insert the numerical values given.	

SOLVE We need to apply Eq. 26.9.	$$V = k\frac{q_1}{r_1} + k\frac{q_2}{r_2} + k\frac{q_3}{r_3}$$
The length of the side of the square is $\ell = 0.243\,\text{m}$. Charges 1 and 3 are each a distance ℓ away and charge 2 is a distance of $\sqrt{2}\,\ell$, the diagonal distance across the square.	$$V = k\frac{q_1}{\ell} + k\frac{q_2}{\sqrt{2}\ell} + k\frac{q_3}{\ell}$$ $$V = \frac{k}{\ell}\left(q_1 + \frac{q_2}{\sqrt{2}} + q_3\right)$$
Now, insert numerical values.	$$V = \frac{8.99\times10^9\,\frac{\text{N·m}^2}{\text{C}^2}}{0.243\,\text{m}}\left(35.0 - \frac{65.0}{\sqrt{2}} + 56.5\right)\times10^{-9}\,\text{C}$$ $$V = \boxed{1.68\times10^3\,\text{V}}$$

CHECK and THINK	
Given the charges and distances from each to the fourth vertex, calculating the total potential is straightforward. There are two positive charges that are closer to the fourth vertex and the further negative charge is smaller than the two positive charges, so it's no surprise that the total potential is positive.	

34. (N) Two identical metal balls of radii 2.50 cm are at a center-to-center distance of 1.00 m from each other (Fig. P26.34). Each ball is charged so that a point at the surface of the first ball has an electric potential of $+1.20 \times 10^3$ V and a point at the surface of the other ball has an electric potential of -1.20×10^3 V. What is the total charge on each ball?

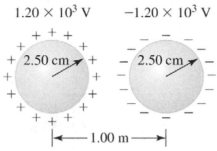

1.20×10^3 V -1.20×10^3 V

2.50 cm   2.50 cm

|← 1.00 m →|

Figure P26.34

INTERPRET and ANTICIPATE First, note that the figure is not to scale. The balls are separated by a large distance compared to their radii. The potential at the location of each sphere is due to its charge as well as the charge on the other sphere, but we expect this contribution to be negligible given their large separation.	
SOLVE The potential on the surface of the left hand sphere is that due to its charge as well as the potential due to the sphere on the right. The spheres are small compared to their separation, so we take the average potential at the sphere on the left due to the sphere in the right to be that due to a charge 1.00 meter away. So, for the positively charged sphere, we write a potential using Eq. 26.9.	$V_+ = \sum \dfrac{kQ}{r} = \dfrac{kQ_+}{0.0250 \text{ m}} + \dfrac{kQ_-}{1 \text{ m}} = 1.20 \times 10^3$ V
Similarly, write the potential for the negatively charged sphere.	$V_- = \sum \dfrac{kQ}{r} = \dfrac{kQ_-}{0.0250 \text{ m}} + \dfrac{kQ_+}{1 \text{ m}} = -1.20 \times 10^3$ V
By symmetry, $Q_+ = Q_-$. We can also show this explicitly by adding these two equations together and solving for this result.	$\dfrac{kQ_+}{0.0250 \text{ m}} + \dfrac{kQ_-}{1 \text{ m}} + \dfrac{kQ_-}{0.0250 \text{ m}} + \dfrac{kQ_+}{1 \text{ m}} = 0$ $41kQ_+ + 41kQ_- = 0$ $Q_- = -Q_+$

Use this relationship in the equation for the potential of the positive sphere from above.	$\dfrac{kQ_+}{0.0250 \text{ m}} - \dfrac{kQ_+}{1 \text{ m}} = 1.20 \times 10^3 \text{ V}$
	$Q_+ = \dfrac{1.20 \times 10^3 \text{ V}}{k\left(\dfrac{1}{0.0250 \text{ m}} - \dfrac{1}{1 \text{ m}}\right)}$
	$Q_+ = \dfrac{1.20 \times 10^3 \text{ V}}{\left(8.99 \times 10^9 \text{ N} \cdot \text{m}^2 / \text{C}\right)\left(39.0 \text{ m}^{-1}\right)}$
	$Q_+ = \boxed{3.42 \times 10^{-9} \text{C}} = 3.42 \text{ nC}$

CHECK and THINK

Indeed, the effect due to the other sphere is small compared to the potential due to the charge on the sphere itself. In fact, if we only had the positively charged sphere and ignored the other completely, then we would have had

$$\frac{kQ_+}{0.0250 \text{ m}} = 1.20 \times 10^3 \text{ V} \quad \rightarrow \quad Q_+ = \frac{(0.0250)(1.20 \times 10^3)}{8.99 \times 10^9} \text{C} = 3.34 \text{ C}$$

which differs by only a few percent.

37. (N) Two charged particles with $q_1 = 5.00 \ \mu C$ and $q_2 = -3.00 \ \mu C$ are placed at two vertices of an equilateral tetrahedron whose edges all have length $s = 4.20$ m (Fig. P26.37). Determine what charge q_3 should be placed at the third vertex so that the total electric potential at the fourth vertex is 2.00 kV.

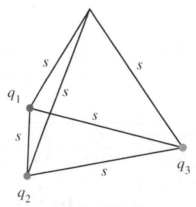

Figure P26.37

INTERPRET and ANTICIPATE

Since potential is a scalar, we can simply add the individual contributions to get the net potential. We also note that all three charges are the same distance from the point at which we want to produce the potential.

SOLVE	
Use Equation 26.8 to find the potential of each charge. Add them to get the desired 2000 V total.	$V = \dfrac{kq_1}{s} + \dfrac{kq_2}{s} + \dfrac{kq_3}{s}$ $V = \dfrac{k}{s}\left(q_1 + q_2 + q_3\right)$
Now solve for the third charge.	$q_3 = \dfrac{sV}{k} - q_1 - q_2$
Finally, insert numerical values.	$q_3 = \dfrac{(4.20)(2000)}{8.99\times 10^9} - 5.00\times 10^{-6} + 3.00\times 10^{-6}$ $q_3 = \boxed{-1.07\times 10^{-6}\ \text{C}} = -1.07\ \mu C$

CHECK and THINK

The third charge is comparable in magnitude to the first two. Although this is a three-dimensional problem, the symmetry of the tetrahedron and the scalar nature of the potential make it relatively easy to solve.

40. (N) A uniformly charged ring with total charge $q = 3.00\ \mu C$ and radius $R = 10.0$ cm is placed with its center at the origin and oriented in the xy plane. What is the difference between the electric potential at the origin and the electric potential at the point $(0, 0, 30.0$ cm$)$?

INTERPRET and ANTICIPATE	
While this is an extended charged object, all parts of the ring are the same distance to the point in question, so we can determine the total potential without needing to perform an integral. We expect the potential to be positive since the charge on the ring is positive, but larger at the center of the ring, so the difference will be negative in moving to a point further from the ring.	

SOLVE	
Since all parts of the ring are the same distance from the point in question in both cases, the total potential is simply Eq. 26.8, where r is	$R_1 = \sqrt{(0.100\ \text{m})^2 + (0.300\ \text{m})^2} = 0.316\ \text{m}$ $R_0 = R$

the distance from each part of the ring to the point in space. First, we determine the distance from any point on the ring to (0, 0, 30.0 cm). At the origin, the distance is simply the radius of the ring $R_0 = R$.	
Now, determine the difference in potential between the two points.	$\Delta V = V_1 - V_0 = \dfrac{kq}{R_1} - \dfrac{kq}{R_0}$
Finally, insert numerical values.	$\Delta V = \left(8.99 \times 10^9 \ \frac{\text{N·m}^2}{\text{C}^2}\right)\left(3 \times 10^{-6} \ \text{C}\right)^2 \left(\dfrac{1}{0.316 \ \text{m}} - \dfrac{1}{0.100 \ \text{m}}\right)$ $\Delta V = \boxed{-1.84 \times 10^5 \ \text{V}}$

CHECK and THINK

Since the potential depends only on the distance of a charge from the point in space, even an extended object is straightforward when the point is equidistant from all parts of the object, as will points along the axis of a ring.

44. (N) Figure P26.44 shows a rod of length $\ell = 1.00$ m aligned with the y axis and oriented so that its lower end is at the origin. The charge density on the rod is given by $\lambda = a + by$, with $a = 2.00 \ \mu\text{C/m}$ and $b = -1.00 \ \mu\text{C/m}^2$. What is the electric potential at point P with coordinates (0, 25.0 cm)? A table of integrals will aid you in solving this problem.

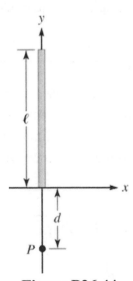

Figure P26.44

Chapter 26 – Electric Potential

INTERPRET and ANTICIPATE Since this is an extended object, we need to integrate over the charge elements dq to add up the potential due to each piece, $k\,dq/r$.	

SOLVE

We can apply Eq. 26.12 to integrate over the charge elements $dq = \lambda\,dx$ located at a distance $d + y$ from point P.

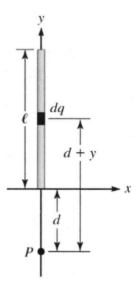

Figure P26.44ANS

Integrate from a distance d to $d+\ell$. That is, integrate $d + y$ from $y = 0$ to ℓ.

$$V = k\int \frac{dq}{r} = k\int \frac{\lambda dx}{r} = k\int_0^\ell \frac{(a+by)\,dy}{d+y}$$

To perform the integral, we write it in two separate terms. A table of integrals is helpful since the integral is not straightforward.

$$V = k\int_0^\ell \left[\frac{a}{d+y} + \frac{by}{d+y}\right]dy$$

$$V = ka\ln(d+y)\Big|_0^\ell + kb\left[y + d\ln\left(\frac{y}{d}+1\right)\right]_0^\ell$$

$$V = ka\ln\left(\frac{d+\ell}{d}\right) + kb\ell + kbd\ln\left(\frac{\ell}{d}+1\right)$$

$$V = k\left[b\ell + (a-bd)\ln\left(1+\frac{\ell}{d}\right)\right]$$

Insert numerical values and solve.

$$V = (8.99\times10^9)\left[(-1.00\times10^{-6})(1.00) + (2.00\times10^{-6} - (0.250)(-1.00\times10^{-6}))\ln\left(1+\frac{1.00}{0.250}\right)\right]$$

$$V = \boxed{2.36\times10^4 \text{ V}}$$

Chapter 26 – Electric Potential

48. (N) A particle with charge 1.60×10^{-19} C enters midway between two charged plates, one positive and the other negative. The initial velocity of the particle is parallel to the plates and along the midline between them (Fig. P26.48). A potential difference of 300.0 V is maintained between the two charged plates. If the lengths of the plates are 10.0 cm and they are separated by 2.00 cm, find the greatest initial velocity for which the particle will not be able to exit the region between the plates. The mass of the particle is 12.0×10^{-24} kg.

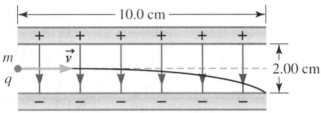

Figure P26.48

INTERPRET and ANTICIPATE
The electric field in the region between the plates is constant. As the particle fails to come out from the plates, the greatest velocity for the charge to not exit (or in other words the minimum speed for which the charge *will* make it out) occurs when the maximum deflection from the middle of the plats is 1.0 cm. We just need to determine the conditions for this maximum displacement.

SOLVE	
The electric field in the region between the plates is constant. It is directed from the positive plate to the negative plate. We find the electric field using Equation 26.18.	$$E = \frac{\Delta V}{\Delta x} = \frac{300 \text{ V}}{0.020 \text{ m}} = 1.5 \times 10^4 \text{ V/m}$$
The only force on the particle is that due to the electric field, $F = qE$, so we can relate the particle's downward acceleration to the electric field.	$$a = \frac{F}{m} = \frac{qE}{m}$$

100

We use kinematics to find the deflection downward. As the particle just fails to come out, the maximum deflection is $y = 0.010$ m. Since there is no initial vertical velocity, $$y = \frac{1}{2}at^2 .$$ In the horizontal direction, the range is $x = 0.10$ m and the particle travels with a constant horizontal velocity v_{x0}. The time of flight of the particle is t.	$$y = \frac{1}{2}at^2 \quad \text{and} \quad x = v_{x0}t$$ $$y = \frac{1}{2}\frac{qE}{m}\left(\frac{x}{v_{x0}}\right)^2$$
Solve for the velocity.	$$v^2 = \frac{1}{2}\frac{qE}{my}x^2$$ $$v = \sqrt{\frac{1}{2}\frac{qE}{my}x^2}$$
Insert numerical values and solve.	$$v = \sqrt{\frac{1}{2}\frac{\left(1.6\times10^{-19}\text{ C}\right)\left(1.5\times10^4\text{ V/m}\right)}{\left(12\times10^{-24}\text{ kg}\right)\left(0.010\text{ m}\right)}\left(0.10\text{ m}\right)^2}$$ $$v = \boxed{1.0\times10^4\text{ m/s}}$$

CHECK and THINK
Charges with speed less than 10,000 m/s will not make it out of the region between the plates. As with trajectory problems, the time for the charge to accelerate and reach the plate is the same, but a faster horizontal velocity would allow the charge to exit this region before crashing into the plate.

53. (N) The function $V = b + cy$ describes the electric potential in the region between $y = -4.00$ m and $y = 4.00$ m, with $b = 12.5$ V and $c = 4.50$ V/m.
a. What is the electric potential at $y = -4.00$ m, at $y = 0$, and at $y = 4.00$ m?
b. What are the magnitude and direction of the electric field at $y = -4.00$ m, at $y = 0$, and at $y = 4.00$ m?

INTERPRET and ANTICIPATE
Given the expression for the potential, we can solve for its value at the three points in question. The electric field is given by the negative of the slope of the potential and points from high to low potential.

Chapter 26 – Electric Potential

SOLVE **a.** To determine the potential, we simply plug in the three y values and solve.	$V(y) = b + cy = 12.5 \text{ V} + (4.50 \text{ V/m})y$ $V(-4.00 \text{ m}) = 12.5 \text{ V} + (4.50 \text{ V/m})(-4.00 \text{ m}) = \boxed{-5.50 \text{ V}}$ $V(0) = 12.5 \text{ V} + (4.50 \text{ V/m})(-4.00 \text{ m}) = \boxed{12.5 \text{ V}}$ $V(+4.00 \text{ m}) = 12.5 \text{ V} + (4.50 \text{ V/m})(-4.00 \text{ m}) = \boxed{30.5 \text{ V}}$
b. The electric field depends on the slope of the potential (Eq. 26.21). The slope is constant, so the electric field points in the negative direction in all three cases.	$E = -\dfrac{dV}{dy} = -c$ $E = -(4.50 \text{ V/m}) = \boxed{4.50 \text{ V/m in the } -y \text{ direction}}$

CHECK and THINK

As anticipated, the electric field points from high to low potential, in this case in the $-y$ direction.

56. (N) The electric potential $V(x, y, z)$ in a region of space is given by $V(x, y, z) = V_0(2x^2 - 3y^2 - z^2)$, where $V_0 = 12.0$ V and $x, y,$ and z are measured in meters. Find the electric field at the point (1.00 m, 1.00 m, 0).

INTERPRET and ANTICIPATE The components of the electric field are given by the negative of the slope of the potential in the $x, y,$ and z directions.	
SOLVE To calculate the electric field, apply Eq. 26.21.	$E_x = -\dfrac{\partial V}{\partial x}, \quad E_y = -\dfrac{\partial V}{\partial y}, \quad E_z = -\dfrac{\partial V}{\partial z}$

Calculate each derivatives.	$E_x = -\dfrac{\partial}{\partial x}\left[V_0\left(2x^2-3y^2-z^2\right)\right] = -4V_0\,x$
	$E_y = -\dfrac{\partial}{\partial y}\left[V_0\left(2x^2-3y^2-z^2\right)\right] = 6V_0\,y$
	$E_z = -\dfrac{\partial}{\partial z}\left[V_0\left(2x^2-3y^2-z^2\right)\right] = 2V_0\,z$
We can express the electric field as a vector using these components.	$\vec{E} = -4V_0\,x\,\hat{\boldsymbol{i}} + 6V_0\,y\,\hat{\boldsymbol{j}} + 2V_0\,z\,\hat{\boldsymbol{k}}$
Finally, plug in numbers to compute a value.	$\vec{E} = -4\left(12.0\text{ V}\right)\left(1.00\text{ m}\right)\hat{\boldsymbol{i}} + 6\left(12.0\text{ V}\right)\left(1.00\text{ m}\right)\hat{\boldsymbol{j}} + 2\left(12.0\text{ V}\right)\left(0\right)\hat{\boldsymbol{k}}$ $\vec{E} = \boxed{\left(-48.0\,\hat{\boldsymbol{i}} + 72.0\,\hat{\boldsymbol{j}}\right)\text{ N/C}}$

CHECK and THINK

The electric field points along the gradient of the potential. Given the potential, we can use the x, y, and z derivatives to determine the electric field, which points along the "downhill" direction along which the potential decreases.

63. (N) A glass sphere with radius 4.00 mm, mass 85.0 g, and total charge 4.00 μC is separated by 150.0 cm from a second glass sphere 2.00 mm in radius, with mass 300.0 g and total charge –5.00 μC. The charge distribution on both spheres is uniform. If the spheres are released from rest, what is the speed of each sphere the instant before they collide?

INTERPRET and ANTICIPATE

Since the spheres are oppositely charged, they attract and will eventually collide. We can determine the initial and final potential energy of the charges and equate this to the final kinetic energy using energy conservation. However, this won't tell us how much kinetic energy (and therefore the speed) of each of the spheres—only the total. The other piece of information we need to use is that momentum is also conserved. This will allow us to relate the final speed of one sphere with the other.

SOLVE	
Consider the two spheres as a system. First, apply conservation of energy (with Eq. 26.3 for the potential energy). Their initial separation is d and final separation	$KE_i + PE_i = KE_f + PE_f$ $0 + \dfrac{k\left(-q_1\right)q_2}{d} = \dfrac{1}{2}m_1v_1^2 + \dfrac{1}{2}m_2v_2^2 + \dfrac{k\left(-q_1\right)q_2}{r_1+r_2}$ (1)

just as they collide is the sum of their radii $r_1 + r_2$. Initially, they are at rest, so the initial kinetic energy is zero. We take m_1 to represent the 85 g sphere and m_2 to be the 300 g sphere. *Note*: in writing (1) this in this way, we have already acknowledged one of the electric charges is negative and included that sign. This means that when we substitute for the charges later, we should substitute their magnitudes.	
Now apply conservation of momentum.	$$0 = m_1 v_1 \hat{i} + m_2 v_2 \left(-\hat{i}\right)$$ $$v_2 = \frac{m_1 v_1}{m_2} \qquad (2)$$
Plug equation (2) into (1) and solve for v_1.	$$\frac{kq_1q_2}{r_1+r_2} - \frac{kq_1q_2}{d} = \frac{1}{2}m_1 v_1^2 + \frac{1}{2}\frac{m_1^2 v_1^2}{m_2}$$ $$kq_1q_2\left(\frac{1}{r_1+r_2} - \frac{1}{d}\right) = \frac{1}{2}\frac{m_1 m_2}{m_2}v_1^2 + \frac{1}{2}\frac{m_1^2 v_1^2}{m_2}$$ $$kq_1q_2\left(\frac{1}{r_1+r_2} - \frac{1}{d}\right) = \left(\frac{m_1\left(m_2+m_1\right)}{2m_2}\right)v_1^2$$ $$v_1 = \sqrt{\frac{2m_2 k_e q_1 q_2}{m_1\left(m_1+m_2\right)}\left(\frac{1}{r_1+r_2} - \frac{1}{d}\right)}$$

Insert numerical values to determine v_1.

$$v_1 = \sqrt{\frac{2(0.300 \text{ kg})(8.99\times10^9 \text{ N}\cdot\text{m}^2/\text{C}^2)(4.00\times10^{-6} \text{ C})(5.00\times10^{-6} \text{ C})}{(0.085 \text{ kg})(0.385 \text{ kg})}\left(\frac{1}{6.00\times10^{-3} \text{ m}} - \frac{1}{1.50 \text{ m}}\right)}$$

$$v_1 = \boxed{23.4 \text{ m/s}}$$

Chapter 26 – Electric Potential

And using equation (2), solve for v_2.	$v_2 = \dfrac{m_1 v_1}{m_2} = \dfrac{0.085 \text{ kg}(23.4 \text{ m/s})}{0.300 \text{ kg}} = \boxed{6.63 \text{ m/s}}$

CHECK and THINK
Note that while energy conservation provides the total final kinetic energy, the momentum conservation is needed. Since the forces between the charged spheres are equal in magnitude, we should not be surprised to see that the heavier sphere (which would have a smaller acceleration given the same force) ends up traveling slower than the lighter sphere.

66. (N) A 5.00-nC charged particle is at point B in a uniform electric field with a magnitude of 625 N/C (Fig. P26.65). What is the change in electric potential experienced by the charge if it is moved from B to A along
a. path 1 and
b. path 2?

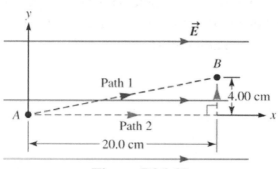

Figure P26.65

INTERPRET and ANTICIPATE
From the special case of planar symmetry, or a region of uniform electric field, the change in electric potential is given by Eq. 26.17, $\Delta V = -E\Delta x$, where Δx is the component of the displacement between two points in the field that is parallel to the field. This displacement component is positive when in the direction of the field and negative when it opposes the field. This displacement component is the same for both paths described, -20.0 cm, so the answers to both (a) and (b) should be the same. The change in potential should turn out to be positive because the electric potential should increase as the particle is moved against the direction of the electric field.

SOLVE	
a. When being moved from B to A, the only component of the displacement that matters is the component parallel to the electric field. In this case, that is -20.0 cm or	$\Delta V = -E\Delta x = -(625 \text{ N/C})(-0.200 \text{ m}) = \boxed{125 \text{ V}}$

0.200 m. The component is negative because it is directed opposite the electric field. Use Eq. 26.17 to calculate the change in electric potential.	
b. Regardless of the path taken from B to A, the parallel component of the displacement is the same.	$\Delta V = -E\Delta x = -(625 \text{ N/C})(-0.200 \text{ m}) = \boxed{125 \text{ V}}$

CHECK and THINK

Electric potential increases as one moves against the direction of an electric field. The difference in electric potential is independent of the electric charge that is being moved; it can be thought of as a property of the space in which the field resides.

68. (N) Figure P26.68 shows three small spheres with identical charges of –3.00 nC placed at the vertices of an equilateral triangle with side $d = 2.50$ cm.

a. Is the electric potential due to the three spheres zero anywhere in the plane that contains the triangle, other than at infinity?

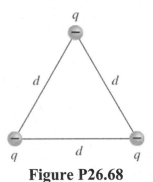

Figure P26.68

Each sphere separately produces a negative potential everywhere. The total potential produced by the three spheres together is then the sum of three negative terms. There is therefore no point at which this total potential is zero (other than infinitely far away, where all three terms approach zero).

b. What is the electric potential at the location of each sphere due to the other two spheres?

INTERPRET and ANTICIPATE

The electric potential is given as the sum of the contributions kq/r from each of the three charges.

SOLVE	
Apply Eq. 26.9. The potential at the location of each charge is due to the other two spheres, so there are two terms in the sum.	$V = \dfrac{kq}{d} + \dfrac{kq}{d} = \dfrac{2kq}{d}$ $V = \dfrac{2\left(8.99 \times 10^9 \ \text{N} \cdot \text{m}^2 / \text{C}^2\right)\left(-3.00 \times 10^{-9} \ \text{C}\right)}{0.0250 \ \text{m}}$ $V = \boxed{-2.16 \times 10^3 \ \text{V}}$

CHECK and THINK
The potential is negative as anticipated in part (a).

75. A long thin wire is used in laser printers to charge the photoreceptor before exposure to light. This is done by applying a large potential difference between the wire and the photoreceptor.

a. (A) Use Equation 26.23,

$$V(r) = \frac{\lambda}{2\pi\varepsilon_0} \ln \frac{R}{r}$$

to determine a relationship between the electric potential V and the magnitude of the electric field E at a distance r from the center of the wire of radius R ($r > R$).

INTERPRET and ANTICIPATE
In Chapter 25 we found an expression for the electric field outside a very long line of charge. We can use this with the expression for potential to relate E to V.

SOLVE	
a. Use the equation for the electric field due to a long wire. We can look this up in Chapter 25, though we can also use Eq. 26.22 since we are given the potential:	$\vec{E} = \dfrac{\lambda}{2\pi\varepsilon_0 r}\hat{r} \quad \rightarrow \quad E_r = \dfrac{\lambda}{2\pi\varepsilon_0 r}$ $\dfrac{\lambda}{2\pi\varepsilon_0} = Er$
$E_r = -\dfrac{\partial V}{\partial r}$ $E_r = -\dfrac{\partial}{\partial r}\left[\dfrac{\lambda}{2\pi\varepsilon_0}\left(\ln R - \ln r\right)\right]$ $E_r = \dfrac{\lambda}{2\pi\varepsilon_0}\dfrac{1}{r}$	

Combining this with the equation given for potential, we can solve for V in terms of E.	$$V(r) = \frac{\lambda}{2\pi\varepsilon_0}\ln\frac{R}{r} = \boxed{Er\ln\frac{R}{r}}$$

CHECK and THINK

This relationship between the magnitude of E and the potential V is valid for points outside the wire, i.e., for $r > R$. We also note that the choice of reference for potential sets $V = 0$ at the surface of the wire.

b. (N) Determine the electric potential at a distance of 2.0 mm from the surface of a wire of radius $R = 0.80$ mm that will produce an electric field of 1.8×10^6 V/m at that point.

INTERPRET and ANTICIPATE

Use the result from part (a).

SOLVE Put in the values given to find the potential. Note that $r = R + 2.0$ mm $= 2.8$ mm.	$$V = \left(1.8\times10^6\right)\left(2.80\times10^{-3}\right)\ln\frac{0.80}{2.0+0.80}$$ $$V = \boxed{-6.3\times10^3 \text{ V}}$$

CHECK and THINK

This negative potential corresponds to a decrease from the value of zero at the surface of the wire and therefore the electric field points away from the wire.

78. (N) An infinite number of charges with $q = 2.0$ μC are placed along the x axis at $x = 1.0$ m, $x = 2.0$ m, $x = 4.0$ m, $x = 8.0$ m, and so on, as shown in Figure P26.78. Determine the electric potential at the point $x = 0$ due to this set of charges. *Hint*: Use the mathematical formula for a geometric series,

$$1 + r + r^2 + r^3 + r^4 + \cdots = \frac{1}{1-r}$$

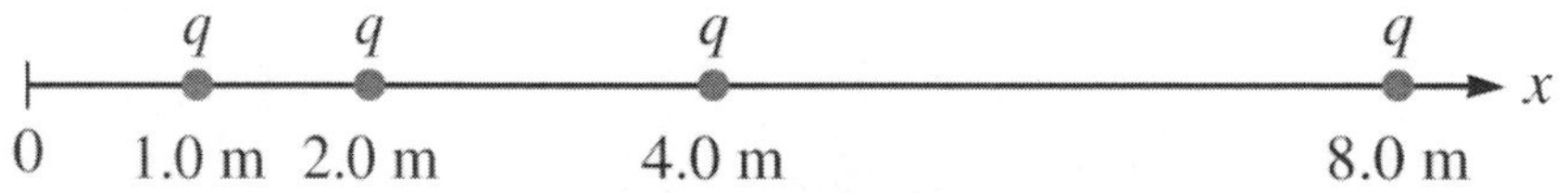

Figure P26.78

INTERPRET and ANTICIPATE

We need to determine the electric potential at $x = 0$ using Equation 26.9. By writing out this sum for an infinite number of charges, we expect to need the geometric series to simplify it.

SOLVE	
The electric potential at $x = 0$ is the sum of the potentials due to the infinite array of charges. Write out the first terms in the series given by Eq. 26.9. Each charge has the same value with distances from the origin that increase as 1, 2, 4, 8, …	$$V(x=0) = k\frac{q}{1} + k\frac{q}{2} + k\frac{q}{4} + k\frac{q}{8} + \cdots$$
This is the geometric series given where $r = 1/2$.	$$V(x=0) = kq\left[1 + \left(\tfrac{1}{2}\right) + \left(\tfrac{1}{2}\right)^2 + \left(\tfrac{1}{2}\right)^3 + \ldots\right]$$ $$V(x=0) = kq\frac{1}{1-\tfrac{1}{2}} = \boxed{2kq}$$

CHECK and THINK

In this case, the potential can be easily be put into the form of the geometric series given to determine the potential at the origin.

27

Capacitors and Batteries

9. (N) A 4.50-μF capacitor is connected to a battery for a long time.
a. If the voltage of the battery is 9.00 V, how much energy is stored in the capacitor?
b. If the voltage of the battery is increased to 24.0 V, how much energy is now stored in the capacitor?

INTERPRET and ANTICIPATE

When a capacitor is connected to a voltage source, it stores charge and energy (electric potential energy equal to the work done in separating the charges). Given the capacitance and voltage, we can determine the energy stored and we expect a larger voltage battery to cause more charge and more energy to be stored.

SOLVE **a.** We can apply Equation 27.3 in both cases.	$U_E = \dfrac{1}{2}CV_C^2$ $U_E = \dfrac{1}{2}\left(4.50\times10^{-6}\text{ F}\right)\left(9.00\text{ V}\right)^2$ $U_E = \boxed{1.82\times10^{-4}\text{ J}}$
b. Now repeat the calculation for the larger voltage.	$U_E = \dfrac{1}{2}CV_C^2$ $U_E = \dfrac{1}{2}\left(4.50\times10^{-6}\text{ F}\right)\left(24.0\text{ V}\right)^2$ $U_E = \boxed{1.30\times10^{-3}\text{ J}}$

CHECK and THINK

As expected, the larger voltage battery leads to an increase in stored energy. Since the energy depends on the square of the voltage, the voltage that is around 2.7 times larger stores about 7 times more energy.

---

12. (N) A capacitor stores 37.5 mJ when the potential difference between its plates is 16.0 V. Then the potential difference is decreased to 8.00 V.
a. After the potential difference is reduced, what is the charge stored by the capacitor?
b. What is its final capacitance?

INTERPRET and ANTICIPATE
The capacitance depends only on the geometry of the capacitor and will not change. With a given capacitance, the charge stored is directly proportional to the voltage across the capacitor.

SOLVE

a. Equation 27.3 relates the energy stored to the capacitance and voltage. Given the energy with 16 V applied, we can determine the capacitance.

$$U_E = \frac{1}{2}CV_C^2 \quad \rightarrow \quad C = \frac{2U_E}{V_C^2}$$

Insert values.

$$C = \frac{2(0.0375\,\text{J})}{(16.0\,\text{V})^2} = 2.93\times10^{-4}\,\text{F}$$

The charge stored depends on the voltage, so it is cut in half (Eq. 27.1).

$$Q = CV$$
$$Q = (2.93\times10^{-4}\,\text{F})(8.00\,\text{V})$$
$$Q = \boxed{2.34\times10^{3}\,\text{C}}$$

b. The capacitance depends on geometry only and therefore does not change when connected to a different voltage source.

$$C = \boxed{2.93\times10^{-4}\,\text{F}}$$

CHECK and THINK
The energy and charge depend on the capacitance and voltage. The capacitance depends on the geometry of the capacitor only and does not change.

16. (N) A 6.50-μF capacitor is connected to a battery. What is the charge on each plate of the capacitor if the voltage of the battery is
a. 10.0 V and
b. 2.00 V?

INTERPRET and ANTICIPATE
The charge on the capacitor depends on the product of the capacitance and voltage.

SOLVE **a.** Apply Equation 27.1 in both cases.	$Q = C\Delta V$ $Q = \left(6.50 \times 10^{-6}\ \text{F}\right)\left(10.0\ \text{V}\right)$ $Q = 6.50 \times 10^{-5}\ \text{C} = \boxed{65.0\ \mu C}$
b.	$Q = C\Delta V$ $Q = \left(6.50 \times 10^{-6}\ \text{F}\right)\left(2.00\ \text{V}\right)$ $Q = 1.30 \times 10^{-5}\ \text{C} = \boxed{13.0\ \mu C}$

CHECK and THINK

In the second case, the voltage is five times smaller, so the charge stored is five times smaller.

19. (N) Two capacitors have capacitances of 6.0 μF and 3.0 μF. Initially, the first one has a potential difference of 18.0 V and the second one 9.0 V. If the capacitors are now connected in parallel, what is their common potential difference?

INTERPRET and ANTICIPATE
The two capacitors are initially charged. When they are connected in parallel, the charges get re-distributed, but are conserved. Since the capacitors are connected in parallel, they will end up with the same potential difference.

SOLVE Using Equation 27.1, we first find the charges Q_1 and Q_2 of the two capacitors.	$Q = CV_C$ $Q_1 = C_1V_1 = \left(6.0 \times 10^{-6}\ \text{F}\right)\left(18\ \text{V}\right)$ $Q_2 = C_2V_2 = \left(3.0 \times 10^{-6}\ \text{F}\right)\left(9.0\ \text{V}\right)$
Assuming they are joined such that the positively charged plates are connected together, the total charge is the sum of the charges on each.	 **Figure P27.19ANS**

	$Q = Q_1 + Q_2$ $Q = \left(6.0 \times 10^{-6}\ \text{F}\right)\left(18\ \text{V}\right) + \left(3.0 \times 10^{-6}\ \text{F}\right)\left(9.0\ \text{V}\right)$
We can now think of this combination as the equivalent capacitance of the two capacitors with charge Q. We can calculate the equivalent capacitance with Eq. 27.8.	$C_{eq} = C_1 + C_2$ $C_{eq} = 6.0\ \mu F + 3.0\ \mu F$ $C_{eq} = 9.0\ \mu F$
Now, apply Eq. 27.1 again. It is also possible to calculate this answer knowing that the voltage across each must be the same since they are in parallel, $V_1 = V_2$.	$V = \dfrac{Q}{C_{eq}}$ $V = \dfrac{\left(6.0 \times 10^{-6}\ \text{F}\right)\left(18\ \text{V}\right) + \left(3.0 \times 10^{-6}\ \text{F}\right)\left(9.0\ \text{V}\right)}{9.0 \times 10^{-6}\ \text{F}}$ $V = \boxed{15\ \text{V}}$

CHECK and THINK

Notice to find the common potential we can use the net charge and the equivalent capacitance of the two parallel conductors.

21. (N) Calculate the equivalent capacitance between points a and b for each of the two networks shown in Figure P27.21. Each capacitor has a capacitance of 1.00 μF.

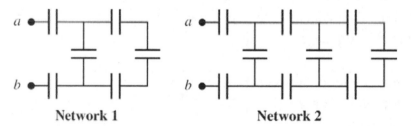

Network 1   Network 2

Figure P27.21

INTERPRET and ANTICIPATE

To find the equivalent capacitance, we need to find sets of capacitors that are either in series or in parallel and replace them with the appropriate equivalent capacitance using Equations 27.7 and 27.8. We'll continue using these rules until we hopefully end up with a single equivalent capacitance.

SOLVE Network 1:	 **Figure P27.21aANS**
In loop ABCD three capacitors are in series. The equivalent capacitance of those three is $C/3$ (Eq. 27.7).	$$\frac{1}{C_{eq}} = \frac{1}{C} + \frac{1}{C} + \frac{1}{C} = \frac{3}{C} \quad \rightarrow \quad C_{eq} = \frac{C}{3}$$
Then in the remaining loop, between points A and D, C and $C/3$ are in parallel. Their equivalent capacitance is $4C/3$ (Eq. 27.8).	$$C_{eq} = C + \frac{C}{3} = \frac{4C}{3}$$
In the final loop between points a and b there are three capacitors in series: C, C, and $4C/3$. Their equivalent capacitance is $(1/C+1/C+3/4C)^{-1} = 4C/11$. This is the equivalent capacitance of the entire network. Each capacitor is $1.00\ \mu F$.	$$\frac{1}{C_{eq}} = \frac{1}{C} + \frac{1}{C} + \frac{3}{4C} = \frac{11}{4C} \quad \rightarrow \quad C_{eq} = \frac{4C}{11}$$ $$C_{eq} = \frac{4C}{11} = \frac{4}{11}(1.00\ \mu F) = \boxed{0.364\ \mu F}$$

Network 2:

The only difference with this network is that there is an extra loop on the end, but the basic procedure is the same. The first three steps are the same: In loop BCDE three capacitors are in series. The equivalent capacitance of those three is $C/3$. Then in the subsequent loop, C and $C/3$ are in parallel. Their equivalent capacitance is $4C/3$. In loop ABEF, three capacitors are in series. The equivalent capacitance of those three is $4C/11$.

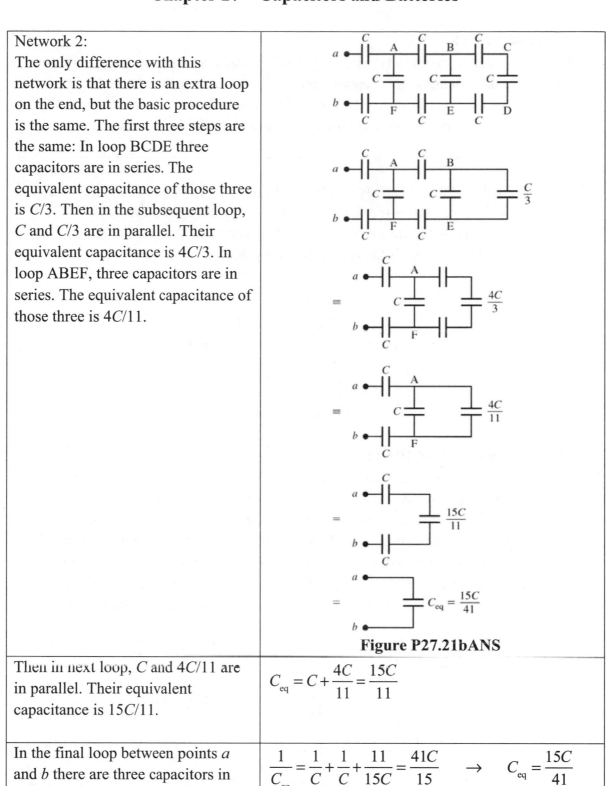

Figure P27.21bANS

Then in next loop, C and $4C/11$ are in parallel. Their equivalent capacitance is $15C/11$.	$$C_{eq} = C + \frac{4C}{11} = \frac{15C}{11}$$

In the final loop between points a and b there are three capacitors in series: C, C, and $15C/11$. Their equivalent capacitance is $(1/C+1/C+11/15C)^{-1} = 15C/41$.	$$\frac{1}{C_{eq}} = \frac{1}{C} + \frac{1}{C} + \frac{11}{15C} = \frac{41C}{15} \quad \rightarrow \quad C_{eq} = \frac{15C}{41}$$ $$C_{eq} = \frac{15C}{41} = \frac{15}{41}(1.00 \ \mu F) = \boxed{0.366 \ \mu F}$$

115

Chapter 27 – Capacitors and Batteries

24. (N) An arrangement of capacitors is shown in Figure P27.23.
a. If $C = 9.70 \times 10^{-5}$ F, what is the equivalent capacitance between points a and b?
b. A battery with a potential difference of 12.00 V is connected to a capacitor with the equivalent capacitance. What is the energy stored by this capacitor?

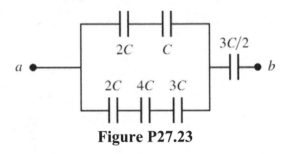

Figure P27.23

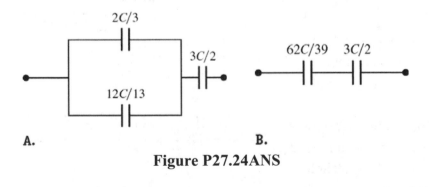

A.   **B.**
Figure P27.24ANS

SOLVE **a.** We first find the equivalent capacitances in the upper and lower branches of the parallel portion of the circuit. There are capacitors in series in the upper and lower branch, so we use Eq. 27.7 and reduce the circuit, as shown in Figure P27.24ANS(A).	$$\frac{1}{C_{up}} = \frac{1}{2C} + \frac{1}{C} = \frac{3}{2C}$$ $$C_{up} = 2C/3$$ $$\frac{1}{C_{down}} = \frac{1}{2C} + \frac{1}{4C} + \frac{1}{3C} = \frac{6}{12C} + \frac{3}{12C} + \frac{4}{12C} = \frac{13}{12C}$$ $$C_{down} = 12C/13$$
Now, the upper and lower branches are in parallel, so use Eq. 27.8 to find the equivalent capacitance of the parallel part of the circuit. This redrawing of the circuit is now shown in Figure P27.24ANS(B).	$$C_{par} = C_{up} + C_{down} = 2C/3 + 12C/13$$ $$C_{par} = 26C/39 + 36C/39$$ $$C_{par} = 62C/39$$
Finally, that result is in series with the last remaining capacitor, so we use Eq. 27.7 again to find the equivalent capacitance of the original circuit.	$$\frac{1}{C_{eq}} = \frac{1}{(62C/39)} + \frac{1}{(3C/2)} = \frac{39}{62C} + \frac{2}{3C}$$ $$\frac{1}{C_{eq}} = \frac{117}{186C} + \frac{124}{186C} = \frac{241}{186C}$$ $$C_{eq} = 186C/241$$
Now that we know the equivalent capacitance, algebraically, we can substitute the known value for C and find the equivalent capacitance numerically.	$$C_{eq} = 186C/241 = (186/241)(9.70 \times 10^{-5} \text{ F})$$ $$C_{eq} = \boxed{7.49 \times 10^{-5} \text{ F}}$$
b. Use Eq. 27.3 to find the energy stored.	$$U_E = \frac{1}{2}CV_C^2 = \frac{1}{2}(7.49 \times 10^{-5} \text{ F})(12.00 \text{ V})^2$$ $$U_E = \boxed{5.39 \times 10^{-3} \text{ J}}$$

CHECK and THINK

We cannot skip steps when reducing a circuit! We must identify combinations of capacitors that are in series, or in parallel, and reduce that portion alone in order to then see how the remaining connections are related. Redrawing the circuit as we find equivalencies and reduce the circuit is a helpful tool in tracking our path through the process of finding an equivalent capacitance.

Chapter 27 – Capacitors and Batteries

27. (N) Find the equivalent capacitance for the network shown in Figure P27.26 if $C_1 = 1.00\ \mu F$, $C_2 = 2.00\ \mu F$, $C_3 = 3.00\ \mu F$, $C_4 = 4.00\ \mu F$, and $C_5 = 5.00\ \mu F$.

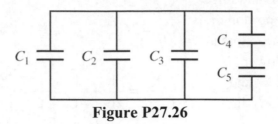

Figure P27.26

INTERPRET and ANTICIPATE	
We can find capacitors that are either in series or parallel and replace them with an equivalent capacitance. By repeating this process, we can reduce this network to a single equivalent capacitance.	
SOLVE	
Capacitors 4 and 5 are in series. Use Eq. 27.7 to replace it with a single capacitor.	$\dfrac{1}{C_{45}} = \dfrac{1}{C_4} + \dfrac{1}{C_5} \quad \rightarrow \quad C_{45} = \dfrac{C_4 C_5}{C_4 + C_5}$
This equivalent capacitance is in parallel with the other three. Use Eq. 27.8 to find the equivalent capacitance of these four capacitances.	$C_{eq} = C_1 + C_2 + C_3 + \dfrac{C_4 C_5}{C_4 + C_5}$
Now, insert numerical values.	$C_{eq} = 1.00\ \mu F + 2.00\ \mu F + 3.00\ \mu F + \dfrac{(4.00)(5.00)}{4.00 + 5.00}\ \mu F$ $C_{eq} = \boxed{8.22\ \mu F}$

CHECK and THINK

Using the formulas for the equivalent capacitance of a series or parallel combination, we can reduce this circuit to a single equivalent capacitance.

30. (N) For the four capacitors in the circuit shown in Figure P27.30, $C_A = 1.00\ \mu F$, $C_B = 4.00\ \mu F$, $C_C = 2.00\ \mu F$, and $C_D = 3.00\ \mu F$. What is the equivalent capacitance between points a and b?

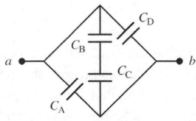

Figure P27.30

118

INTERPRET and ANTICIPATE We can find capacitors that are either in series or parallel and replace them with an equivalent capacitance. By repeating this process, we can reduce this network to a single equivalent capacitance.	
SOLVE Capacitors B and C are in series. Use Eq. 27.7 to replace it with a single capacitor.	$$\frac{1}{C_{BC}} = \frac{1}{4.00 \ \mu F} + \frac{1}{2.00 \ \mu F} = \frac{3}{4.00 \ \mu F}$$ $$C_{BC} = 1.33 \ \mu F$$
This equivalent capacitance is in parallel with the other three.	 **Figure P27.30ANS**
Use Eq. 27.8 to find the equivalent capacitance of these four capacitances.	$$C_{eq} = C_A + C_{BC} + C_D$$ $$C_{eq} = 1.00 \ \mu F + 1.33 \ \mu F + 3.00 \ \mu F = \boxed{5.33 \ \mu F}$$
CHECK and THINK Using the formulas for the equivalent capacitance of a series or parallel combination, we can reduce this circuit to a single equivalent capacitance.	

36. (N) A variable capacitor like the one shown in Figure P27.36 is often used as a tuning element in a high-frequency radio circuit. Suppose the air gap between a pair of plates is 0.300 mm and the pair behaves like a parallel-plate capacitor.
a. Determine the effective area of the plates when they are rotated to be fully meshed (overlapped), where the capacitance of the pair has a maximum value of 100.0 pF.
b. Determine the effective area of the plates when they are rotated to the point where the pair of plates has the minimum capacitance of 2.00 pF.

Chapter 27 – Capacitors and Batteries

Figure P27.36

INTERPRET and ANTICIPATE	
The capacitance of a parallel plate capacitor is proportional to the area of the plates and inversely proportional to their separation. Given the gap and capacitance, we can determine the effective area of the plates.	
SOLVE **a.** Apply Eq. 27.10.	$C = \dfrac{\varepsilon_0 A}{d} \quad \rightarrow \quad A = \dfrac{dC}{\varepsilon_0}$
Determine the effective area for the maximum capacitance. This corresponds to the maximum effective area.	$A_{max} = \dfrac{dC_{max}}{\varepsilon_0}$ $A_{max} = \dfrac{(0.30 \times 10^{-3})(100 \times 10^{-12})}{8.85 \times 10^{-12}}$ $A_{max} = \boxed{3.39 \times 10^{-3} \ \text{m}^2}$
b. Now calculate using the minimum capacitance.	$A_{min} = \dfrac{dC_{min}}{\varepsilon_0}$ $A_{min} = \dfrac{(0.30 \times 10^{-3})(2.0 \times 10^{-12})}{8.85 \times 10^{-12}}$ $A_{min} = \boxed{6.78 \times 10^{-5} \ \text{m}^2}$
CHECK and THINK	
Since capacitance is proportional to effective area, the area for the maximum capacitance is 50 times larger than for the minimum capacitance, equal to the capacitance ratio of 100/2.	

41. (N) The capacitance of most commonly used capacitors is small (~ 1 pF $< C < \sim 1$ μF). A capacitance of 1 F is very large. In this problem, we estimate the size of a 1-F capacitor using parallel-plate geometry. If the plates of a 1.0-F, air-filled, parallel-plate capacitor are separated by 50.0 μm (about the diameter of a human hair), what is the area of each plate?

INTERPRET and ANTICIPATE
The capacitance of a parallel plate capacitor is proportional to the area of the plates and inversely proportional to their separation. Given the gap and capacitance, we can determine the effective area of the plates. Since we're told that 1 F is a large capacitance, we'd expect that the area of the plates will be large.

SOLVE	
Apply Eq. 27.10.	$$C = \frac{\varepsilon_0 A}{d} \quad \rightarrow \quad A = \frac{Cd}{\varepsilon_0}$$ $$A = \frac{(1.0\text{ F})(50.0 \times 10^{-6}\text{ m})}{8.85 \times 10^{-12}} = \boxed{5.6 \times 10^6 \text{ m}^2}$$

CHECK and THINK
One Farad *must* be a very large capacitance… the area is *really* large—almost six square kilometers!

43. (N) A 5.69-pF spherical capacitor carries a charge of 1.54 μC.
a. What is the potential difference across the capacitor?
b. If the radial separation between the two spherical shells is 6.52×10^{-3} m, what are the inner and outer radii of the spherical conductors? *Hint*: See Problem 32.

INTERPRET and ANTICIPATE
When we look at Problem 32, we find a formula for the capacitance of this arrangement of spherical shells. This is still a capacitor, and so the capacitance must also be defined by Eq. 27.1, as well. We can find the potential difference by using Eq. 27.1 first. Then, we must try to use the equation from Problem 32 to create an equation involving the inner and outer radii. Knowing the separation can be written as, $r_{out} - r_{in} = 6.52 \times 10^{-3}$ m, we will then have two equations with two unknowns, which means we should have a solvable system of equations.

SOLVE	
a. Use Eq. 27.1 to find the potential difference.	$$V = Q/C = (1.54 \times 10^{-6}\text{ C})/(5.69 \times 10^{-12}\text{ F})$$ $$V = \boxed{2.71 \times 10^5 \text{ V}}$$
b. First, we note we have one known relationship between the inner and outer	$$r_{out} - r_{in} = 6.52 \times 10^{-3}\text{ m}$$

radii of the capacitor (their separation). Once we create another, we can solve the system of equations using substitution.	
Now, write the equation from Problem 32.	$C = 4\pi\varepsilon_0 \dfrac{r_{in} r_{out}}{r_{out} - r_{in}}$
Let's solve this equation for the outer radius.	$r_{in} r_{out} = \dfrac{(r_{out} - r_{in})C}{4\pi\varepsilon_0}$ $r_{in} r_{out} = \dfrac{(6.52\times10^{-3}\ \text{m})(5.69\times10^{-12}\ \text{F})}{4\pi(8.85\times10^{-12}\ \text{F/m})}$ $r_{out} = \dfrac{(6.52\times10^{-3}\ \text{m})(5.69\times10^{-12}\ \text{F})}{4\pi(8.85\times10^{-12}\ \text{F/m})r_{in}}$
Now, substitute this into the equation for the separation of the radii to solve for the inner radius.	$6.52\times10^{-3}\ \text{m} = r_{out} - r_{in}$ $6.52\times10^{-3}\ \text{m} = \dfrac{(6.52\times10^{-3}\ \text{m})(5.69\times10^{-12}\ \text{F})}{4\pi(8.85\times10^{-12}\ \text{F/m})r_{in}} - r_{in}$ $(6.52\times10^{-3}\ \text{m})r_{in} = \dfrac{(6.52\times10^{-3}\ \text{m})(5.69\times10^{-12}\ \text{F})}{4\pi(8.85\times10^{-12}\ \text{F/m})} - r_{in}^{\ 2}$ $r_{in}^{\ 2} + (6.52\times10^{-3}\ \text{m})r_{in} - \dfrac{(6.52\times10^{-3}\ \text{m})(5.69\times10^{-12}\ \text{F})}{4\pi(8.85\times10^{-12}\ \text{F/m})} = 0$
The result we get is a quadratic equation, which can be solved using the quadratic formula. Realize that nothing should be rounded until we achieve the final answer. In this case, one of the results is negative and the other is positive. The physically meaningful result is the positive answer in this case, since we are looking for a radius of a sphere.	$ar_{in}^2 + br_{in} + c = 0$ $r_{in} = \dfrac{-b \pm \sqrt{b^2 - 4ac}}{2a}$ Here: $a = 1$ $b = 6.52\times10^{-3}\ \text{m}$ $c = -\dfrac{(6.52\times10^{-3}\ \text{m})(5.69\times10^{-12}\ \text{F})}{4\pi(8.85\times10^{-12}\ \text{F/m})}$ $r_{in} = \boxed{1.53\times10^{-2}\ \text{m}}$

Then, go back to the separation of the radii to find the outer radius.	$r_{out} - 1.53 \times 10^{-2}$ m $= 6.52 \times 10^{-3}$ m $r_{out} = \boxed{2.18 \times 10^{-2} \text{ m}}$
CHECK and THINK The outer radius is bigger than the inner radius, which makes sense. Note that Eq. 27.1 can be applied, even in these odder scenarios or arrangements of oppositely-charged conductors.	

47. (N) The plates of an air-filled parallel-plate capacitor with a plate area of 16.0 cm^2 and a separation of 9.00 mm are charged to a 145-V potential difference. After the plates are disconnected from the source, a porcelain dielectric with $\kappa = 6.5$ is inserted between the plates of the capacitor.
a. What is the charge on the capacitor before and after the dielectric is inserted?
b. What is the capacitance of the capacitor after the dielectric is inserted?
c. What is the potential difference between the plates of the capacitor after the dielectric is inserted?
d. What is the magnitude of the change in the energy stored in the capacitor after the dielectric is inserted?

INTERPRET and ANTICIPATE	
When a dielectric is inserted into a capacitor, the capacitance increases. Since the charged capacitor is disconnected from the source, the charge has nowhere to go and must remain constant. Using these, we can then determine the voltage and energy stored in the capacitor.	

SOLVE **a.** The initial capacitance of a parallel plate capacitor can be calculated using Eq. 27.10.	$C_i = \dfrac{\varepsilon_0 A}{d}$
With Eq. 27.1, we can determine the charge on the capacitor. The charged capacitor is disconnected from the source, so the two sides of the capacitor are electrically isolated and the charge must be constant. That is, $Q_i = Q_f = Q$.	$Q = C_i V_i = \dfrac{\varepsilon_0 A V_i}{d}$ $Q = \dfrac{\left(8.85 \times 10^{-12} \frac{\text{C}^2}{\text{N} \cdot \text{m}^2}\right)\left(16.0 \times 10^{-4} \text{ m}^2\right)\left(145 \text{ V}\right)}{\left(9.00 \times 10^{-3} \text{ m}\right)}$ $Q = \boxed{2.28 \times 10^{-10} \text{ C}}$

b. When a dielectric is inserted, the capacitance increases by a factor equal to the dielectric constant as in Eq. 27.18.	$C_f = \dfrac{\kappa \varepsilon_0 A}{d}$ $C_f = \dfrac{6.50\left(8.85\times10^{-12}\ \frac{C^2}{N\cdot m^2}\right)\left(16.0\times10^{-4}\ m^2\right)}{\left(9.00\times10^{-3}\ m\right)}$ $C_f = \boxed{1.02\times10^{-11}\ F}$
c. Based on Equation 27.1, given a fixed charge, the voltage decreases with increasing capacitance. The voltage decreases by a factor equal to the dielectric constant, 145 V / 6.5 = 22.3 V.	$V_f = \dfrac{Q}{C_f} = \dfrac{228\ pC}{1.02\times10^{-11}\ F} = \boxed{22.3\ V}$
d. We can calculate the initial energy with Eq. 27.3.	$U_i = \dfrac{1}{2}C_i V_i^2 = \dfrac{1}{2}\dfrac{\varepsilon_0 A}{d}V_i^2$ $U_i = \dfrac{1}{2}\dfrac{\left(8.85\times10^{-12}\ \frac{C^2}{N\cdot m^2}\right)\left(16.0\times10^{-4}\ m^2\right)}{\left(9.00\times10^{-3}\ m\right)}\left(145\ V\right)^2$ $U_i = 1.654\times10^{-8}\ J$
Now, calculate the final energy.	$U_f = \dfrac{1}{2}C_f V_f^2 = \dfrac{1}{2}\left(1.02\times10^{-11}\ F\right)\left(22.3\ V\right)^2$ $U_f = 2.54\times10^{-9}\ J$
Finally, calculate the difference and quote the magnitude of the change.	$\Delta U = U_f - U_i = -1.40\times10^{-8}\ J$ $\left|\Delta U\right| = \boxed{1.40\times10^{-8}\ J}$

CHECK and THINK

When a charged capacitor is disconnected from the voltage source and a dielectric is inserted, the charge must remain constant. Since the capacitance increases, the voltage and energy stored both decrease.

54. (A) A parallel-plate capacitor with an air gap has capacitance C_0. It is connected to a battery with potential V_0 that gives it charge Q_0 and stored energy U_0. After the capacitor is disconnected from the battery, a dielectric with constant $\kappa = 3$ is inserted into the air gap, completely filling it. In terms of the initial values, find the new capacitance C, charge Q, potential V, and stored energy U.

INTERPRET and ANTICIPATE The capacitance increase by a factor of κ. The charge has no path to enter or leave the capacitor plates, so it is fixed. The potential then must decrease (Eq. 27.1). We calculate the energy (Eq. 27.3) and find that it decreases.	

SOLVE When a dielectric is inserted, the capacitance increases by a factor equal to the dielectric constant (Eq. 27.18).	$C = \kappa C_0 = \boxed{3C_0}$
Since the capacitor is not connected and the plates are electrically isolated, the charge remains unchanged.	$Q = \boxed{Q_0}$
We can determine the voltage with Eq. 27.1.	$V = \dfrac{Q}{C} = \dfrac{Q_0}{3C_0} = \boxed{\dfrac{V_0}{3}}$
Finally, calculate the energy with Eq. 27.3.	$U = \tfrac{1}{2}CV^2 = \tfrac{1}{2}\left(3C_0\right)\left(\dfrac{V_0}{3}\right)^2 = \boxed{\dfrac{U_0}{3}}$

CHECK and THINK When a charged capacitor is disconnected from the voltage source and a dielectric is inserted, the charge remains constant. Since the capacitance increases, the voltage and energy stored both decrease as well.

57. (N) Five hundred 8.00-μF capacitors are connected in parallel and then charged to a potential of 25.0 kV. For how long will the stored energy light a 100.0-W bulb until no energy remains in the capacitors?

INTERPRET and ANTICIPATE The equivalent capacitance for capacitors in parallel is the sum of all the capacitances. Given the potential difference, we can calculate the total energy stored and then how long we can light the bulb.

SOLVE The equivalent capacitance for capacitors in parallel is given by Eq. 27.8, so all 500 capacitances add together.	$C_{eq} = \left(500\right)\left(8.0\ \mu\text{F}\right) = 4.0 \times 10^3\ \mu\text{F}$

The energy stored can be calculated with Eq. 27.3.	$U_C = \dfrac{1}{2} C_{eq} V^2$ $U_C = \dfrac{1}{2}\left(4.0 \times 10^{-3}\ \text{F}\right)\left(25 \times 10^3\ \text{V}\right)^2$ $U_C = 1.25 \times 10^6\ \text{J}$
This can be equated with the energy used by the light bulb (equal to the power of the bulb times time).	$t = \dfrac{1.25 \times 10^6\ \text{J}}{100\ \text{W}} = \boxed{1.25 \times 10^4\ \text{s}} = 3\,\text{h}\ 28\,\text{m}\ 20\,\text{s}$

CHECK and THINK

In this case, with 500 capacitors charged with a pretty substantial voltage, we can run the light bulb for about three and a half hours.

62. (A) An air-filled parallel-plate capacitor is charged to a certain potential difference. A dielectric is then inserted in the capacitor to completely fill the space between the plates. Then the charge on the plates is increased by a factor of three to restore the original potential difference. Determine the dielectric constant.

INTERPRET and ANTICIPATE

The capacitance increases by a factor equal to the dielectric constant. We can relate the given voltage and charge to capacitance and then determine the dielectric constant.

SOLVE Equation 27.1 relates capacitance, voltage, and charge on a capacitor.	$V = \dfrac{Q}{C}$
With a dielectric, we can apply Eq. 27.17, assuming that the charge has increased by a factor of three in order to have the same voltage as above. That is, when the dielectric κ is inserted, $Q \rightarrow 3Q$ and V remains unchanged.	$V = \dfrac{3Q}{\kappa C}$
Combining these, we find that the dielectric constant must be 3.	$\kappa = \boxed{3}$

Chapter 27 – Capacitors and Batteries

CHECK and THINK

In this case, the fact that the charge increases by a factor of three at the same voltage tells is that the dielectric constant is three (because the capacitance must increase by a factor of three for the charge to increase by three at the given voltage).

65. (N) Nerve cells in the human body and in other animals are modeled as very long cylindrical capacitors. Portions of some nerves are covered with a layer of fat known as myelin, which functions as the dielectric ($\kappa = 7$) between two plates in the cylindrical capacitor model. The potential difference between the inner and outer walls of myelin in resting nerve cells is roughly $V_{inner} - V_{outer} = 270$ mV. Find the linear charge density on the inner (positive) plate. *Hint*: Use the result of Example 27.8.

INTERPRET and ANTICIPATE

We will use the capacitance per unit length determined in Example 27.8. Given the capacitance per unit length and voltage, we can determine the charge per unit length.

SOLVE We can use the capacitance per unit length calculated in Example 27.8.	$\dfrac{C}{L} = 3.1 \times 10^{-10}$ F/m
Equation 27.1 relates charge to capacitance and voltage. If we divide both sides by a length L, we can use the quantity above.	$Q = CV \quad \rightarrow \quad \dfrac{Q}{L} = \dfrac{C}{L} V$
Insert numerical values.	$\dfrac{Q}{L} = \left(3.1 \times 10^{-10} \text{ F/m}\right)\left(70 \times 10^{-3} \text{ V}\right) = \boxed{2.2 \times 10^{-11} \text{ C/m}}$

CHECK and THINK

The charge per unit length is 22 pC/m. For a capacitor, Q is the magnitude of charge on both plates, so this is the magnitude of the charge per unit length for both the positive and the negative plate.

70. (N) Three capacitors with capacitances 2.00×10^{-3} μF, 4.00×10^{-3} μF, and 6.00×10^{-3} μF are connected in series. Is it possible to apply a potential difference of 11.00×10^3 V across the set if the breakdown voltage of each capacitor is 4.00×10^3 V?

Chapter 27 – Capacitors and Batteries

INTERPRET and ANTICIPATE

We need to determine if the voltage across any one of the capacitors is above the breakdown voltage.

SOLVE	
For capacitors in series, they each have the same charge, which we can find by first determining the equivalent capacitance (Eq. 27.7).	$$\frac{1}{C_{eq}} = \frac{1}{0.002\ \mu F} + \frac{1}{0.004\ \mu F} + \frac{1}{0.006\ \mu F} = \frac{916.7}{\mu F}$$ $$C_{eq} = 1.09 \times 10^{-3}\ \mu F$$
The charge is then found with Eq. 27.1.	$$Q = C_{eq}V = \left(1.09 \times 10^{-3}\ \mu F\right)\left(1.1 \times 10^4\ V\right) = 12\ \mu C$$
The voltage across the smallest capacitor is then found to be 6000 V, which is above the breakdown voltage, so it is not possible.	$$V_1 = \frac{Q}{C_1} = \frac{12\ \mu C}{0.002\ \mu F} = 6000\ V$$

CHECK and THINK

The total voltage across the three capacitors in series is 11000 V, but is not the same for each. The charge for capacitors in series is the same, so the smallest capacitance has the largest voltage across it. In this case, it's found to be above the breakdown voltage, so the capacitor will fail.

73. (N, C) In a laboratory, you find a 9.00-V battery and a 12.0-V battery. You also find a 30.0-μF capacitor and a 45.0-μF capacitor. Your challenge is to store the maximum possible energy. You may use as much of this equipment as you wish. Describe your solution, and draw a schematic diagram of your network. How much energy is stored by the capacitor(s)?

INTERPRET and ANTICIPATE

The energy stored depends on the voltage across the capacitors and the capacitance. We will try to create a circuit to maximize these quantities.

SOLVE	
We can combine the capacitors in parallel to produce the largest equivalent capacitance (either capacitor individually will have a smaller capacitance and the	$$C_{eq} = 30.0\ \mu F + 45.0\ \mu F = 75.0\ \mu F$$

combination in series will have an equivalent capacitance smaller than either of the individual values).	
The largest voltage will be produced by using the two batteries in series.	$V_{tot} = 9.00 \text{ V} + 12.0 \text{ V} = 21.0 \text{ V}$
A sketch of the circuit is shown.	 **Figure P27.73ANS**
Calculate the energy with Eq. 27.3.	$U_E = \frac{1}{2}C_{eq}V_{tot}^2 = \frac{1}{2}(75.0 \ \mu\text{F})(21.0 \text{ V})^2 = \boxed{16.5 \text{ mJ}}$

CHECK and THINK
If we maximize the equivalent capacitance and voltage across the network, we can maximize the energy stored.

79. (N) When connected in series, two capacitors have an equivalent capacitance of 3.00 μF. The same two capacitors have an equivalent capacitance of 13.0 μF when connected in parallel. What is the capacitance of each of the capacitors?

INTERPRET and ANTICIPATE
We can write expressions for the equivalent capacitance in both cases and solve for the individual capacitances.

SOLVE	
Use Eq. 27.7 for the equivalent capacitance for the series combination.	$\dfrac{1}{C_s} = \dfrac{1}{C_1} + \dfrac{1}{C_2} = \dfrac{1}{3.00}$
Now, use Eq. 27.8 for the same capacitors in parallel.	$C_p = C_1 + C_2 = 13.0$

Solve this for C_2 and substitute into the equation above.	$$\frac{1}{3.00} = \frac{1}{C_1} + \frac{1}{13.0 - C_1} = \frac{13.0}{C_1(13.0 - C_1)}$$
Solve for C_1.	$$C_1^2 - 13.0C_1 + 39.0 = 0$$ $$C_1 = \frac{13.0 \pm \sqrt{(13.0)^2 - 4(39.0)}}{2} = 8.30 \ \mu F \text{ or } 4.70 \ \mu F$$
The two solutions turn out to be the two values of capacitance. For instance, if we arbitrarily choose the first one, we can solve for C_2 using the equivalent parallel capacitance from above and find that we recover the second value.	$$C_1 = \boxed{8.30 \ \mu F}$$ $$C_2 = C_p - C_1$$ $$C_2 = 13.0 - 8.30 = \boxed{4.70 \ \mu F}$$

CHECK and THINK

Given the formulas for equivalent capacitance, we are able to write two equations with two unknowns. It is then only a matter of algebra to solve for each capacitance.

82. (N) Consider an infinitely long network with identical capacitors arranged as shown in Figure P27.82. Determine the equivalent capacitance of such a network. Each capacitor has a capacitance of 1.00 μF.

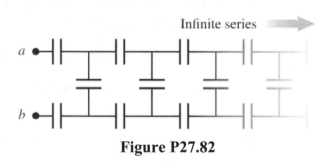

Figure P27.82

INTERPRET and ANTICIPATE

We assume this infinite network has an equivalent capacitance. In this case, if we add one more unit to the left, the equivalent capacitance of *this* network must *also* be equal to the equivalent capacitance of the infinite network.

SOLVE Looking at the figure, we replace the infinite sequence with the equivalent capacitance and add one more unit on the left. Now have in loop ACDB two capacitors C and C_{eq} in parallel for which we can obtain the equivalent capacitance $C + C_{eq}$. This again has to be combined with two other capacitors C and C in series across terminals a and b.	 **Figure P27.82ANS**
The final equivalent capacitance of *this* network must *also* be equal to C_{eq} (since an infinite network and a network with infinite plus one unit is the same). Apply the formula for capacitors in series (Eq. 27.7).	$$\frac{1}{C_{eq}} = \frac{1}{C} + \frac{1}{C} + \frac{1}{C+C_{eq}}$$
Solve for C_{eq}.	$$\frac{(C+C_{eq})-C_{eq}}{C_{eq}(C+C_{eq})} = \frac{2}{C}$$ $$2C_{eq}^{\,2} + 2C_{eq}C - C^2 = 0$$ $$C_{eq} = \frac{-2C+\sqrt{4C^2+8C^2}}{4} = \frac{\sqrt{3}-1}{2}C = \boxed{0.366C}$$

CHECK and THINK

This problem relies on the fact that if an infinite arrangement of capacitors has an equivalent capacitance, adding one more unit can't change this limiting value. With this in mind, we can write the capacitance with an additional unit added and solve for the equivalent capacitance.

28

Current and Resistance

6. (N) In a low-energy transmission electron microscope (TEM), the typical average beam current is about $-1.00\ \mu A$. If a material sample is exposed to this electron current for 10.0 min, how many electrons impact the material during this time

INTERPRET and ANTICIPATE	
The amount of charge can be found by using Eq. 28.1. Since the current is constant, we can write Eq. 28.1 as $I = \Delta q/\Delta t$. Using the fact that 1 A = 1 C/s, we can find the amount of charge that impact the material in 10.0 min. We must then use the fundamental electric charge to find the number of electrons this total charge represents. Because the fundamental electric charge is so small, we expect this to be a very large number of electrons, especially with the beam running for 10.0 min.	

SOLVE The total time can be expressed in seconds. Then, use Eq. 28.1 to find the total amount of charge that impacts the material in that time.	$\Delta t = (10.0\ \text{min}) \cdot \left(\dfrac{60\ \text{s}}{1\ \text{min}}\right) = 6.00 \times 10^2\ \text{s}$ $\Delta q = I\Delta t = \left(-1.00 \times 10^{-6}\ \text{A}\right)\left(6.00 \times 10^2\ \text{s}\right)$ $\Delta q = -6.00 \times 10^{-4}\ \text{C}$
Then, use the electric charge of one electron to find their number. This isn't so much an equation as it is the idea of dividing a large collection of something by the smallest part, to find the number of those parts present.	$N = \Delta q/-e = \left(-6.00 \times 10^{-4}\ \text{C}\right)/\left(-1.60 \times 10^{-19}\ \text{C}\right)$ $N = \boxed{3.75 \times 10^{15}\ \text{electrons}}$

CHECK and THINK Even with a micro-amp current, the number of electrons deposited on the material by the beam is quite large. We could divide this total number by the 600 s to see how many impact the material each second: 6.25×10^{12} !

9. (A) Positively charged ions move along a single axis, forming a beam. The amount of charge that passes through a cross section of the beam is given by $q = q_0 \cos \omega t$, where q_0 is the amount of charge that passes through at $t = 0$ and ω is the angular frequency. Find an expression for the current as a function of time.

INTERPRET and ANTICIPATE
The current is the rate at which charge flows per unit time. This is the rate of charge flow, or the derivative with respect to time of the charge $q(t)$. Given the expression for the charge, it is straightforward to take the derivative to find the answer.

SOLVE	
Use Equation 28.1 to find the answer.	$I = \dfrac{dq}{dt} = \dfrac{d}{dt}\left(q_0 \cos \omega t\right) = \boxed{-q_0 \omega \sin \omega t}$

CHECK and THINK

Since the angular frequency has units of 1/s, notice that the factor $q_0\omega$ has units of C/s or A, as we would hope. If we sketch the two solutions, it becomes even clearer. The charge starts at a maximum and decreases, so the current is negative (decreasing the charge) and reaches a maximum value when the charge is changing most rapidly.

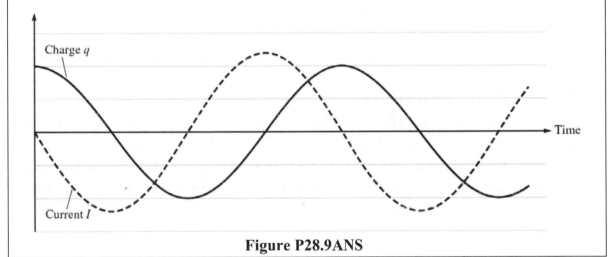

Charge q

Current I

Time

Figure P28.9ANS

15. (N) The current in a wire varies with time (measured in seconds) as $I = 24\text{ A} - (0.12 \text{ A/s}^2)\, t^2$. Determine the amount of charge that flows through a cross-sectional area of the wire between $t = 0$ and $t = 12$ s.

INTERPRET and ANTICIPATE
The current is equal to the rate that charge flows per unit time, dq/dt. We can integrate the expression for current with respect to time to determine the total amount of charge

that flows through the wire.	
SOLVE Use the definition given by Equation 28.1.	$I = \dfrac{dq}{dt} \quad \rightarrow \quad dq = I\,dt$
Next, we integrate with respect to time to obtain the charge that flows through a cross-sectional area of the wire between $t = 0$ s and $t = 12$ s.	$\displaystyle\int_0^Q dq = \int_{0\,s}^{12\,s} I\,dt = \int_{0\,s}^{12\,s}\left[24\text{ A} - \left(0.12\text{ A/s}^2\right)t^2\right]dt$ $Q = \left[24\,t - 0.12\dfrac{t^3}{3}\right]\Bigg\vert_{0\,s}^{12\,s}$ $Q = (288 - 69)\text{C}$ $Q = \boxed{2.2\times10^2\text{ C}}$

CHECK and THINK
Notice that the current is not constant as it varies with time, so we had to integrate to find the total flow of charge in 12 s.

18. (N) A copper wire that has a cross-sectional diameter of 3.500 mm has a current of 1.241 A.
a. What is the current density?
b. What is the total charge that passes by a certain location along the wire in 2.000 s?

INTERPRET and ANTICIPATE We can use the diameter to define the cross-sectional radius of the wire ($d/2$) and then get the cross-sectional area. Eq. 28.3 can then be used to find the current density. As for finding the total charge that passes by in 2.000 s, Eq. 28.1 should be sufficient, knowing the current in the wire. Since 1 A = 1 C/s, and we are going to find the total charge that passes by in 2.000 s, we expect the numerical value to be double the numerical value of the current.	
SOLVE **a.** The radius is half the diameter, so using the area of a circle for the cross-sectional area of the wire, we find the result shown to the right.	$A = \pi\left(d/2\right)^2$
Then, Eq. 28.3 is used to find the current density.	$J = I/A = \left(1.241\text{ A}\right)\Big/\left[\pi\left(3.500\times10^{-3}\text{ m}/2\right)^2\right]$ $J = \boxed{1.290\times10^5\text{ A/m}^2}$

b. Using the current and Eq. 28.1, we find the charge that passes by in the time of 2.000 s.	$\Delta q = I\Delta t = (1.241 \text{ A})(2.000 \text{ s}) = \boxed{2.482 \text{ C}}$

CHECK and THINK

As expected, the numerical value of part (b) is double that of the current. Every second, 1.241 C of charge flow past that location. Incidentally, that is a lot of electric charge passing by each second! How many fundamental electric charges (or electrons) does that represent?

23. (N) A copper wire that is 2.00 mm in radius with density 8.94 g/cm^3 has a current of 8.00 A. The molar mass of copper is 63.5463, and each copper atom contributes one free electron. What is the drift speed of the electrons in the copper wire?

INTERPRET and ANTICIPATE

The current, which is given, depends on quantities such as the drift speed of the electrons and the number density of the charges. We can use these expressions to solve for the drift speed.

SOLVE We can use Equations 28.3 and 28.5 relate the current to the drift speed. (This is actually just Eq. 28.6.) Solve for the drift speed.	$J = \dfrac{I}{A} = nev_d$ $v_d = \dfrac{I}{neA}$
We are given the current ($I = 8.00$ A) and the fundamental charge of the electron e is a constant. The cross-sectional area of a circular wire is the area of a circle, which we can calculate.	$A = \pi r^2 = \pi (2.00 \times 10^{-3} \text{ m})^2$

The number density n is the number of charges per unit volume. From the density, we know that there are 8.94 grams of copper atoms per cm^3. The molar mass tells us the mass of an Avagadro's number of atoms. We also know there is 1 free electron per atom.

$$n = 8.94\frac{\text{g}}{\text{cm}^3}\left(\frac{6.022 \times 10^{23} \text{ atoms}}{63.5463 \text{ g}}\right)\left(\frac{1 \text{ free electron}}{\text{atom}}\right)\left(\frac{100 \text{ cm}}{\text{m}}\right)^3 = 8.47 \times 10^{28} \frac{\text{electrons}}{\text{m}^3}$$

This value has the right units and is very close to that found in Table 28.2. It would also be possible to determine the volume that each atom occupies since this quantity is equivalent to how many atoms there are per cubic meter.

Now, insert numerical values and solve.

$$v_d = \frac{8.00 \text{ A}}{\left(8.47 \times 10^{28} \text{ atoms/m}^3\right)\left(1.6 \times 10^{-19} \text{ C}\right)\pi\left(2.00 \times 10^{-3} \text{ m}\right)^2} = \boxed{4.70 \times 10^{-5} \text{ m/s}}$$

CHECK and THINK

The value is small, but this is consistent with Example 28.2.

---

26. (N) Consider the nanotube described in Problem 25. The electron number density is 6.57×10^{28} m^{-3}. What is the mean free time for the electrons flowing in a current along the carbon nanotube?

INTERPRET and ANTICIPATE

Note that the resistivity of the nanotube was given in Problem 25, $\rho = 3.40 \times 10^{-8}$ $\Omega \cdot$m.

Knowing the electron number density, the mass of an electron, m_e, and the fundamental electric charge, e, we can find the mean free time using Eq. 28.18. Most mean free times are very short (picoseconds), though note that the nanotube may represent more of an extreme case. We are certainly expecting a small numerical value.

SOLVE	
We have all of the necessary value to make use of Eq. 28.18. We solve for the mean free time.	$\rho = \dfrac{m_e}{ne^2 t_{mf}}$ $t_{mf} = \dfrac{m_e}{ne^2\rho} = \dfrac{9.109 \times 10^{-31} \text{ kg}}{\left(6.57 \times 10^{28} \text{ m}^{-3}\right)\left(1.6 \times 10^{-19} \text{ C}\right)^2\left(3.40 \times 10^{-8} \text{ } \Omega \cdot \text{m}\right)}$ $t_{mf} = \boxed{1.59 \times 10^{-14} \text{ s}}$

CHECK and THINK

The mean free time is very short here. This means that the electrons travel less time before interacting with another charge, than in a typical material.

---

33. (N) Two concentric, metal spherical shells of radii $a = 4.0$ cm and $b = 8.0$ cm are separated by aluminum as shown in Figure P28.33. The inner sphere has a total charge Q at any time. If the two spheres are maintained at a potential difference of 2.0 V via an external source, calculate the current from one sphere to the other.

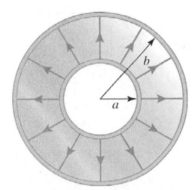

Figure P28.33

INTERPRET and ANTICIPATE Between the two shells, there is an electric field that depends on the charge of the inner shell. (This is something that we know from Gauss's law in Chapter 25.) The electric field drives a current, which we will try to calculate, assuming that the inner sphere is somehow continuously maintained at charge Q. The electric field is also directly related to the potential difference between the two spheres, which is given.	
SOLVE The electric field determines the current density (Eq. 28.16), which can be integrated to determine the current (Eq. 28.3).	$I = \int \vec{J} \cdot d\vec{A} = \sigma \int \vec{E} \cdot d\vec{A}$
The electric field between the two spherical shells can be found using Gauss's law and is given by Equation 25.14, where $q = Q$ is the charge of the inner sphere. Notice that the field is in the radial direction.	$\vec{E} = \dfrac{1}{4\pi\varepsilon_0} \dfrac{q}{r^2} \hat{r}$
Imagine a spherical surface between the two shells. The area vector points radially outward in the same direction as the electric field. The area of the spherical shell is $4\pi r^2$. We can use this to determine the total current from the inner sphere, through the spherical surface, to the outer sphere.	$I = \sigma \int \dfrac{1}{4\pi\varepsilon_0} \dfrac{Q}{r^2} \hat{r} \cdot dA\hat{r}$ $I = \sigma \dfrac{Q}{4\pi\varepsilon_0} \int \dfrac{1}{r^2} dA$ $I = \sigma \dfrac{Q}{4\pi\varepsilon_0 r^2} \left(4\pi r^2 \right)$ $I = \dfrac{\sigma Q}{\varepsilon_0}$ (1)

While this expression would allow us to calculate the current, we actually don't know the charge on the inner sphere. Since we are given the potential difference, let's use the electric field to determine the potential difference between the two shells. Remember that the voltage is the integral of the electric field along a path from one point to another. This will allow us to relate the charge to the potential difference. We use a and b to indicate the radii of the spheres.	$$V = -\int_a^b \vec{E} \cdot d\vec{r}$$ $$V = -\frac{Q}{4\pi\varepsilon_0}\int_a^b \frac{1}{r^2}dr$$ $$V = \frac{Q}{4\pi\varepsilon_0}\left(\frac{1}{a} - \frac{1}{b}\right) \qquad (2)$$
We now use Equation (1) to solve for Q and then substitute it in Eq. (2).	$$I = \frac{4\pi\sigma\,ab}{(b-a)}V$$
Now insert numerical values for the radii of the shells ($a = 0.04$ m, $b = 0.08$ m), the potential difference ($V = 2.0$ V), and the conductivity of aluminum from Table 28.2 ($\sigma = 3.767 \times 10^7\ \Omega^{-1}\text{m}^{-1}$).	$$I = \frac{4\pi\left(3.767\times10^7\ \Omega^{-1}\text{m}^{-1}\right)(0.08\text{ m})(0.04\text{ m})}{0.08\text{ m} - 0.04\text{ m}}(2.0\text{ V})$$ $$I = \boxed{7.6\times10^7\text{ A}}$$

CHECK and THINK

The electric field drives a current and creates a voltage between the two concentric shells. Using these facts, we were able to relate the voltage to the current. Notice that the specific results depend on the geometry of the problem (for instance that they are spherical conductors of a given size) and the material between them (as the conductivity of this medium affects the current flow).

37. (N) A copper wire has a diameter of 0.400 mm. What length of the wire has a resistance of 2.00 Ω?

INTERPRET and ANTICIPATE

The resistance is proportional to length and inversely proportional to cross sectional area. Given the material (which determines the resistivity) and the desired resistance, we can determine the length of the wire needed.

SOLVE Use Eq. 28.24.	$R = \rho \dfrac{\ell}{A}$
The cross sectional area is circular. The radius is half the given diameter, or 0.200 mm.	$A = \pi r^2 = \pi \left(0.200 \times 10^{-3} \text{ m}\right)^2 = 1.26 \times 10^{-7} \text{ m}^2$
Solve for the length and insert values. The resistivity can be found in Table 28.2.	$\ell = \dfrac{RA}{\rho}$ $\ell = \dfrac{\left(2.00 \ \Omega\right)\left(1.26 \times 10^{-7} \text{ m}^2\right)}{1.678 \times 10^{-8} \ \Omega \cdot \text{m}} = \boxed{15.0 \text{ m}}$

CHECK and THINK

The resistance of a wire is determined by the material (and its resistivity specifically), length, and cross-sectional area. Given three of these quantities, we can easily calculate the fourth.

42. (N) A long, thin cylindrical conductor is made of silver. Its resistance is 78.5 Ω, and its mass is 0.0250 kg. Find the length and radius of the conductor.

INTERPRET and ANTICIPATE

The resistance of a wire depends on its resistivity (which depends on the material), length, and cross sectional area of the wire. The mass will allow us to determine the total volume of the wire, which equals the length times the area.

SOLVE The resistance of a wire is given by Eq. 28.24. The resistivity for silver can be found in Table 28.2.	$R = \rho \dfrac{\ell}{A}$ $\rho = 1.586 \times 10^{-8} \ \Omega \cdot \text{m}$
We can determine the ratio of the length to area.	$\dfrac{\ell}{A} = \dfrac{R}{\rho} = \dfrac{78.5 \ \Omega}{1.586 \times 10^{-8} \ \Omega \cdot \text{m}} = 4.95 \times 10^9 \text{m}^{-1}$ (1)

The volume of the wire (a cylinder) is equal to its length times its cross-sectional area.	$V = \ell A$
One way to calculate the volume of 0.0250 kg of silver is to look up its density, 10.49 g/cm^3 = 10490 kg/m^3, which equals mass over volume.	$V = \dfrac{0.0250\ \text{kg}}{10490\ \text{kg/m}^3} = 2.38 \times 10^{-6}\ \text{m}^3$ $\ell A = 2.38 \times 10^{-6}\ \text{m}^3 \qquad (2)$

Another way to determine the volume is to use the data in Table 28.2. The number density of conduction electrons $n = 5.86 \times 10^{28}$ m^{-3} tells us how many electrons there are per unit volume. Assuming one electron per atom and using the molar mass of 107.87 g/mol, we can determine the volume of the 25.0 g wire. (Think of this as a unit conversion where we convert the mass in grams to volume using known relationships.)

$$25.0\,\text{g}\left(\frac{\text{mole}}{107.87\ \text{g}}\right)\left(\frac{6.022 \times 10^{23}\ \text{atoms}}{1\ \text{mole}}\right)\left(\frac{1e^-}{1\ \text{atom}}\right)\left(\frac{\text{m}^3}{5.86 \times 10^{28}\,e^-}\right) = 2.38 \times 10^{-6}\ \text{m}^3$$

Now, combine Equations (1) and (2). For instance, solve (1) for length and substitute into (2) to determine the cross sectional area.	$\ell A = \left(4.95 \times 10^9\ \text{m}^{-1}\right) A^2 = 2.38 \times 10^{-6}\ \text{m}^3$ $A = \sqrt{\dfrac{2.38 \times 10^{-6}\ \text{m}^3}{4.95 \times 10^9\ \text{m}^{-1}}} = 2.19 \times 10^{-8}\ \text{m}^2 \qquad (3)$
The area of the circular cross-section allows us to determine the radius of the wire.	$A = \pi r^2$ $r = \sqrt{\dfrac{A}{\pi}} = \sqrt{\dfrac{2.19 \times 10^{-8}\ \text{m}^2}{\pi}} = 8.35 \times 10^{-5}\ \text{m}$
Finally, plug Eq. (3) back into Equation (1) or (2) to determine the length.	$\ell = \left(4.95 \times 10^9\ \text{m}^{-1}\right)\left(2.19 \times 10^{-8}\ \text{m}^2\right) = \boxed{108\ \text{m}}$

CHECK and THINK

The wire turns out to be over 100 meters in length! The fixed amount of material determines the cross sectional area of the wire as we draw the wire to greater lengths. Since these are the factors that determine the resistance of the wire, we can use this to determine the dimensions of our wire.

44. (A) Two wires with different resistivities, ρ_1 and ρ_2, are supposed to have the same resistance. If the radii of the wires are r_1 and r_2, find a function for the length of the second wire in terms of the length of the first wire.

INTERPRET and ANTICIPATE	
We can use Eq. 28.24, written for each resistor, to relate the properties of each and find an expression for the length of the new wire. We expect to also see ratios of the given resistivities and radii in our final expression.	
SOLVE Write Eq. 28.24 for each resistor.	$$R_1 = \rho_1 \frac{\ell_1}{A_1}$$ $$R_2 = \rho_2 \frac{\ell_2}{A_2}$$
Now, equate the two resistances, use the fact that the cross-sectional area can be expressed as the area of a circle, and solve for the length of the second wire in terms of the first.	$$\rho_2 \frac{\ell_2}{A_2} = \rho_1 \frac{\ell_1}{A_1}$$ $$\rho_2 \frac{\ell_2}{\pi r_2^2} = \rho_1 \frac{\ell_1}{\pi r_1^2}$$ $$\boxed{\ell_2 = \frac{\rho_1 r_2^2}{\rho_2 r_1^2} \ell_1}$$

CHECK and THINK
The relationship is a linear one, where the slope depends on the resistivities and the radii of the wires.

47. (N) A 75.0-cm length of a cylindrical silver wire with a radius of 0.150 mm is extended horizontally between two leads. The potential at the left end of the wire is 3.20 V, and the potential at the right end is zero. The resistivity of silver is 1.586×10^{-8} $\Omega \cdot$m.

a. What are the magnitude and direction of the electric field in the wire?
b. What is the resistance of the wire?
c. What are the magnitude and direction of the current in the wire?
d. What is the current density in the wire?

INTERPRET and ANTICIPATE
This problem requires us to relate a variety of properties of a current through a wire.

SOLVE **a.** The magnitude of the electric field is equal to the slope or gradient of the potential and points from high to low potential.	$$E_x = -\frac{dV}{dx} = -\frac{\Delta V}{\Delta x}$$ $$E_x = -\frac{(0 - 3.20 \text{ V})}{0.750 \text{ m}}$$ $$E_x = \boxed{4.27 \text{ V/m from left to right}}$$
b. The resistance depends on the material and geometry of the wire and can be calculated with Equation 28.24.	$$R = \rho \frac{\ell}{A}$$ $$R = \frac{\left(1.586 \times 10^{-8} \ \Omega \cdot \text{m}\right)\left(0.750 \text{ m}\right)}{\pi \left(0.150 \times 10^{-3} \text{ m}\right)^2}$$ $$R = \boxed{0.168 \ \Omega}$$
c. The current can be found using Ohm's law (Eq. 28.25) and travels in the direction of the electric field.	$$I = \frac{\Delta V}{R} = \frac{3.20 \text{ V}}{0.168 \ \Omega}$$ $$I = \boxed{19.0 \text{ A from left to right}}$$
d. The current density is current per cross-sectional area, $A = \pi r^2$ (Eq. 28.3).	$$J = \frac{I}{A}$$ $$J = \frac{19.0 \text{ A}}{\pi \left(0.150 \times 10^{-3} \text{ m}\right)^2}$$ $$J = \boxed{2.69 \times 10^8 \text{ A/m}^2}$$

CHECK and THINK

Given information about an experimental set-up, we are able to use the relationships from the chapter to determine a variety of quantities related to the resistance of the wire and current through it. All of these values sound like typical values we might encounter in a physics lab.

55. (N) A two-slice bread toaster consumes 850.0 W of power when plugged into a 120.0-V source.

a. What is the current in the toaster?

b. What is the resistance of the coils in the toaster?

INTERPRET and ANTICIPATE	
The power drawn by an electrical device depends on the current times voltage. With Ohm's law, we can also determine the resistance.	

SOLVE **a.** The power drawn by an electrical device depends on current and voltage as given by Equation 28.33.	$P = I\Delta V \quad \rightarrow \quad I = \dfrac{P}{\Delta V}$
Insert numerical values to determine the current.	$I = \dfrac{850.0 \text{ W}}{120.0 \text{ V}} = \boxed{7.083 \text{ A}}$
b. We can now use Ohm's law, i.e. the voltage across a resistor equals current times resistance (Eq. 28.25).	$\Delta V = IR \quad \rightarrow \quad R = \dfrac{\Delta V}{I}$
Plug in values.	$R = \dfrac{120.0 \text{ V}}{7.083 \text{ A}} = \boxed{16.94 \ \Omega}$

CHECK and THINK	
Given the formulas for power and Ohm's law, this is a straightforward calculation.	

59. (A) A resistor is connected to a variable power supply. Initially, the power supply's terminal potential is V_0 and the current through the resistor is I_0. The terminal potential is then doubled.

Find an expression for the new
a. current through the resistor,
b. power delivered by the power supply, and
c. power used by the resistor in terms of the initial parameters.

INTERPRET and ANTICIPATE	
The current driven by a voltage depends on resistance according to Ohm's law ($V = IR$). The power delivered by a voltage source or dissipated by a resistor is given by $P = IV$.	

SOLVE **a.** Use Ohm's law, Equation 28.25, for the initial situation.	$V_0 = I_0 R$

The final terminal potential (voltage) is double the initial value. Apply Eq. 28.25 again.	$V_f = 2V_0$ $I_f = \dfrac{V_f}{R} = \dfrac{2V_0}{R} = \boxed{2I_0}$
b. The power is current times terminal potential (Eq. 28.33).	$P_0 = I_0 V_0$ $P_f = I_f V_f = (2I_0)(2V_0) = \boxed{4V_0 I_0} = 4P_0$
c. The power dissipated by a resistor is I^2R as given in Eq. 28.34.	$P_f = I_f^2 R = (2I_0)^2 R = 4I_0^2 R$ $P_f = 4I_0^2(V_0/I_0) = \boxed{4V_0 I_0} = 4P_0$

CHECK and THINK

The power supplied by the power supply is dissipated by the resistor, so the answers for parts (b) and (c) are identical. Both the current and voltage double, so the power increases by a factor of four.

62. (N) High-voltage transmission lines carrying electricity from a generating station to a local switching station 125 km away carry a current of 850 A. How much power is lost because of resistance in each wire during transmission from the generating station to the local station if the wire's resistance is 0.240 Ω/km?

INTERPRET and ANTICIPATE

The power dissipated by a circuit element can be calculated as $P = I\Delta V$ or $P = I^2R$. We are given the current and can easily calculate the resistance of the 125 km wire.

SOLVE The resistance of a wire is the resistance per kilometer times the length.	$R = \left(\dfrac{0.240\ \Omega}{\text{km}}\right)(125\ \text{km}) = 30.0\ \Omega$
The power dissipated is then found using Eq. 28.34.	$P = I^2 R = (850\ \text{A})^2(30.0\ \Omega)$ $P = \boxed{2.17 \times 10^7\ \text{W}} = 21.7\ \text{MW}$

CHECK and THINK

This is a pretty significant amount of power lost to heat.

66. (N) A beam of electrons incident on a target carries −0.325 mA of current. How many electrons strike the target each minute?

INTERPRET and ANTICIPATE	
A current describes an amount of charge flowing per unit time. We can determine how much charge is transferred in one minute and then how many electrons comprise this amount of charge. Since the electron charge is *so* small, we expect a very large number of electrons even for currents of milliamps.	

SOLVE	
Current is charge transferred per unit time (Eq. 28.1) and the charge is equal to the number of charges N times the fundamental charge of the electron e. Solve for the number of charges N.	$I = \dfrac{q}{\Delta t} = \dfrac{Ne}{\Delta t}$ $N = \dfrac{I\Delta t}{e}$
Insert numerical values. We note that it must actually be a negative charge transferred if it's a beam of electrons.	$N = \dfrac{\left(-0.325\times10^{-3}\ \text{C/s}\right)\left(60.0\ \text{s}\right)}{-1.60\times10^{-19}\ \text{C/electron}} = \boxed{1.22\times10^{17}\ \text{electrons}}$

CHECK and THINK	
As expected, this is a very large number indeed (10^{17})!	

69. (A) If the electric field in a uniform conductor is E and v_d is the corresponding drift velocity of the free electrons in the conductor, obtain an expression for the drift speed as a function of the electric field E. Assume the cross-sectional area of the conductor is A, the number density of electrons is n, and ρ is the resistivity of the conductor.

INTERPRET and ANTICIPATE	
Both the electric field and the drift velocity can be related to the current density, so we will use these expressions to express the drift velocity as a function of the electric field.	

SOLVE	
Equation 28.5 relates the current density to the drift velocity. This can be plugged into Equation 28.19, which relates the electric field to current density.	$J = nev_d$ $E = \rho J = \rho nev_d$

Solve for the drift velocity.	$v_d = \dfrac{E}{\rho n e}$

CHECK and THINK

The drift velocity increases linearly with electric field, so the larger the field, the larger the rate at which charges are driven through the wire. We also see that a larger resistivity leads to a smaller drift velocity, which makes sense.

---

71. (N, C) Two separate 60.0-V batteries are each connected across separate resistors. One is connected across a 250.0-Ω resistor, and the other is connected across a 350.0-Ω resistor.

a. How much power is delivered by the battery in each case?

b. Which resistor consumes 1000 J of energy in the shortest time?

INTERPRET and ANTICIPATE

In both cases, we can determine the power delivered by the battery using Eq. 28.31. We will need Eq. 28.25 to find the current through the battery. We expect that the power delivered by the battery would be greater for the 250.0-Ω resistor, since the current will be greater with less resistance. Likewise, it should be the case that more energy is being delivered to the 250.0-Ω resistor each second, and thus it should consume 1000 W faster than the other resistor.

SOLVE **a.** Use Eq. 28.25 to write an expression for the current in each case.	$I_1 = \Delta V / R_1 = (60.0\text{ V})/(250.0\ \Omega)$ $I_2 = \Delta V / R_2 = (60.0\text{ V})/(350.0\ \Omega)$
Then, Use Eq. 28.31 to find the power delivered by the battery in each case.	$P_{bat,\,1} = (60.0\text{ V})(60.0\text{ V})/(250.0\ \Omega) = \boxed{14.4\text{ W}}$ $P_{bat,\,2} = (60.0\text{ V})(60.0\text{ V})/(350.0\ \Omega) = \boxed{10.3\text{ W}}$

b. $\boxed{\text{The 250.0-}\Omega\text{ resistor}}$ will receive the 1000 W first because more energy is delivered per second by the battery in its case (14.4 W). Remember that a W is a J/s!

CHECK and THINK

The current drawn through the battery will depend on the load placed across it. With less resistance, the current will be higher and the energy delivered by the battery each second (power) will be greater.

74. (N) The resistance of a copper wire is measured to be 3.50 Ω at 22.0°C. What is the resistance of this wire at 43.0°C? The temperature coefficient for copper is $\alpha = 3.9 \times 10^{-3}$ C^{-1}.

INTERPRET and ANTICIPATE Temperature affects the resistivity of the material. For most materials, a higher temperature leads to a larger resistivity, so we expect the resistance to be above 3.50 Ω.	
SOLVE The resistance of a wire depends on its resistivity (i.e. the material that it's made of), which is a temperature dependent quantity, length, and cross-sectional area according to Equation 28.24. The resistivity varies with temperature according to Equation 28.20. We will assume that the length and area of the wire are not affected by the temperature change, though in practice, thermal expansion can occur as well (see Problem 75).	$R = \rho \dfrac{\ell}{A}$ $\rho = \rho_0 \left[1 + \alpha\left(T - T_0\right)\right]$
Combining these, and defining the resistance at the reference temperature of 22.0°C to be $R_0 = \rho_0 \dfrac{\ell}{A}$, we can express the resistance as a function of temperature. (This is actually the goal of Problem 36.)	$R = R_0 \left[1 + \alpha\left(T - T_0\right)\right]$
Now, plug in numbers and solve for the resistance at 43.0°C.	$R = \left(3.50\ \Omega\right)\left[1 + \left(3.93 \times 10^{-3}\ \text{°C}^{-1}\right)\left(43.0\text{°C} - 22.0\text{°C}\right)\right]$ $R = \boxed{3.79\ \Omega}$

CHECK and THINK

The resistance is higher at the higher temperature, as is typically the case. While not a dramatic difference, it is significant enough that a careful measurement could be thrown off by such a variation.

77. (A) Consider the situation described in Problem 76. Find an expression for the drift velocity of the electrons flowing in the wire as a function of I, the number density n, and the two radii r_a and r_b.

INTERPRET and ANTICIPATE

In order to solve this problem, we must solve Problem 76 first and find an expression for the current density in the wire. Once we have that, however, we can find the drift velocity using Eq. 28.5. In order to find the current density, we will need to find an expression for the cross-sectional area of the wire and make use of Eq. 28.3.

SOLVE	
The cross-sectional area of the wire can be found by using the inner and outer radius shown in Fig. P28.76. The area is the difference between the circle mapped out by the outer radius and the circle mapped out by the inner radius. The difference between the area of those two circles is the area we see shaded in the figure.	$A = \pi r_b^2 - \pi r_a^2 = \pi \left(r_b^2 - r_a^2 \right)$
Now, use Eq. 28.3 to write an expression for the current density in the wire.	$J = \dfrac{I}{A} = \dfrac{I}{\pi \left(r_b^2 - r_a^2 \right)}$
Lastly, use Eq. 28.5 and solve for the drift velocity. Really, this is more of a drift speed since we don't have a coordinate system to use with vectors.	$J = nev_d$ $v_d = \dfrac{J}{ne} = \left(\dfrac{I}{\pi \left(r_b^2 - r_a^2 \right)} \right) \Big/ ne$ $\boxed{v_d = \dfrac{I}{ne\pi \left(r_b^2 - r_a^2 \right)}}$

CHECK and THINK
Note that as the inner radius decreases, the drift velocity will decrease. With more cross-sectional area for the charges to move through, but with the same current, the rate of movement must go down.

81. (N) A generating station and a switching station 140.0 km away are connected by a high-voltage aluminum transmission line that has a current of 850.0 A. Each wire of the transmission line has a cross-sectional area of 750.0 mm^2, and the density of free electrons in aluminum is 6.022×10^{28} electrons/m^3. How much time does it take for one electron to travel from the generating station to the switching station in this wire?

INTERPRET and ANTICIPATE
To determine the time for a particular electron to travel the 140 km, we need to first calculate the drift speed. We expect this speed to be very slow (a fraction of a meter per hour) so the electron should take a long time to travel this distance.

SOLVE
The drift velocity is related to the current density (Eq. 28.3) according to Equation 28.5.

$$J = \frac{I}{A} = nqv_d$$

Solve this for the drift speed.

$$v_d = \frac{I}{nqA}$$

The drift speed, as it sounds, is the average distance an electron travels per unit time. Use this to solve for the time interval.

$$v_d = \frac{d}{\Delta t} \quad \rightarrow \quad \Delta t = \frac{d}{v_d} = \frac{dnqA}{I}$$

We now insert numerical values.

$$\Delta t = \frac{\left(1.40\times10^5 \text{ m}\right)\left(6.022\times10^{28} \text{ m}^{-3}\right)\left(1.60\times10^{-19} \text{ C}\right)\left(750\times10^{-6} \text{ m}^2\right)}{850 \text{ A}}$$

$$\Delta t = \boxed{1.19\times10^9 \text{ s}} = 37.7 \text{ yr}$$

CHECK and THINK
This is in fact a *very* long time—nearly 38 years to travel less than 100 miles!!

29

Direct Current (DC) Circuits

6. (N) If the switch in Figure P29.4 is closed, use the values given in the figure to find the potential difference between the points

a. $V_b - V_a$,

b. $V_c - V_b$,

c. $V_d - V_c$, and

d. $V_a - V_d$.

If you worked Problem 5, compare your answers in each case.

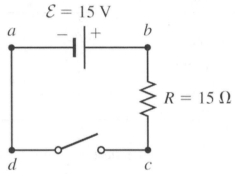

Figure P29.4

INTERPRET and ANTICIPATE
The potential difference can be found between any two points in a circuit. An ideal emf source (without an internal resistance) maintains a constant *potential difference* between its terminals. An ideal wire maintains a constant *potential* (zero potential difference) between any two points. For a closed loop, Kirchoff's loop tells us that the sum of all the potential increases must equal the sum of all the potential decreases (or the potential difference around a closed loop adds up to zero).

SOLVE	
a. $V_b - V_a$ is the voltage across the battery. Point b is at a higher potential than point a, which we know because the battery indicates that the right side (+) is at high potential and the left side (−) is at low potential.	**a.** $V_b - V_a = \boxed{+15 \text{ V}}$

150

Chapter 29 – Direct Current (DC) Circuits

c., d. Let's skip part (b) for a moment. Notice that points d and a are connected by an ideal wire. When the switch is closed, the same is true between points c and d. An ideal wire forces both sides to be at the same potential and therefore a zero potential difference.	**c.** $V_d - V_c = \boxed{0}$ **d.** $V_a - V_d = \boxed{0}$
b. Let's return to part (b). We know from Kirchoff's loop rule that the total potential difference around a closed loop is zero, so there must be a 15 V drop between points b and c (i.e. the battery increases potential by 15 volts, so there must be a 15 volt drop across the resistor). A second way to see this is that the current, given by Ohm's law ($I = V/R$) is 1 A, so the voltage across it is 15 V. The current goes clockwise through the circuit and from high to low potential in the resistor, so it is a potential drop.	**b.** $V_c - V_b = \boxed{-15 \text{ V}}$

CHECK and THINK
Using our knowledge of emf devices, ideal wires, and resistors, we are able to determine the voltage that would be read by a voltmeter between any two points in the circuit. In problem 5 ($V_b - V_a = +15$ V, $V_c - V_b = 0$, $V_d - V_c = -15$ V, $V_a - V_d = 0$), the sum of potential changes still satisfies Ohm's law, but as the circuit is open, there is no current through the circuit and the potential drop is across the open switch. The battery maintains a 15 V potential difference between points a and b in either case.

8. (A) Two circuits made up of identical ideal emf devices and resistors are shown in Figure P29.8. What is the potential difference $V_b - V_a$
a. for circuit 1 and
b. for circuit 2?
c. Find expressions for both the current in the resistor in circuit 1 and the current in the resistor in circuit 2, and compare them.

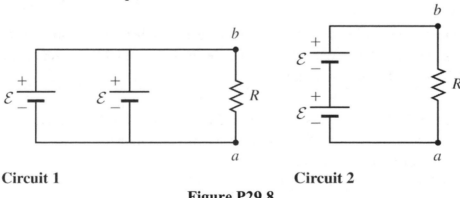

Circuit 1   Circuit 2
Figure P29.8

151

INTERPRET and ANTICIPATE Given an arrangement of emf devices, we can determine the total potential difference across the resistor.	

SOLVE **a.** In circuit 1, the two emf devices are in parallel. The ideal wire across the bottom is all at the same potential (imagine for instance that it's at 0 volts). Each battery produces a potential difference such that the wire at the top is at a voltage of $+\mathcal{E}$ (so each battery connects between wires at 0 and $+\mathcal{E}$). In general, circuit elements in parallel have the same potential difference across them.	$\Delta V_1 = V_b - V_a = \boxed{+\mathcal{E}}$
b. These emf devices are in series. Again imagining the bottom wire to be at 0 volts, the battery on the bottom produces a potential difference such that the wire between the batteries is at $+\mathcal{E}$. The battery on the top similarly produces a potential difference such that the wire on top is at $+\mathcal{E}$ above the potential between the batteries, or $+2\mathcal{E}$. In general, the total voltage across circuit elements in series is the sum of the voltages across each.	$\Delta V_2 = V_b - V_a = \boxed{+2\mathcal{E}}$
c. In both cases, we use Ohm's law with the potential difference we determined above.	$I_1 = \dfrac{\Delta V_1}{R} = \boxed{\dfrac{\mathcal{E}}{R}}$ $I_2 = \dfrac{\Delta V_1}{R} = \boxed{\dfrac{2\mathcal{E}}{R}}$

CHECK and THINK Circuit elements in parallel have the same voltage across them while those in series have voltages that add together. Two batteries in series can drive a larger current than either a single battery or batteries in parallel.

11. (N) The terminal voltage of a real battery that delivers 15.0 W of power to a load resistor is 13.4 V, and its emf is 16.0 V. What is the resistance of
a. the load resistor in this circuit and
b. the internal resistor of the battery?

INTERPRET and ANTICIPATE

The power dissipated by the load resistor depends on the voltage across it and the resistance ($P = \Delta V^2/R$). With the emf and terminal voltage of the battery, we can determine the internal resistance.

SOLVE **a.** The power delivered to the resistor can be determined with the voltage across it and its resistance (or as $P = I^2R$ if we know the current).	$$P = \frac{(\Delta V)^2}{R} \quad \rightarrow \quad R = \frac{(\Delta V)^2}{P}$$
Substitute the given values.	$$R = \frac{(13.4 \text{ V})^2}{15.0 \text{ W}} = \boxed{12.0 \; \Omega}$$
b. The terminal voltage (the 13.4 volts of the "real battery") depends on the emf of the battery and the internal resistance as given by Eq. 29.3. Solve for the internal resistance.	$$\Delta V = \mathcal{E} - Ir$$ $$r = \frac{\mathcal{E} - \Delta V}{I}$$
We need to determine the current I. Since we know the terminal voltage and the resistance of the load resistor from part (a), we can use Ohm's law to determine the current through the resistor, which is the current through the circuit.	$$\Delta V = IR \quad \rightarrow \quad I = \frac{\Delta V}{R}$$ $$I = \frac{13.4 \text{ V}}{12.0 \; \Omega} = 1.12 \text{ A}$$
Now, insert values to determine the internal resistance r.	$$r = \frac{16.0 \text{ V} - 13.4 \text{ V}}{1.12 \text{ A}} = \boxed{2.32 \; \Omega}$$

CHECK and THINK

The real battery has a terminal voltage a bit lower than the emf of the ideal battery. The internal resistance is relatively small, as we might expect.

13. (N) Eight real batteries, each with an emf of 5.00 V and an internal resistance of 0.200 V, are connected end to end in a loop as in Figure P29.13. What is the terminal voltage across one of the batteries between points *a* and *b*?

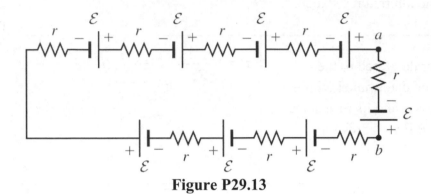

Figure P29.13

INTERPRET and ANTICIPATE	
The circuit shown contains elements that are all in series with each other, so the voltages of the batteries and the internal resistances add together. We can determine the total voltage of all the batteries and the total internal resistance and treat the entire network as if it was a single battery.	

SOLVE	
There are eight batteries in series. According to Kirchoff's loop rule, the total voltage increase due to these batteries is the sum of all eight. Similarly, the resistors are in series, so use Equation 29.6 to find the equivalent resistance.	$\mathcal{E}_{tot} = 8(5.00 \text{ V}) = 40.0 \text{ V}$ $R_{tot} = 8(0.200 \text{ }\Omega) = 1.60 \text{ }\Omega$
We can now determine the current in the circuit.	$I = \dfrac{\mathcal{E}_{tot}}{r_{tot}} = \dfrac{40.0 \text{ V}}{1.60 \text{ }\Omega} = 25.0 \text{ A}$
Calculate the terminal potential across this single battery with Eq. 29.3.	$V_{ab} = \mathcal{E} - Ir$ $V_{ab} = 5 \text{ V} - (25 \text{ A})(0.2 \text{ }\Omega) = \boxed{0 \text{ V}}$

CHECK and THINK	

In this case, there is nothing else connected to the batteries, so a large current is produed as we're basically shorting out the batteries. We know from Kirchoff's loop rule that the voltage change around the entire loop must be zero, so all of the voltage increases from

Chapter 29 – Direct Current (DC) Circuits

the batteries must equal the voltage decreases across the resistors. Since each battery is identical, that means the voltage of each battery must be equal to the voltage drop across its internal resistance. We'd get the same result by simply shorting out a single battery.

16. (N) In Figure P29.15, three resistors are connected to an ideal emf device. The resistances are $R_1 = 13.4\ \Omega$, $R_2 = 20.5\ \Omega$, and $R_3 = 9.8\ \Omega$. The current through the last resistor is 7.55 mA.

a. What is the current through the other two resistors?

b. What is the terminal potential of the emf device?

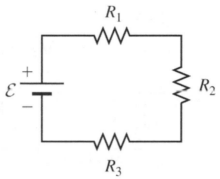

Figure P29.15

INTERPRET and ANTICIPATE We are given a series circuit. There is a single current through the entire circuit and the voltage across each can be determined with Ohm's law. The total change in potential around the closed circuit must be zero according to Kirchoff's loop rule.	

SOLVE **a.** The current is the same for any circuit elements in series. That is, there's a single current going around this loop.	$I_1 = I_2 = I_3 = 7.55$ mA
b. According to Kirchoff's loop rule, the total change in potential around a closed loop is zero. In other words, the sum of all the potential increases (the emf) equals the sum of all potential decreases (due to the resistors).	$\mathcal{E} - IR_1 - IR_2 - IR_3 = 0$ $\mathcal{E} = I(R_1 + R_2 + R_3)$ $\mathcal{E} = (7.55 \times 10^{-3}\ \text{A})(13.4\ \Omega + 20.5\ \Omega + 9.8\ \Omega)$ $\mathcal{E} = \boxed{0.330\ \text{V}}$

CHECK and THINK Since this is a series circuit, the current is the same through each resistor and the emf must equal the total potential drop across all the resistors.

19. (N) An ideal emf device with $\mathcal{E} = 9.00$ V is connected to two resistors in series. One of the resistors has a resistance of 145 Ω, and the other has unknown resistance R. If the current through the emf device is 0.0155 A, what is the resistance R?

INTERPRET and ANTICIPATE

We can use Kirchoff's loop rule in order to create an equation that involves the unknown resistance and relates it to the other unknown quantities. This circuit has only one loop and we choose to follow the current as we create the equation. We sketch the circuit and choose to orient the emf device such that the current travels clockwise around the circuit.

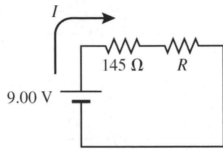

Figure P29.19ANS

SOLVE

Beginning with the emf device, the potential goes up by 9.00 V as we cross from the low potential to the high potential side, traveling clockwise around the circuit. Then, the potential drops based on Ohm's Law, $\Delta V = IR$, across each resistor.	$9.00\ \text{V} - I(145\ \Omega) - IR = 0$ $9.00\ \text{V} - (0.0155\ \text{A})(145\ \Omega) - (0.0155\ \text{A})R = 0$
Now, solve for the unknown resistance.	$R = \dfrac{9.00\ \text{V} - (0.0155\ \text{A})(145\ \Omega)}{(0.0155\ \text{A})} = \boxed{436\ \Omega}$

CHECK and THINK

When beginning this problem, one of the things that we could say was certainly true was that the sum of the changes in potential about a closed, non-repeating, loop in the circuit must be zero (Kirchoff's loop rule). We expressed this truth and examined it to see if it allowed us to solve for the unknown resistance. It did in this case!

---

24. The emf devices and lightbulbs in Figure P29.24 are identical.

a. (A) Find an expression for the current in each bulb.

b. (C) List the bulbs in order from brightest to dimmest. Explain your answer.

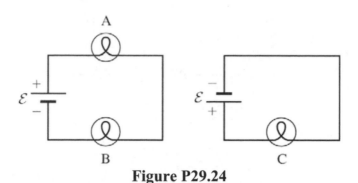

Figure P29.24

INTERPRET and ANTICIPATE	
We can use Ohm's law with the voltage and total resistance of the circuit.	

SOLVE	Loop with bulbs A and B:
a. Each bulb is assumed to have the same resistance R. For resistors in series, the equivalent resistance is the sum of the resistances (Eq. 29.6). We can then use Ohm's law to relate the emf to the current. Since each circuit is a single loop, the current through bulbs A and B must be the same.	$R_{eq} = R + R = 2R$ $$I_A = I_B = \boxed{\dfrac{\mathcal{E}}{2R}}$$ Loop with bulb C: $$I_C = \boxed{\dfrac{\mathcal{E}}{R}}$$
b. The brightness of the bulb increases with the current. (Roughly speaking, the brightness depends on the power.) Since bulbs A and B have the same current which is half that in C, they should be at the same brightness but be dimmer than bulb C.	Brightness: $\boxed{A = B < C}$

CHECK and THINK	
With Ohm's law, we are able to determine the current (which is related to the brightness) for each bulb.	

27. (N) Determine the currents through the resistors R_2, R_5, R_6, and R_7 in the set of junctions and branches shown in Figure P29.27. *Hint*: Use Kirchhoff's junction rule, be sure to consider the branches where a current is shown, and assume the branches that appear disconnected are connected to other parts of the circuit.

15.0 A   3.0 A

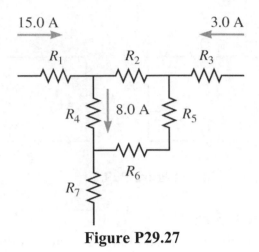

Figure P29.27

INTERPRET and ANTICIPATE Kirchoff's junction rule says the sum of all the currents into a junction equals the sum of all currents out of the junction. We'll apply this at each junction.	

SOLVE There are three junctions that we identify in the circuit diagram.	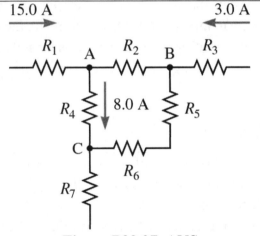 **Figure P29.27aANS**
Consider junction A first. We assume current I_2 (the current through resistor) is to the right, from junction A to B. Using the junction rule, we see that the current must be 7 A. Note that if we assumed the other direction on current I_2, we would just end up with a minus sign in front, which would indicate that the current was in the opposite direction of what we guessed.	Junction A: current in = current out $I_1 = I_2 + I_4$ $15.0 \text{ A} = I_2 + 8.0 \text{ A}$ $\boxed{I_2 = 7.0 \text{ A}}$

At junction B, the current I_2 is into the junction as is the 3.0 A current through R_3. Since resistors R_5 and R_6 are in series, the current must be the same through both, so I_5 = I_6 and we assume this current is out of junction B (current from junctions B to C).	Junction B: current in = current out $I_2 + I_3 = I_5$ $I_2 + 3\,\text{A} = I_5$ $\boxed{I_5 = I_6 = 10.0\,\text{A}}$
Finally, at junction C, I_4 and I_6 both go into the junction, so we assume I_7 is out of the junction (going down out of this part of the circuit).	Junction C: current in = current out $I_4 + I_6 = I_7$ $8.0\,\text{A} + 10.0\,\text{A} = I_7$ $\boxed{I_7 = 18.0\,\text{A}}$
We can sketch the final circuit to confirm that everything makes sense.	 **Figure P29.27bANS**

CHECK and THINK

In each case, charge conservation tells is that the current into each junction must equal the current out of the junction.

31. (A) Six resistors with resistances $7R$, $6R$, $2R$, R, $R/2$, and $R/4$ are connected in parallel. What is the equivalent resistance of this combination?

INTERPRET and ANTICIPATE

Because all of the resistors are in parallel, we can find the equivalent resistance using Eq. 29.7. The equivalent resistance should be less than any of the individual resistances because this is true when determining equivalent resistance of resistors in parallel.

SOLVE Write Eq. 29.7, using each of the resistances in the problem statement.	$$\frac{1}{R_{eq}} = \sum_{i=1}^{6} \frac{1}{R_i} = \frac{1}{7R} + \frac{1}{6R} + \frac{1}{2R} + \frac{1}{R} + \frac{1}{R/2} + \frac{1}{R/4}$$ $$\frac{1}{R_{eq}} = \frac{6}{42R} + \frac{7}{42R} + \frac{21}{42R} + \frac{42}{42R} + \frac{84}{42R} + \frac{168}{42R}$$
Now, solve for the equivalent resistance.	$$\frac{1}{R_{eq}} = \frac{6}{42R} + \frac{7}{42R} + \frac{21}{42R} + \frac{42}{42R} + \frac{84}{42R} + \frac{168}{42R} = \frac{328}{42R}$$ $$\frac{1}{R_{eq}} = \frac{164}{21R}$$ $$\boxed{R_{eq} = \frac{21R}{164}}$$

CHECK and THINK
Note that the equivalent resistance is less than any of the individual resistances. If it is more comfortable, you could combine two resistors at a time, where each iterative equivalent resistance would be in parallel with the remaining resistances. You will eventually end up with one final equivalent resistance from that process that will be the same as the result here.

36. (A) Each resistor shown in Figure P29.36 has resistance R. An ideal emf device ($\mathcal{E}$) is connected to points a and b via two leads (not shown in the figure). Find an expression for the current through the emf device.

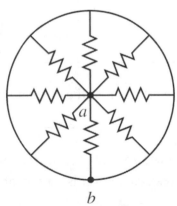

b
Figure P29.36

INTERPRET and ANTICIPATE
Note that all points around the outside of the circuit in Figure P29.36 are electrically equivalent, or at the same electric potential. When connecting the ideal emf device to the center and the outside, each of the resistors will be in parallel with the emf device. We

can redraw the circuit as shown in Figure P29.36ANS. This means that each of the resistors will be in parallel with each other. We can find the equivalent resistance of this arrangement to model the circuit as an emf device connected to a single resistor, and then use Ohm's Law to find the current through the emf device.

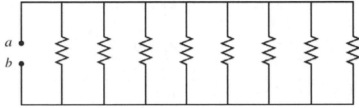

Figure P29.36ANS

SOLVE	
Use Eq. 29.7 to find the equivalent resistance of eight identical resistors in parallel.	$$\frac{1}{R_{eq}} = \sum_{i=1}^{8} \frac{1}{R_i} = \frac{1}{R} + \frac{1}{R} + \frac{1}{R} + \frac{1}{R} + \frac{1}{R} + \frac{1}{R} + \frac{1}{R} + \frac{1}{R}$$ $$\frac{1}{R_{eq}} = \frac{8}{R}$$ $$R_{eq} = R/8$$
Now, use Ohm's law to find the current that must pass through the emf device, connected at a and b, with the single equivalent resistance connected to it.	$$I = \frac{\mathcal{E}}{R_{eq}} = \frac{\mathcal{E}}{R/8} = \boxed{8\mathcal{E}/R}$$

CHECK and THINK

When the arrangement of resistors might seem odd, follow the branches, or paths current can follow, in order to aid in redrawing the circuit in a more familiar form. Just be sure to preserve all connections as they are, and not create new ones that did not exist originally. Ohm's law can be applied to the equivalent circuit in order to determine the current drawn through the emf device when connected to the arrangement of resistors.

40. (N) The emf in Figure P29.40 is 4.54 V. The resistances are $R_1 = 13.0 \ \Omega$, $R_2 = 26.0 \ \Omega$, and $R_3 = 39.0 \ \Omega$. Find
a. the current in each resistor,
b. the power consumed by each resistor, and
c. the power supplied by the emf device.

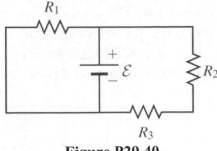

Figure P29.40

INTERPRET and ANTICIPATE Given a circuit, we can use Kirchoff's laws to relate voltages and currents in different parts of the circuit. Resistor 1 is the smallest resistance and not in series with any others, so we might guess that the largest current is I_1.	

SOLVE **a.** We first label currents in the circuit, anticipating that they leave the positive terminal of the battery.	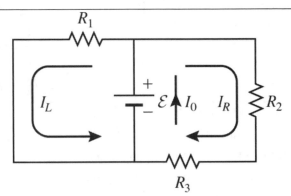 **Figure P29.40aANS**
Write Kirchoff's loop rule going counter-clockwise around the loop on the left and solve for I_L. The current flows from high to low potential, so the voltage across the resistor is negative. We find the current through resistor 1 is $I_1 = 0.349$ A.	$\mathcal{E} - I_L R_1 = 0$ $I_L = \dfrac{\mathcal{E}}{R_1} = \dfrac{4.54 \text{ V}}{13.0 \text{ } \Omega} = \boxed{0.349 \text{ A}}$
Repeat for the loop on the right. The current through resistors 1 and 2 must be the same since they are sin series. We find $I_2 = I_3 = 0.0698$ A.	$\mathcal{E} - I_R R_2 - I_R R_3 = 0$ $I_R = \dfrac{\mathcal{E}}{R_2 + R_3} = \dfrac{4.54 \text{ V}}{(26.0 + 39.0) \text{ } \Omega} = \boxed{0.0698 \text{ A}}$

b. The power consumed by a resistor is $P = I^2 R$.	$P_1 = I_1^2 R_1 = (0.349 \text{ A})^2 (13.0 \ \Omega) = \boxed{1.58 \text{ W}}$ $P_2 = I_2^2 R_2 = (0.0698 \text{ A})^2 (26.0 \ \Omega) = \boxed{0.127 \text{ W}}$ $P_3 = I_3^2 R_3 = (0.0698 \text{ A})^2 (39.0 \ \Omega) = \boxed{0.190 \text{ W}}$
c. The power delivered by the emf device is $P = I\mathcal{E}$. The current through the battery can be found using Kirchoff's junction rule at the junction on the top branch. The sum of the currents into the junction (in this case I_0) must equal the current leaving the junction (I_L and I_R).	$P = I_0 \mathcal{E}$ $I_0 = I_R + I_L = 0.349 \text{ A} + 0.0698 \text{ A} = 0.419 \text{ A}$
Calculate the power. Note that the total current drawn by the resistors, $P_1 + P_2 + P_3 = 1.90$ W as well, which makes sense—the power supplied by the emf device is dissipated by the resistors.	$P = I_0 \mathcal{E} = (0.419 \text{ A})(4.54 \text{ V}) = \boxed{1.90 \text{ W}}$

CHECK and THINK

We're able to analyze this circuit using Kirchoff's rules. Notice that this circuit can actually be drawn in a simpler form as shown, with R_2 and R_3 (in series) in parallel with R_1. Since these branches each have the emf of the battery across it, we could use Ohm's law directly. We could also find the equivalent resistance of this network and the total current drawn I_0.

$$\frac{1}{R_{eq}} = \frac{1}{R_1} + \frac{1}{R_2 + R_3} = \frac{1}{13} + \frac{1}{26 + 39} = 0.923 \ \Omega^{-1}$$

$$R_{eq} = 10.83 \ \Omega$$

$$I_0 = \frac{\mathcal{E}}{R_{eq}} = \frac{4.54 \text{ V}}{10.83 \ \Omega} = 0.419 \text{ A} \text{ , which agrees with what we found above.}$$

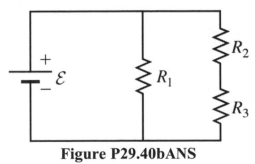

Figure P29.40bANS

Chapter 29 – Direct Current (DC) Circuits

43. (N) The emfs in Figure P29.43 are $\mathcal{E}_1 = 6.00$ V and $\mathcal{E}_2 = 12.0$ V. The resistances are $R_1 = 15.0\ \Omega$, $R_2 = 30.0\ \Omega$, $R_3 = 45.0\ \Omega$, and $R_4 = 60.0\ \Omega$. Find the current in each resistor when the switch is
a. open and
b. closed.

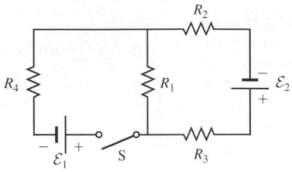

Figure P29.43

INTERPRET and ANTICIPATE When the switch is open, the branch on the left is open and does not factor into the circuit. With the switch closed, we have two emf devices which are not in series or parallel, so we will need to use Kirchoff's rules.	

SOLVE **a.** With the switch open, the branch on the left will not have a current through it, as it is not a closed loop.	$\boxed{I_4 = 0}$
The loop on the right consists of three resistors in series and an emf device. We can find the Equivalent resistance (Eq. 29.6) and then apply Ohm's law to find the current. The current is the same through all circuit elements in series.	$R_{eq} = R_1 + R_2 + R_3$ $R_{eq} = \left(15.0 + 30.0 + 45.0\right)\ \Omega = 90.0\ \Omega$ $I = \dfrac{\mathcal{E}_2}{R_{eq}} = \dfrac{12.0\text{ V}}{90.0\ \Omega} = 0.133\text{ A}$ $\boxed{I_1 = I_2 = I_3 = 0.133\text{ A}}$

164

© 2016 Cengage Learning. All Rights Reserved. May not be scanned, copied or duplicated, or posted to a publicly accessible website, in whole or in part.

b. Now, we imagine closing the switch. First, we label each branch with currents.	 **Figure P29.43ANS**
Write Kirchoff's loop rule for the loop on the left. The voltage decreases through a resistor in the direction of the current and increases in the direction opposite the current. Starting at the bottom left corner, we go counter-clockwise through the loop: increase in potential (– to +) through the emf device, increase in potential through resistor 1, and decrease in potential through resistor 4. This determines the sign of each term.	$$\mathcal{E}_1 + I_1 R_1 - I_4 R_4 = 0 \qquad (1)$$
Write Kirchoff's loop rule for the loop on the right. Starting at the upper right and going clockwise: increase in potential (– to +) through the emf device, decrease in potential through resistor 3, increase in potential through resistor 1, and decrease in potential through resistor 2. Notice that current I_2 goes through both R_2 and R_3.	$$\mathcal{E}_2 - I_2 R_3 + I_1 R_1 - I_2 R_2 = 0$$ $$\mathcal{E}_2 + I_1 R_1 - I_2 \left(R_2 + R_3 \right) = 0 \qquad (2)$$
We can also use Kirchoff's junction rule at either the top or bottom junction.	$$I_4 + I_1 + I_2 = 0 \qquad (3)$$

Chapter 29 – Direct Current (DC) Circuits

We have three unknown currents and three equations, so it's just a matter of algebra now. For instance, we can try to eliminate I_1 from a couple equations to end up with two equations and two unknown currents. Solve Eq. (3) for I_1 and substitute into (2).	$\mathcal{E}_2 + \left(-I_2 - I_4\right)R_1 - I_2\left(R_2 + R_3\right) = 0$ $\mathcal{E}_2 - I_4 R_1 - I_2\left(R_1 + R_2 + R_3\right) = 0 \qquad (4)$
Now, solve Eq. (3) for I_1 and insert it into (1).	$\mathcal{E}_1 + \left(-I_2 - I_4\right)R_1 - I_4 R_4 = 0$ $\mathcal{E}_1 - I_2 R_1 - I_4\left(R_1 + R_4\right) = 0 \qquad (5)$
Notice that with Eq. (4) and (5) we have two equations in terms of only two unknowns, I_2 and I_4. We might solve Eq. (5) for I_2 and substitute it into Eq. (4).	$I_2 = \dfrac{\mathcal{E}_1 - I_4\left(R_1 + R_4\right)}{R_1}$ $\mathcal{E}_2 - I_4 R_1 - \left[\dfrac{\mathcal{E}_1 - I_4\left(R_1 + R_4\right)}{R_1}\right]\left(R_1 + R_2 + R_3\right) = 0$
Let's insert values and see what we get (dropping units and extra digits for clarity). The resulting negative sign simply means that our initial guess for the direction of current 4 was the opposite of its actual direction. That is, current 4 really goes upward through resistor 4.	$12.0 - I_4\left(15.0\right) - \left[\dfrac{6.00 - I_4\left(75.0\right)}{15.0}\right]\left(90.0\right) = 0$ $12.0 - \left(15.0\right)I_4 - 36.0 + \left(450.0\right)I_4 = 0$ $I_4 = \dfrac{24.0}{435.0} = \boxed{0.0552\ \text{A}} \qquad (6)$
Ok, that was a lot of work, but once we have one current, life gets easier. Plug I_4 (Eq. (6)) into Eq. (5) to get I_2.	$6.00 - I_2\left(15.0\right) - \left(0.0552\right)\left(75.0\right) = 0$ $I_2 = \dfrac{1.86}{15.0} = \boxed{0.124\ \text{A}}$
R_2 and R_3 are both on the same branch, so they must have the same current.	$I_3 = I_2 = \boxed{0.124\ \text{A}}$
Finally, use the loop rule (Eq. (3)) to get I_1.	$I_1 = -I_2 - I_4 = \boxed{-0.179\ \text{A}}$

CHECK and THINK

Part (b) was significantly more work given that the circuit was no longer a simple voltage

166

Chapter 29 – Direct Current (DC) Circuits

source and equivalent resistance, but armed with Kirchoff's rules and algebra, we can determine the currents in every branch. The key is to be careful and proceed in an organized way—one minus sign or dropped parentheses can easily cause all of your answers to change! Notice also that the currents in part (b) are not at all obvious after doing part (a). With multiple batteries and multiple branches, it's not always intuitive how the currents will work out.

46. (N) Figure P29.46 shows a circuit with a 12.0-V battery connected to four resistors. How much power is delivered to each resistor?

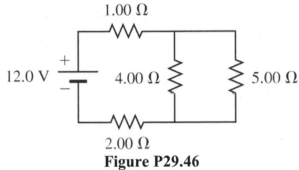

Figure P29.46

INTERPRET and ANTICIPATE

To determine the power dissipated by each resistor, we need to determine either the potential difference across the resistor or the current through the resistor $\left(P = \dfrac{\Delta V^2}{R} = I^2 R\right)$. We can start by finding the equivalent resistance, which will allow us to determine the total current drawn from the battery.

SOLVE We start by determining the equivalent resistance of the circuit and then the total current through the battery. The two resistors on the right side are in parallel, so we use Equation 29.7.	$\dfrac{1}{R_p} = \dfrac{1}{4.00\ \Omega} + \dfrac{1}{5.00\ \Omega} = \dfrac{9}{20.0\ \Omega}$ $R_p = \dfrac{20.0\ \Omega}{9} = 2.22\ \Omega$
This resultant is then in series with the other two resistors, so use Eq. 29.6.	$R_{eq} = 1.00\ \Omega + 2.22\ \Omega + 2.00\ \Omega = 5.22\ \Omega$

The current supplied by the battery can be found using Ohm's law.	$I_{battery} = \dfrac{\Delta V}{R_{eq}} = \dfrac{12.0\ \text{V}}{5.22\ \Omega} = 2.30\ \text{A}$
The power delivered to the 1 Ω and 2 Ω resistors is given by $P = I^2R$. Since they are in series with the battery, the current for both is the same as that through the battery.	$P_{1.00\ \Omega} = (2.30\ \text{A})^2 (1.00\ \Omega) = \boxed{5.28\ \text{W}}$ $P_{2.00\ \Omega} = (2.30\ \text{A})^2 (2.00\ \Omega) = \boxed{10.6\ \text{W}}$
The voltage drop across the 1 Ω and 2 Ω resistors plus the voltage across the parallel combination must add up to the voltage of the battery. This is a consequence of Kirchoff's loop rule. We can calculate the voltage across the 1 Ω and 2 Ω resistors using Ohm's law, $\Delta V = IR$.	$\Delta V_{1.00\ \Omega} = (2.30\ \text{A})(1.00\ \Omega) = 2.30\ \text{V}$ $\Delta V_{2.00\ \Omega} = (2.30\ \text{A})(2.00\ \Omega) = 4.60\ \text{V}$
The voltage across the two parallel resistors can now be determined. This is the voltage across both the 4 Ω and 5 Ω resistors.	$\Delta V_p = 12.0\ \text{V} - 2.30\ \text{V} - 4.60\ \text{V}$ $\Delta V_p = 5.10\ \text{V} = \Delta V_{4.00\ \Omega} = \Delta V_{5.00\ \Omega}$
Finally, calculate the power dissipated by each of these using $P = \Delta V^2/R$.	$P_{4.00\ \Omega} = \dfrac{\Delta V^2}{R} = \dfrac{(5.10\ \text{V})^2}{(4.00\ \Omega)} = \boxed{6.50\ \text{W}}$ $P_{5.00\ \Omega} = \dfrac{\Delta V^2}{R} = \dfrac{(5.10\ \text{V})^2}{(5.00\ \Omega)} = \boxed{5.20\ \text{W}}$

CHECK and THINK

We can double-check this by determining the power drawn by the equivalent circuit,

$P_{tot} = \dfrac{\Delta V^2}{R} = \dfrac{(12\ \text{V})^2}{5.22\ \Omega} = 27.6\ \text{W}$. This is indeed equal to the power drawn by all four

resistors added together (5.28 W + 10.6 W + 6.50 W + 5.20 W).

49. (N) Three resistors with resistances $R_1 = R/2$ and $R_2 = R_3 = R$ are connected as shown, and a potential difference of 225 V is applied across terminals a and b (Fig. P29.49).
a. If the resistor R_1 dissipates 75.0 W of power, what is the value of R?

b. What is the total power supplied to the circuit by the emf?

c. What is the potential difference across each of the three resistors?

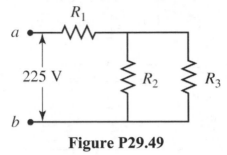

Figure P29.49

INTERPRET and ANTICIPATE	
The power dissipated by a resistor can be expressed as $P = I^2 R = \dfrac{\Delta V^2}{R}$. We can use these relationships to connect the given and desired information.	

SOLVE	
a. Note that R_2 and R_3 are in parallel, so according to Equation 29.6, their equivalent resistance R_{23} is $R/2$, the same resistance as R_1.	$\dfrac{1}{R_{23}} = \dfrac{1}{R_2} + \dfrac{1}{R_3} = \dfrac{2}{R} \quad \rightarrow \quad R_{23} = \dfrac{1}{2} R$
Since R_1 and the equivalent resistance R_{23} are the same, the potential difference across them is the same, or each has 250/2 V = 112.5 V across it.	$\Delta V_1 = \dfrac{225 \text{ V}}{2} = 112.5 \text{ V}$ $\Delta V_2 = \Delta V_3 = \Delta V_1$
We can now determine the resistance of R_1 using the power and voltage given.	$P_1 = \dfrac{\Delta V_1^2}{R_1} = \dfrac{\Delta V_1^2}{R/2} \quad \rightarrow \quad R = \dfrac{2 \Delta V_1^2}{P_1}$ $R = \dfrac{2(112.5 \text{ V})^2}{75.0 \text{ W}} = \boxed{338 \ \Omega}$
b. For resistors 2 and 3, the potential difference is the same as for resistor 1, but the resistance is R instead of $R/2$, therefore the power dissipated by either resistor 2 or 3 is half that of resistor 1.	$P_2 = \dfrac{\Delta V_{23}^2}{R_2} = \dfrac{(112.5 \text{ V})^2}{338 \ \Omega} = 37.5 \text{ W} = P_3$

The total power is the sum of these three.	$P_{tot} = P_1 + P_2 + P_3$ $P_{tot} = 75.0 \text{ W} + 37.5 \text{ W} + 37.5 \text{ W}$ $P_{tot} = \boxed{1.50 \times 10^2 \text{ W}}$
c. We determined the potential difference in part (a). We write it here to three significant figures.	$\Delta V_1 = \Delta V_2 = \Delta V_3 = \boxed{113 \text{ V}}$

CHECK and THINK

Electrical power can be related to voltage and resistance to relate the quantities needed. We see that the arrangement of resistors ($R/2$ in series with a parallel combination of R and R) means that the voltage across each is the same. However, the power (which is inversely proportional to resistance but proportional to voltage squared) is not the same for all three resistors.

55. (N) A 650.0-Ω resistor is connected across the terminals of a 12.0-nF capacitor that carries an initial charge of 7.40 μC.
a. What is the magnitude of the maximum current in the resistor?
b. What is the current in the resistor 5.00 μs after the circuit is completed and the capacitor begins to discharge through the resistor?
c. How much charge remains in the capacitor 5.00 μs after the circuit is completed?

INTERPRET and ANTICIPATE

This is a discharging RC circuit. As the charge on the capacitor decreases from its initial value, the current will initially be greatest and will exponentially decrease to zero with a time constant of $\tau = RC$.

SOLVE **a.** The current in a discharging capacitor decreases according to Equation 29.17 where $I_{max} = Q_{max}/RC$ and the time constant is $\tau = RC$ (Eq. 29.13).	$I(t) = -I_{max} e^{-t/\tau}$ $I_{max} = \dfrac{Q_{max}}{RC} = \dfrac{7.40 \times 10^{-6} \text{ C}}{(650 \ \Omega)(12.0 \times 10^{-9} \text{ F})} = \boxed{0.949 \text{ A}}$

b. Using Eq. 29.17 above, we can determine the current at this particular time.	$I(t) = -I_{max} e^{-t/RC}$ $I(t) = -(0.949 \text{ A}) \exp\left[-\dfrac{5.00 \times 10^{-6} \text{ s}}{(650 \ \Omega)(12.0 \times 10^{-9} \text{ F})} \right]$ $I(t) = \boxed{-0.500 \text{ A}}$
c. The charge is given by Eq. 29.16.	$q(t) = Q_{max} e^{-t/RC}$ $q(t) = (7.40 \ \mu\text{C}) \exp\left[-\dfrac{5.00 \times 10^{-6} \text{ s}}{(650 \ \Omega)(12.0 \times 10^{-9} \text{ F})} \right]$ $q(t) = \boxed{3.90 \ \mu\text{C}}$

CHECK and THINK

For a discharging RC circuit, both the charge and current decrease exponentially. Given the resistance, capacitance, and initial charge, we can calculate both the charge and current at any time as it discharges.

58. (A) A real battery with internal resistance r and emf $\mathcal{E}$ is used to charge a capacitor with capacitance C. A resistor R is put in series with the battery and the capacitor when charging.

a. Write an expression for the time constant of this circuit.

b. Find an expression for the time when the capacitor has reached half its maximum charge.

c. Find an expression for the current through the capacitor at this time.

INTERPRET and ANTICIPATE

If we find the equivalent resistance in the circuit, we can make use of Eq. 29.13, 29.14, and 29.15 to model the charging behavior of the RC circuit. The internal resistance of the battery is in series with the battery and will thus be in series with the resistor, R, and the capacitor, C. Thus, we can use Eq. 29.6 to find the equivalent resistance. Note that the resistance will be higher than if we used an ideal emf device in our calculations, which is modeled with no internal resistance.

SOLVE	
a. Use Eq. 29.6 to find the equivalent resistance.	$R_{eq} = r + R$

Now, use Eq. 29.13 to find the time constant for the circuit.	$\tau = R_{eq}C = \boxed{(r+R)C}$
b. We can express the charge as a function of time, while charging, by using Eq. 29.14. Note that we want to find the time when the charge is half the maximum charge, so we set q equal to $C\mathcal{E}/2$.	$q = C\mathcal{E}\left(1 - e^{-\frac{t}{\tau}}\right)$ $\dfrac{C\mathcal{E}}{2} = C\mathcal{E}\left(1 - e^{-\frac{t}{(r+R)C}}\right)$
Now, solve for the time, t, when the capacitor is charged to half its maximum value.	$1/2 = 1 - e^{-\frac{t}{(r+R)C}}$ $e^{-\frac{t}{(r+R)C}} = 1/2$ $\ln\left[e^{-\frac{t}{(r+R)C}}\right] = \ln(1/2)$ $-\dfrac{t}{(r+R)C} = \ln(1/2)$ $t = \boxed{-\ln(1/2)(r+R)C}$
c. Use the time from part (b) with Eq. 29.15 to find the current while charging.	$I = \dfrac{\mathcal{E}}{R_{eq}}e^{-\frac{t}{\tau}} = \dfrac{\mathcal{E}}{r+R}e^{-\frac{-\ln(1/2)(r+R)C}{(r+R)C}}$ $I = \dfrac{\mathcal{E}}{r+R}e^{\ln(1/2)} = \boxed{\dfrac{\mathcal{E}}{2(r+R)}}$

CHECK and THINK

We can see that if the internal resistance were not included in the model, the time constant would be less, the capacitor would reach its half-charge state in less time, and the current at that time would be greater. These are all ways of saying the same thing: With more resistance in the RC circuit, it will take longer to charge, the maximum current will be less (and thus the current when it is half-charged), and the time constant reflects both these facts in that it is larger.

65. (N) The reading on the ammeter in Figure P29.65 is 3.00 A, and the current runs from right to left through the ammeter. What is the value of current

a. I_1 and

b. I_2?

Chapter 29 – Direct Current (DC) Circuits

c. What is the emf $\mathcal{E}$?

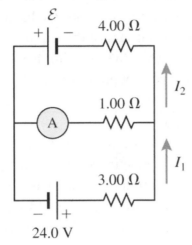

Figure P29.65

INTERPRET and ANTICIPATE The ammeter reading tells us the current through the middle branch. We can apply Kirchoff's junction and loop rules to determine the desired quantities.	

SOLVE **a.** Apply Kirchhoff's junction rule to the junction on the right, assuming the current through the ammeter $I_3 = 3.00$ A, is flowing from right to left (out of the junction).	current in = current out of junction $I_1 = I_2 + 3.00$ A
Apply Kirchhoff's loop rule to the bottom loop and solve for I_1.	$24 \text{ V} - (3 \ \Omega)I_1 - (1 \ \Omega)(3.00 \text{ A}) = 0$ $I_1 = \dfrac{21}{3} = \boxed{7.00 \text{ A}}$
b. Using the junction rule above, we can determine I_2.	$I_2 = I_1 - 3.00 \text{ A} = 7.00 \text{ A} - 3.00 \text{ A} = \boxed{4.00 \text{ A}}$
c. Apply Kirchhoff's loop rule to the top loop.	$\mathcal{E} + (1 \ \Omega)(3.00 \text{ A}) - (4 \ \Omega)I_2 = 0$ $\mathcal{E} = 4(4.00) - 1(3.00) = \boxed{13.0 \text{ V}}$

CHECK and THINK Using Kirchoff's rules (both the junction and loop rule), we are able to relate the currents and voltages in different parts of the circuit.

74. (A) A capacitor with capacitance C is charged to a total charge Q. The capacitor is then connected in parallel to two resistors (R_1 and R_2) that are also connected in parallel.
a. Write an expression for the time constant of this circuit.
b. Find an expression for the time when the capacitor has lost half of its charge.
c. Find an expression for the current through the capacitor at this time.

INTERPRET and ANTICIPATE	
If we find the equivalent resistance in the circuit, we can make use of Eq. 29.13, 29.16, and 29.17 to model the discharging behavior of the RC circuit. The two resistors are in parallel, thus, we can use Eq. 29.7 to find the equivalent resistance of the circuit. Note that the equivalent resistance will be less than either individual resistor, which should make the time constant less, and cause the capacitor to discharge in less time than if it were connected to either resistor alone.	
SOLVE **a.** Use Eq. 29.7 to find the equivalent resistance.	$\dfrac{1}{R_{eq}} = \dfrac{1}{R_1} + \dfrac{1}{R_2} = \dfrac{R_1 + R_2}{R_1 R_2}$ $R_{eq} = \dfrac{R_1 R_2}{R_1 + R_2}$
Now, use Eq. 29.13 to find the time constant for the circuit.	$\tau = R_{eq}C = \boxed{\dfrac{R_1 R_2}{R_1 + R_2}C}$
b. We can express the charge as a function of time, while discharging, by using Eq. 29.16. Note that we want to find the time when the charge is half the maximum charge, so we set q equal to $q_{max}/2$.	$q = q_{max}e^{-\frac{t}{\tau}}$ $\dfrac{q_{max}}{2} = q_{max}e^{-\frac{t}{\tau}}$
Now, solve for the time, t, when the capacitor is discharged to half its maximum value.	$1/2 = e^{-\frac{t}{\tau}}$ $\ln(1/2) = \ln\left(e^{-\frac{t}{\tau}}\right)$ $-\dfrac{t}{\tau} = \ln(1/2)$ $t = -\tau\ln(1/2) = \boxed{-\dfrac{R_1 R_2}{R_1 + R_2}C\ln(1/2)}$

c. Use the time from part (b) with Eq. 29.17 to find the current while discharging.	$I = I_0 e^{-\frac{t}{\tau}} = \dfrac{Q/C}{R_1 R_2/(R_1 + R_2)} e^{-\frac{\frac{R_1 R_2}{R_1 + R_2} C \ln(1/2)}{\frac{R_1 R_2}{R_1 + R_2} C}}$
	$I = \dfrac{Q(R_1 + R_2)}{R_1 R_2 C} e^{\ln(1/2)} = \boxed{\dfrac{Q(R_1 + R_2)}{2 R_1 R_2 C}}$

CHECK and THINK

We can see that if the equivalent resistance is less, the time constant would be less, the capacitor would reach its half-charge state in less time, and the current at that time would be greater. These are all ways of saying the same thing: With less resistance in the RC circuit, it will take less time to discharge, the maximum current will be greater (and thus the current when it is half-charged), and the time constant reflects both these facts in that it is smaller.

80. (A) Calculate the equivalent resistance between points P and Q of the electrical network shown in Figure P29.80.

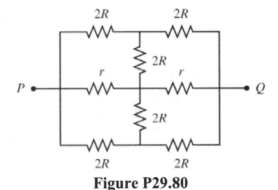

Figure P29.80

INTERPRET and ANTICIPATE

The network looks pretty complicated, but we notice that it is symmetric which can help us simplify the situation. In particular, the drawing is symmetric on the left and right sides. Therefore, if a voltage is applied across points P and Q, points a, b, and c, much each be at exactly half of the total voltage drop. That is points a, b, and c are at exactly the same potential.

For instance, we know that if we applied a voltage across points P and Q, the current through each of the resistors r must be the same (i.e. because it's symmetric left/right). If the current is the same, the voltage drop must be the same and $V_{Pb} = V_{Qb}$. The same story holds for the two resistors $2R$ on both the top and bottom of the network.

Since points a, b, and c are at exactly the same potential, there is no current through the vertical resistors and we can remove these resistors from the network without affecting its behavior at all. The remaining horizontal branches each consist of two resistors in series, which can be added to find the equivalent resistance (Eq. 29.6). That allows us to draw the simplified, equivalent circuit diagram shown.

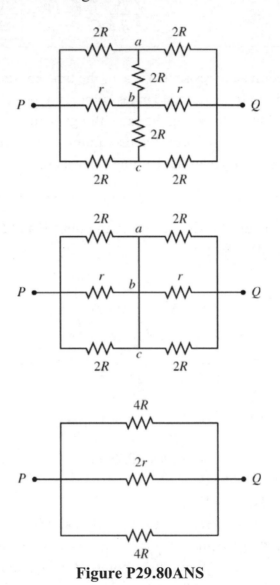

Figure P29.80ANS

SOLVE	
We now have three resistors $4R$, $2r$, and $4R$ in parallel and can use Eq. 29.7 to determine the equivalent resistance.	$\dfrac{1}{R_{eq}} = \dfrac{1}{4R} + \dfrac{1}{2r} + \dfrac{1}{4R}$ $\dfrac{1}{R_{eq}} = \dfrac{1}{2R} + \dfrac{1}{2r}$ $\dfrac{1}{R_{eq}} = \dfrac{R+r}{2Rr}$ $R_{eq} = \boxed{\dfrac{2rR}{R+r}}$

CHECK and THINK

The challenging part of this problem is realizing that we can simplify the network based on symmetry. If we did not realize this, we might imagine connecting the circuit to a battery and writing Kirchoff's rules to determine the total current drawn by the battery. With Ohm's law, we could then find the equivalent resistance of the network. Noting and taking advantage of the symmetry makes our lives much easier though!

88. (A) Figure P29.88 shows a circuit that consists of two identical emf devices. If $R_1 = R_2 = R$ and the switch is open, find an expression (in terms of R and $\mathcal{E}$) for the current I_1 that is in resistor 1.

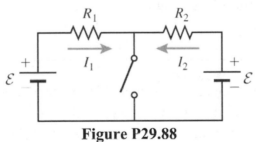

Figure P29.88

INTERPRET and ANTICIPATE	
The circuit contains two batteries and a couple resistors. We can use Kirchoff's loop rule to relate the voltages to the current.	

SOLVE	
The simplest solution is to notice that the emf devices are in series and oriented in the opposite direction. Therefore, the total voltage of these two emf devices in series is $\mathcal{E} - \mathcal{E} = 0$. There is no net voltage to drive a current, so the current must be zero!	$I = 0$

We can also confirm this result by going through Kirchoff's rules: This is a single loop and therefore there must be a single current that travels around the loop. Let's assume that I_1 is the actual current and replace I_2 with current I_1 traveling to the right.

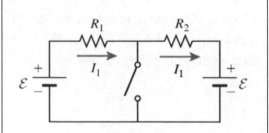

Figure P29.88ANS

Apply Kirchoff's loop rule starting at the bottom left and moving around clockwise. As we go through the first emf device on the left, from the negative to positive terminal, the potential increases by $+\mathcal{E}$. Since current flows from high to low potential through a resistor, moving along the direction of current through R_1 and R_2 leads to voltage drops of $-I_1R_1$ and $-I_1R_2$. Finally, going down through the battery on the right, from the positive to negative terminal, corresponds to a decrease in potential of $-\mathcal{E}$. Solving for the current, we see that it is zero as we found above.

$$+\mathcal{E} - I_1R_1 - I_1R_2 - \mathcal{E} = 0$$

$$I_1(R_1 + R_2) = 0$$

$$I_1 = \boxed{0}$$

CHECK and THINK

In this case, the two voltage sources are oriented in opposite directions and effectively cancel each other out. In fact, if you have an electronic device with two batteries and flip one of them around, you will indeed find that it will stop working!

30

Magnetic Fields and Forces

9. (C) Figure P30.9 shows very long current-carrying wires. Using the coordinate system indicated (with the z axis out of the page), state the direction of the magnetic field at point P in each case.

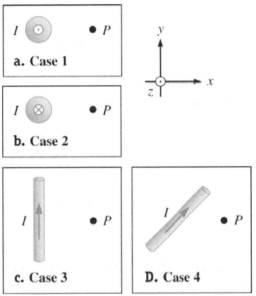

Figure P30.9

The direction in each case can be found using the simple right hand rule. Point your thumb along the direction of the current and wrap your fingers around to indicate the direction of the magnetic field lines, which form closed loops circling the current.

a. The current comes out of the page, so your fingers wrap around counter-clockwise. That is, the magnetic field forms closed loops pointing counter-clockwise around the current. At point P, the field points up towards the top of the page in the $\boxed{+\hat{j}}$ direction.

b. The current goes into the page, so your fingers wrap around clockwise. So the magnetic field forms closed loops pointing clockwise around the current. At point P, the field points down towards the bottom of the page in the $\boxed{-\hat{j}}$ direction.

c. Point your thumb up along the current and wrap your fingers around to determine that the magnetic field points into the page, in the $\boxed{-\hat{k}}$ direction.

d. Just as in part (c), the magnetic field points into the page in the $\boxed{-\hat{k}}$ direction.

a. Case 1

c. Case 3

b. Case 2

d. Case 4

Figure P30.9ANS

11. (C) Figure P30.11 shows three configurations of wires and the resultant magnetic fields due to current in the wires. What is the direction of the current that gives the resultant magnetic field shown in each case?

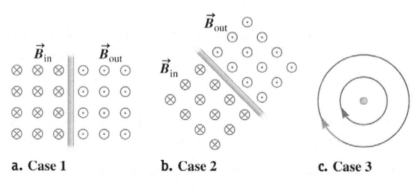

a. Case 1

b. Case 2

c. Case 3

Figure P30.11

The simple right hand rule relates the direction of current to the magnetic field, Specifically, if you point your thumb along the current, your fingers wrap around the wire in the direction of the magnetic field. So in each case, we simply need to try the simple right hand rule with the current going through the wire in the two possible directions to see which is consistent with the field lines indicated. The current in each as is:

a. Downward, along the wire from the top toward the bottom.

b. The current should flow along the wire from the upper left to lower right.

c. The current should flow along the wire into the page.

In each case, if the current was in the opposite direction, the field lines would simply be in the opposite direction.

12. (N) Review A proton is accelerated from rest through a 5.00-V potential difference.
a. What is the proton's speed after it has been accelerated?
b. What is the maximum magnetic field that this proton produces at a point that is 1.00 m from the proton?

INTERPRET and ANTICIPATE

Given a charge accelerated through a potential difference, we can easily calculate the change in energy and use energy conservation to determine its speed. Given its speed, we can use the Biot-Savart law to determine the magnetic field produced.

SOLVE	
a. Use energy conservation to equate the change in potential energy to the change in kinetic energy. Remember that the energy change for a charge q moving through a potential difference of V is $q\Delta V$.	$q\Delta V = \dfrac{1}{2}mv^2 \quad \rightarrow \quad v = \sqrt{\dfrac{2q\Delta V}{m}}$
Insert values to calculate the speed. The charge and mass of a proton are $q = 1.6 \times 10^{-19}$ C and $m = 1.67 \times 10^{-27}$ kg.	$v = \sqrt{\dfrac{2\left(1.6\times10^{-19}\text{ C}\right)\left(5.00\text{ V}\right)}{1.67\times10^{-27}\text{ kg}}} = \boxed{3.10\times10^{4}\text{ m/s}}$
b. The magnetic field in the vicinity of a moving charge is given by the Biot-Savart law (Equation 30.1).	$\vec{B} = \left(\dfrac{\mu_0}{4\pi}\right)q\dfrac{\vec{v}\times\vec{r}}{r^3}$

Chapter 30 – Magnetic Fields and Forces

<table>
<tr><td>The maximum value occurs when the angle between the velocity and the radial vector is 90 degrees and $\left|\vec{v}\times\vec{r}\right| = vr\sin\varphi = vr$. That is, at a point to the side of the charge, perpendicular to its velocity.</td><td>$B_{max} = \left(\dfrac{\mu_0}{4\pi}\right)\dfrac{qv}{r^2}$</td></tr>
<tr><td>Finally, insert numerical values.</td><td>$B_{max} = \left(10^{-7}\ \tfrac{\text{T·m}}{\text{A}}\right)\left(1.60\times10^{-19}\ \text{C}\right)\dfrac{\left(3.10\times10^4\ \tfrac{\text{m}}{\text{s}}\right)}{\left(1.00\ \text{m}\right)^2}$

$B_{max} = \boxed{4.96\times10^{-22}\ \text{T}}$</td></tr>
</table>

CHECK and THINK

The field is quite small, but this is not to surprising for a point that is one meter away from a single proton. Typical magnetic fields arise as a net field due to the motion of very many charges.

15. (N) For the proton in Problem 14, what are the magnitude and direction of the magnetic field at point A?

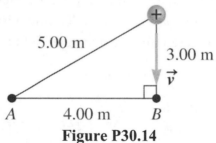

Figure P30.14

INTERPRET and ANTICIPATE

The magnetic field around a moving charge can be found using the Biot-Savart law. The right hand rule for the cross product can be used to find the direction of the field.

<table>
<tr><td>**SOLVE**
The Biot-Savart law is given by Equation 30.1.</td><td>$\vec{B} = \left(\dfrac{\mu_0}{4\pi}\right)q\dfrac{\vec{v}\times\vec{r}}{r^3}$</td></tr>
<tr><td>The direction can be found using the right hand rule. The velocity vector $\vec{v}$ points downward and $\vec{r}$ points from the proton to point A.</td><td>B points into the page</td></tr>
</table>

Apply the right hand rule by pointing your fingers downward along the velocity and then curl (or "push") your fingers towards the $\vec{r}$ vector, in which case your thumb points into the page, in the direction of the magnetic field.	
The magnitude can now be found from the Biot-Savart law. The charge of the proton is $q = 1.60 \times 10^{-19}$ C. According to Equation 30.3, the magnitude of the cross product depends on the sine of the angle between the two vectors, which is 53.1 degrees. The distance to point A is $r = 5$ m. Using our typical coordinate system, with the z axis pointing out of the page, we might express this as $\vec{B} = -5.12 \times 10^{-21} \hat{k}$ T .	$\left\| \vec{v} \times \vec{r} \right\| = vr\sin\varphi$ $B = \left(\dfrac{\mu_0}{4\pi} \right) q \dfrac{v\sin\varphi}{r^2}$ $B = \left(10^{-7} \; \frac{\text{T·m}}{\text{A}} \right) \left(1.60 \times 10^{-19} \text{ C} \right) \dfrac{\left(1.00 \times 10^7 \; \frac{\text{m}}{\text{s}} \right) \left(\sin 53.1° \right)}{\left(5.00 \text{ m} \right)^2}$ $B = \boxed{5.12 \times 10^{-21} \text{ T into the page}}$

CHECK and THINK

We found the magnetic field using the Biot-Savart law, which allows us to find both the magnitude and the direction. The field due to this single proton (even moving as fast as it is moving) is quite weak compared to the magnetic field of the Earth.

18. (A) Two long, straight, parallel wires are shown in Figure P30.18. The current in the wire on the left is double the current in the wire on the right. Find an expression for the magnetic field at points A and B. Use the indicated coordinate system to write your answer in component form.

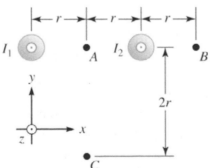

Figure P30.18

183

Chapter 30 – Magnetic Fields and Forces

INTERPRET and ANTICIPATE The magnetic field at a given location due to a long current-carrying wire depends on the current as well as the distance from the wire to the point. The direction can be found using the simple right hand rule. In this case, with two currents, we can find the field due to each wire and use vector addition to find the sum.	

SOLVE For a single wire, the field can be calculated using Equation 30.9. We are told that $I_1 = 2I_2$. The direction of the field can be found using the simple right hand rule (SRHR): for each wire, point your thumb out of the page along the direction of the current and your fingers wrap around counter-clockwise indicating the direction of the magnetic field.	$B = \dfrac{\mu_0 I}{2\pi r}$
Point A is a distance r from each wire. At this location, using the SRHR, the field due to wire 1 points upward $\left(+\hat{j}\right)$ and the field due to wire 2 points downward $\left(-\hat{j}\right)$. We can now determine the net force.	$\vec{B}_A = \dfrac{\mu_0 I_1}{2\pi r_1}\hat{j} - \dfrac{\mu_0 I_2}{2\pi r_2}\hat{j}$ $\vec{B}_A = \dfrac{\mu_0 2I_2}{2\pi r}\hat{j} - \dfrac{\mu_0 I_2}{2\pi r}\hat{j}$ $\boxed{\vec{B}_A = \dfrac{\mu_0 I_2}{2\pi r}\hat{j}}$
Point B is a distance r from wire 2 and a distance $3r$ from wire 1. Applying the SRHR, the field at point B points upwards $\left(+\hat{j}\right)$ for both currents.	$\vec{B}_B = \dfrac{\mu_0 I_1}{2\pi r_1}\hat{j} + \dfrac{\mu_0 I_2}{2\pi r_2}\hat{j}$ $\vec{B}_B = \dfrac{\mu_0(2I_2)}{2\pi(3r)}\hat{j} + \dfrac{\mu_0 I_2}{2\pi r}\hat{j}$ $\boxed{\vec{B}_B = \dfrac{5\mu_0 I_2}{6\pi r}\hat{j}}$

CHECK and THINK

In both cases, we determine the magnetic field due to each wire and add them to determine the total magnetic field. The fields due to each current point in opposite directions at point A, so the field is actually smaller than at point B.

24. (A) A wire is bent in the form of a square loop with sides of length L (Fig. P30.24). If a steady current I flows in the loop, determine the magnitude of the magnetic field at point P in the center of the square.

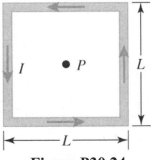

Figure P30.24

INTERPRET and ANTICIPATE
The square loop consists of four equal segments of current. We can use the formula for the magnetic field due to a finite wire segment for each and multiply by four to get the total field.

SOLVE

First, notice that this is a symmetric situation in that the point P at the center is the same distance ($L/2$) from the center of each of the sides. Consider the side on the right: use the simple right hand rule for this wire segment to find that the magnetic field inside the square loop points out of the page. You can also imagine a little length vector along this current (pointing upward) and an r vector pointing from this current element to point P (pointing to the left) as a second way to verify that the magnetic field direction—$d\vec{\ell} \times \vec{r}$ points out of the page indicating the direction of the magnetic field. You can repeat this process to confirm that the magnetic field due to each of the four sides is outward.

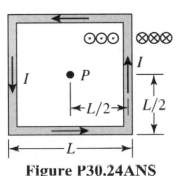

Figure P30.24ANS

Calculate the magnetic field due to one side (a finite length wire) using Equation 30.8 (derived in Example 30.1). Here, R is the distance to the center of the

$$B_{side} = \frac{\mu_0 I}{2\pi R}\left[\frac{L}{\left(L^2 + 4R^2\right)^{1/2}}\right]$$

segment ($L/2$) and L is the length of the side.	$$B_{side} = \frac{\mu_0 I}{2\pi(L/2)} \left[\frac{L}{\left(L^2 + 4(L/2)^2\right)^{1/2}} \right]$$ $$B_{side} = \frac{\mu_0 I}{\pi L} \left[\frac{L}{\left(2L^2\right)^{1/2}} \right]$$ $$B_{side} = \frac{\mu_0 I}{\sqrt{2}\pi R} = \frac{\sqrt{2}\mu_0 I}{2\pi R}$$
The total field due to all four edges of the square is four times that due to one side.	$$B_{total} = 4B_{side} = \boxed{\frac{2\sqrt{2}\mu_0 I}{\pi R}} \text{ out of the page}$$

CHECK and THINK

It looks similar to the expression for the magnetic field at the center of a loop,

$B_{loop} = \dfrac{\mu_0 I}{2R}$ (Eq. 30.10), though the numerical factor is a bit different. This makes sense, since this *is* a loop that is just a different shape. The field at the center depends on the current and size of the loop in the same way though. For instance, a larger current would produce a larger field while making the square larger would lead to a smaller field at the center.

29. (N) Figure P30.29 shows two current-carrying loops with $I_1 = 4.00$ A clockwise and $I_2 = 9.00$ A counterclockwise, placed with their centers at the origin of the xy plane.
a. If $r_1 = 10.0$ cm and $r_2 = 16.0$ cm, what are the magnitude and direction of the magnetic field due to the two loops at the origin?
b. If r_1 is held constant at 10.0 cm, what would r_2 have to be for the magnetic field at the origin to be 0?

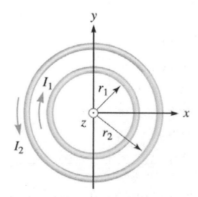

Figure P30.29

INTERPRET and ANTICIPATE

A current loop is one of the special cases described in Example 30.2. We can use the formula from that calculation and the right hand rule to determine the magnetic field due to each current loop and then add these fields as vectors.

SOLVE **a.** The magnetic field along a line going through the center of a current loop is given by Equation 30.10. R is the radius of the loop with current I. y is the distance along the center axis from the center of the loop, so the center of the loop is given by $y = 0$.	$$B = \frac{\mu_0 I R^2}{2\left(R^2 + y^2\right)^{3/2}}$$ $$B_{center}\left(y=0\right) = \frac{\mu_0 I}{2R} \qquad (1)$$
To determine the direction, we can use the simple right hand rule. For instance, for I_1, the current on the left side of the loop is going upward. Point your thumb upward toward the top of the page and wrap your fingers around the wire. The magnetic field is pointing into the page $\left(-\hat{k}\right)$ inside the loop (and therefore at the center of the loop). Use Eq. (1) above to calculate the magnitude.	$$\vec{B}_1 = -\frac{\mu_0 I_1}{2r_1}\hat{k}$$
Similarly, using the simple right hand rule for I_2, the magnetic field points out of the page $\left(+\hat{k}\right)$.	$$\vec{B}_2 = \frac{\mu_0 I_1}{2r_1}\hat{k}$$
The net field is the vector sum of these two. Equivalently, we can think of it as "the field due to current 2 out of the page	$$\vec{B}_{net} = \left(\frac{\mu_0 I_1}{2r_1} - \frac{\mu_0 I_1}{2r_1}\right)\hat{k} = \frac{\mu_0}{2}\left(\frac{I_2}{r_2} - \frac{I_1}{r_1}\right)\hat{k}$$

187

minus the field due to current 1 into the page equals the total field out of the page."	$B_{net} = \dfrac{\mu_0}{2}\left(\dfrac{I_2}{r_2} - \dfrac{I_1}{r_1}\right)$ out of the page $\qquad(2)$
Now, plug in numbers	$\vec{B}_{net} = \dfrac{\left(4\pi \times 10^{-7}\ \frac{T \cdot m}{A}\right)}{2}\left(\dfrac{9.00\ A}{16.0 \times 10^{-2}\ m} - \dfrac{4.00\ A}{10.0 \times 10^{-2}\ m}\right)\hat{k}$ $\boxed{\vec{B}_{net} = 1.02 \times 10^{-5}\,\hat{k}\ T\ \text{(out of the page)}}$ $B_{net} = 1.02 \times 10^{-5}\ T$ out of the page
b. We want the net magnetic field to equal zero, so set Equation (2) equal to zero and solve for the value of r_2 that satisfies this criterion.	$\dfrac{\mu_0}{2}\left(\dfrac{I_2}{r_2} - \dfrac{I_1}{r_1}\right) = 0 \quad \rightarrow \quad \dfrac{I_2}{r_2} - \dfrac{I_1}{r_1} = 0$ $r_2 = \left(\dfrac{I_2}{I_1}\right)r_1$ $r_2 = \left(\dfrac{9.00\ A}{4.00\ A}\right)(10.0\ cm) = 22.5\ cm = \boxed{0.225\ m}$

CHECK and THINK

Given the special case of a loop of current discussed in Example 30.2 and the simple right hand rule, we can determine the magnetic field magnitude and direction at the center of a loop. With two loops, we simply need to do this twice and find the net field as the vector sum. From Equation (1) above, we see that the field at the center of a loop is proportional to the current and inversely proportional to its radius. For the two fields to cancel and result in no field at the center, the currents must go in opposite directions around the loop and the ratio of the radii needs to be the same as the ratio of the currents.

33. (N) A square loop of wire with side length 0.205 m carries a current of 1.50 A. What is the magnitude of the magnetic field along the axis of the square loop, 2.50 m from the center?

INTERPRET and ANTICIPATE

We can use Eq. 30.12 to find the magnitude of the magnetic moment of the loop of current. Then, because we are looking for the magnitude of the magnetic field at a large distance, compared to the size of the loop, along the axis of the magnetic dipole, we can use Eq. 30.15 to find the magnitude of the magnetic field.

SOLVE The magnetic moment can be found using the area of the loop (a square in this case) and the current in the loop.	$\mu = IA = (1.50\ \text{A})(0.205\ \text{m})^2$
Now, use this magnetic moment, Eq. 30.15, and the other given info to find the approximate magnitude of the magnetic field 2.50 m along the axis of the dipole.	$B \approx \left(\dfrac{\mu_0}{2\pi}\right)\dfrac{\mu}{y^3} = \left(\dfrac{4\pi \times 10^{-7}\ \text{T·m/A}}{2\pi}\right)\dfrac{(1.50\ \text{A})(0.205\ \text{m})^2}{(2.50\ \text{m})^3}$ $B \approx \boxed{8.07 \times 10^{-10}\ \text{T}}$

CHECK and THINK
When the location of interest is very far from the current loop (very large compared to the dimensions of the loop), Eq. 30.15 is a good approximation for the magnetic field due to the loop. Note that for an exact calculation, we would use the Biot-Savart law.

42. (N) The velocity vector of a singly charged helium ion ($m_{\text{He}} = 6.64 \times 10^{-27}$ kg) is given by $\vec{v} = 4.50 \times 10^5 \hat{i}$ m/s. The acceleration of the ion in a region of space with a uniform magnetic field is 8.50×10^{12} m/s^2 in the positive y direction. The velocity is perpendicular to the field direction. What are the magnitude and direction of the magnetic field in this region?

INTERPRET and ANTICIPATE The magnetic force depends on the charge, velocity, and magnetic field, as well as the angle between the velocity and the field. With the quantities given, we can determine the magnitude and direction of the field needed to produce the observed force.	
SOLVE The magnetic force is given by Eq. 30.18. From Newton's second law, we also know $F = ma$, so we can determine the required magnetic force to produce this acceleration.	$F = |q|vB \sin \varphi = ma$
We're told that the velocity is perpendicular to the field direction, so $\varphi = 90°$ and $\sin 90° = 1$. Solve this expression for magnetic field.	$B = \dfrac{ma}{|q|v}$

Now, plug in values. Since the ion is singly charged, the magnitude of the charge is $q = e = 1.6 \times 10^{-19}$ C. The resultant field is in the metric unit Tesla.	$B = \dfrac{\left(6.64 \times 10^{-27} \text{ kg}\right)\left(8.50 \times 10^{12} \text{ m/s}^2\right)}{\left(1.60 \times 10^{-19} \text{ C}\right)\left(4.50 \times 10^{5} \text{ m/s}\right)} = 0.784$ T
To get the direction, use the right hand rule. If you point your fingers to the right in the direction of the velocity ($+x$) and curl them into the page ($-z$), your thumb points towards the top of the page ($+y$) in the direction of the force we need. We're told that the velocity and field are perpendicular and the force is always perpendicular to the velocity and the field, so this makes sense.	$B = 0.784$ T into the page or $\vec{B} = \boxed{-0.784 \, \hat{k} \text{ T}}$ 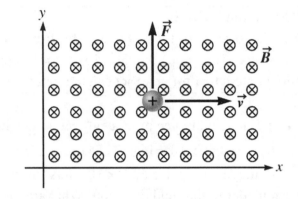 **Figure P30.42ANS**

CHECK and THINK

With the magnetic force law and the right hand rule, we can relate the field and velocity to the resultant force.

49. (A) A proton and a helium nucleus (consisting of two protons and two neutrons) pass through a velocity selector and into a mass spectrometer. The radius of the proton's circular path is r_p. Find an expression for the radius r of the helium nucleus's path in terms of r_p. (You may assume the mass of a proton is roughly equal to the mass of a neutron, and the helium nucleus has the same speed as the proton.)

INTERPRET and ANTICIPATE

A velocity selector has electric and magnetic fields that exert forces in opposite directions on a moving charge. Since the magnetic force depends on velocity, this means that charges of only a specific velocity have zero net force on them and are allowed to pass through undeflected. This velocity and charge then determines the radius of the circular orbit in the mass spectrometer.

SOLVE The velocity selector is described in Example 30.5, which derives a formula for the velocity of a charged particle that can pass through undeflected. The electric and magnetic fields in the velocity selector (E_{vs} and B_{vs}) are the same for both the proton and helium nucleus, so their velocities are indeed the same as they enter the mass spectrometer as suggested.	$$v = \frac{E_{vs}}{B_{vs}}$$
In the mass spectrometer, there is no electric field and only a magnetic field, B_{ms}. The magnetic force keeps the charge in the circular path with radius given by Eq. 30.23.	$$r = \frac{mv}{qB}$$
We can now calculate the ratio and then express the radius of the helium nucleus in terms of the proton. Compared to a single proton, the helium nucleus has twice the charge (two protons) and four times the mass (two protons plus two neutrons). The velocity out of the velocity selector and the magnetic field in the mass spectrometer are the same for both particles.	$$\frac{r_{He}}{r_p} = \frac{\dfrac{m_{He}v}{q_{He}B}}{\dfrac{m_p v}{q_p B}} = \frac{m_{He}}{m_p}\frac{q_p}{q_{He}} = \frac{4}{1}\cdot\frac{1}{2} = 2$$ $$r_{He} = \boxed{2r_p}$$

CHECK and THINK
We are able to determine the radius of the helium nucleus as asked. While it has the same speed as the proton exiting the velocity selector, it travels with a different radius orbit in the mass spectrometer. Indeed, this is what the mass spectrometer does—separating different charged particles, typically in order to determine which particles are in the incoming beam.

53. (N) A rectangular silver strip is 2.50 cm wide and 0.050 cm thick. It is in a magnetic field perpendicular to its surface (Fig. 30.41, page 957). The magnetic field is uniform, with a magnitude of 1.75 T. The strip carries a current of 6.45 A. According to Table 28.2, the number density of charge carriers in silver is 5.86×10^{28} m^{-3}. Find the Hall voltage for this strip.

INTERPRET and ANTICIPATE	
The Hall voltage depends on the quantities given. For instance, a larger number of charges or current traveling through a larger magnetic field will lead to a larger amount of charge deflected across the metal strip to produce a larger Hall voltage.	

SOLVE	
The Hall voltage is expressed in Eq. 30.30, which depends on the current I, magnetic field B, electron charge e, thickness of the strip w, and number density of charge carriers n.	$n = \dfrac{IB}{ew\Delta V_{Hall}} \quad \rightarrow \quad V_{Hall} = \dfrac{IB}{ewn}$
Now, insert values.	$V_{Hall} = \dfrac{(6.45\text{ A})(1.75\text{ T})}{(1.6\times10^{-19}\text{ C})(0.00050\text{ m})(5.86\times10^{28}\text{ m}^3)}$ $V_{Hall} = \boxed{2.41\times10^{-6}\text{ V}}$

CHECK and THINK	
The Hall voltage is small, but significant as it depends on the nature (and the sign specifically!) of the charge carriers.	

59. (N) A current of 5.64 A flows along a wire with $\vec{\ell} = \left(1.382\hat{i} - 2.095\hat{j}\right)$m. The wire resides in a uniform magnetic field $\vec{B} = \left(-0.300\hat{j} + 0.750\hat{k}\right)$ T. What is the magnetic force acting on the wire?

INTERPRET and ANTICIPATE
The magnetic force can be found using Eq. 30.32. We must be able to find a vector cross product using the component forms of the current direction and magnetic field. A vector cross product in three dimensions can be calculated in the following way:
$\vec{A} \times \vec{B} = \begin{vmatrix} \hat{i} & \hat{j} & \hat{k} \\ A_x & A_y & A_z \\ B_x & B_y & B_z \end{vmatrix} = \left(A_y B_z - A_z B_y\right)\hat{i} - \left(A_x B_z - A_z B_x\right)\hat{j} + \left(A_x B_y - A_y B_x\right)\hat{k}$

Given the known x and y components for the current direction and magnetic field, we expect that the magnetic force will have each component, when we compare the cross product above to Eq. 30.32.

SOLVE	
Writing Eq. 30.32, and using the two vectors from the problem statement, we find an expression for the magnetic force on the wire.	$\vec{F} = I\vec{\ell} \times \vec{B} = (5.64 \text{ A}) \begin{vmatrix} \hat{i} & \hat{j} & \hat{k} \\ 1.382 \text{ m} & -2.095 \text{ m} & 0 \\ 0 & -0.300 \text{ T} & 0.750 \text{ T} \end{vmatrix}$ $\vec{F} = (5.64 \text{ A}) \begin{bmatrix} (-2.095 \text{ m})(0.750 \text{ T})\hat{i} \\ -(1.382 \text{ m})(0.750 \text{ T})\hat{j} \\ +(1.382 \text{ m})(-0.300 \text{ T})\hat{k} \end{bmatrix}$ $\boxed{\vec{F} = \left(-8.86\hat{i} - 5.85\hat{j} - 2.34\hat{k}\right) \text{ N}}$

CHECK and THINK

As expected, the magnetic force points in all three component directions. Note that one of the features of any mathematical vector cross product is that the resultant vector must be perpendicular to the other two. If desired, we could now find the magnitude of the resultant force and the direction.

60. (N) A wire with a current of $I = 8.00$ A directed along the positive y axis is embedded in a uniform magnetic field perpendicular to the current in the wire. If the wire experiences a magnetic force of 50.0 mN/m in the positive z direction, what are the magnitude and direction of the magnetic field in this region?

INTERPRET and ANTICIPATE	
A current in a magnetic field experiences a force. With the quantities given and the right hand rule, we can determine the magnitude and direction of the magnetic field required to produce the given force.	

SOLVE	
The magnitude of the magnetic force on a current is given by Equation 30.33.	$F = BI\ell \sin\varphi$
Solve for the magnetic field. F/ℓ represents the force per unit length on the wire, which is given. The angle is that between the wire and the magnetic field, which we are told are perpendicular, so $\varphi = 90°$.	$B = \dfrac{F/\ell}{I \sin\varphi}$

Insert numerical values.	$B = \dfrac{0.0500 \text{ N/m}}{(8.00 \text{ A})\sin 90°} = 6.25 \times 10^{-3} \text{ T}$
To determine the direction, use the right hand rule for the cross product $\vec{F} = I(\vec{\ell} \times \vec{B})$ (Eq. 30.32). If you point your fingers along the direction of the current towards the top of the page ($+y$) and wrap them to the left ($-x$) in the direction of the field, the force will be in the direction your thumb points, or out of the page ($+z$), as needed.	$B = 6.25 \times 10^{-3}$ T in the $-x$ direction or $\boxed{B = \boxed{-6.25 \times 10^{-3} \hat{i} \text{ T}}}$

CHECK and THINK

We are able to determine the magnitude and direction of the magnetic field required to produce the given force with the equation for the magnetic force on a wire.

66. Two long, straight, parallel wires carry identical currents of 6.30 A in the positive x direction. The separation between the wires is 8.50 cm.

a. (C) Is the force between the wires attractive or repulsive?

Imagine the two wires lying side by side on a table with the two currents flowing toward you, with wire 1 on the left and wire 2 on the right. The right-hand rule that relates current to field direction shows the magnetic field due to wire 1 at the location of wire 2 is directed vertically upward. Then, the right-hand rule that relates current and magnetic field to force gives the direction of the force experienced by wire 2 as being to the left, toward wire 1.

Similarly, the field due to wire 2 points downward at the location of wire 1 and the force on wire 1 due to this field is to the right, towards wire 2.

Therefore, the force between the wires is $\boxed{attractive}$. (For two parallel wires with currents in opposite directions, the force would be repulsive.)

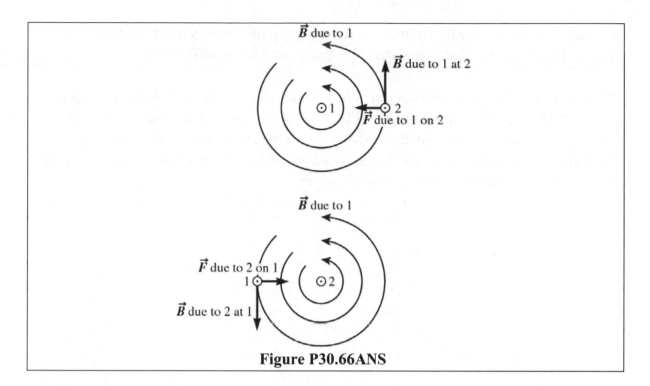

Figure P30.66ANS

b. (N) What is the magnitude of the force per unit length that each wire exerts on the other?

INTERPRET and ANTICIPATE
The force per unit length depends on the magnitude of each current, I_1 and I_2, and the distance separating the two parallel wires r. We can insert these values and determine the attractive force.

SOLVE	
Equation 30.37 represents the force per unit length that one wire exerts on the other.	$$f = \frac{\mu_0 I_1 I_2}{2\pi r}$$
Insert numerical values.	$$f = \frac{\left(4\pi \times 10^{-7}\ \text{T}\cdot\text{m/A}\right)\left(6.30\ \text{A}\right)^2}{2\pi\left(8.50 \times 10^{-2}\ \text{m}\right)}$$ $$f = \boxed{9.34 \times 10^{-5}\ \text{N/m}}$$

CHECK and THINK
The force per unit length increases with the currents and decreases with the separation. This makes sense given what we saw in part (a)—the larger current 1, for instance, or the closer the wires are spaced, the larger the field at the location of current 2 and therefore

> the larger the force. We find a force per unit length since longer wires would mean a greater length of wire that experiences this force and a larger total force.

72. (N) A straight conductor between points O and P has a mass of 0.50 kg and a length of 1.0 m and carries a current of 12 A (Fig. P30.72). It is hinged at O and is placed in a plane perpendicular to a magnetic field of 2.0 T. If the conductor begins from rest, determine the angular acceleration of the conductor due to the magnetic force. Ignore the effect of the gravitational force on the conductor.

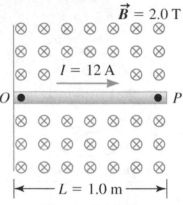

Figure P30.72

INTERPRET and ANTICIPATE
The current experiences a magnetic force in the plane that causes the wire to rotate. We can determine the force along the wire to calculate the total torque and then use rotational kinematics to find the angular acceleration.

SOLVE

We can find the force on a tiny section of the current as shown. Use the right hand rule to determine the direction: point your fingers to the right along the current and wrap them into the page along the field direction such that your thumb points towards the top of the page indicating the direction of the force.

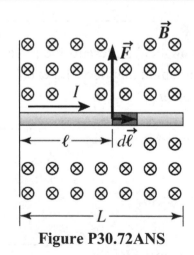

Figure P30.72ANS

The magnitude of the force on this section is given by Equation 30.32. The angle between the current direction and the magnetic field is 90 degrees.	$\left\lvert d\vec{F}_B \right\rvert = \left\lvert I \left(d\vec{\ell} \times \vec{B} \right) \right\rvert = I\,d\ell\,B\sin 90°$ $dF_B = IBd\ell$
While the force is the same for each current element, the torque due to each $d\ell$ depends on distance from the rotation point O, so we must integrate over the entire length of the wire (from $r = 0$ to L) to determine the total torque. The angle between the radial vector from the rotation axis and the force (perpendicular to the rod) is 90 degrees.	$d\tau_B = \left\lvert \vec{\ell} \times d\vec{F}_B \right\rvert = \ell\,dF_B \sin 90° = IB\ell\,d\ell$ $\tau_B = IB \int_0^L \ell\,d\ell = \dfrac{1}{2} IBL^2$
The torque produces an angular acceleration according to Newton's second law for rotational dynamics. We indicate the rotational moment of inertia as I_{rot} to avoid confusion with current I. I_{rot} equals $ML^2/3$ for a thin rod rotating about its end.	$\tau = I_{rot}\alpha$ $\alpha = \dfrac{\tau}{I_{rot}} = \dfrac{\frac{1}{2}BL^2}{\frac{1}{3}ML^2} = \dfrac{3IB}{2M}$
Finally we substitute the numbers to find the counter-clockwise angular acceleration.	$\alpha = \dfrac{3(12\text{ A})(2.0\text{ T})}{2(0.05\text{ kg})}$ $\alpha = \boxed{7.2 \times 10^2 \text{ rad/s}^2}$

CHECK and THINK

Notice that since there is a force along the length of the wire at different distances from the rotation axis, we needed to integrate to find the total torque due to the force on all of these current elements. The final angular acceleration of around 100 revolutions per second squared is pretty substantial!

77. (N) Figure P30.77 shows two long current-carrying wires in the xy plane with $I_1 = 2.00$ A to the left and $I_2 = 7.00$ A downward. The wires are not in contact with each other. What are the magnitude and direction of the magnetic field due to the two wires **a.** at point A, a distance $a = 60.0$ cm and $b = -50.0$ cm in the x and y directions from the intersection of the wires, and **b.** at $z = -50.0$ cm behind the intersection of the wires?

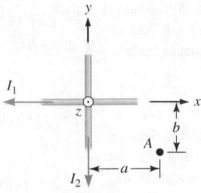

Figure P30.77

INTERPRET and ANTICIPATE
The magnetic field at a given location is the vector sum of the magnetic fields due to each of the wires. The magnetic field due to a wire is $B = \mu_0 I/2\pi r$, where r is the perpendicular distance to the wire and the direction is given by the simple right hand rule.

SOLVE	
a. Let's first determine the directions of the field at point A using the simple right hand rule. For current I_1, point your thumb along the direction of the current to the left. Your fingers wrap around in the direction of the magnetic field, so it points out of the page at A. (Your fingers should wrap into the page above the wire, down behind, out of the page below the wire, and up in front.) Similarly, for wire I_2, the field points out of the page at point A. (Thumb down along the current and fingers wrap out of the page to the right of the wire.) Therefore, the total field is the sum of the fields due to wires 1 and 2 and points out of the page.	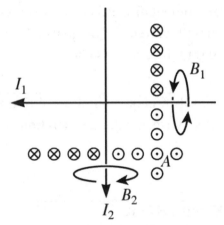 **Figure P30.77ANS**

Calculate the field due to each wire using Equation 30.9, where r is the distance to the wire for each of the currents.	$$B = B_1 + B_2 = \frac{\mu_0 I_1}{2\pi r_1} + \frac{\mu_0 I_2}{2\pi r_2}$$ $$B = \frac{\mu_0}{2\pi}\left(\frac{I_1}{r_1} + \frac{I_2}{r_2}\right)$$
Plug in numerical values.	$$B = \frac{\left(4\pi\times10^{-7}\ \text{T}\cdot\text{m/A}\right)}{2\pi}\left(\frac{2.00\ \text{A}}{0.500\ \text{m}} + \frac{7.00\ \text{A}}{0.600\ \text{m}}\right)$$ $$B = \boxed{3.13\times10^{-6}\ \text{T out of the page}}$$ or $\vec{B} = \boxed{3.13\times10^{-6}\,\hat{k}\ \text{T}}$
b. Using the simple right hand rule again as above at a point *behind* the wires, the field due to wire 1 points downward $\left(-\hat{j}\right)$ and the field due to wire 2 points to the right $\left(+\hat{i}\right)$. Calculate the field due to each using Eq. 30.9 again.	$$B_1 = \frac{\mu_0 I_1}{2\pi r} = \frac{\left(4\pi\times10^{-7}\ \text{T}\cdot\text{m/A}\right)\left(2.00\ \text{A}\right)}{2\pi\left(0.500\ \text{m}\right)}$$ $B_1 = 0.800\times10^{-6}\ \text{T}$ (in the $-y$ direction) $\vec{B}_1 = -0.800\times10^{-6}\,\hat{j}\ \text{T}$ $$B_2 = \frac{\mu_0 I_2}{2\pi r} = \frac{\left(4\pi\times10^{-7}\ \text{T}\cdot\text{m/A}\right)\left(7.00\ \text{A}\right)}{2\pi\left(0.500\ \text{m}\right)}$$ $B_2 = 2.80\times10^{-6}\ \text{T}$ (in $+x$ direction) $\vec{B}_2 = 2.80\times10^{-6}\,\hat{i}\ \text{T}$
The total field is the vector sum.	$$\vec{B}_{\text{tot}} = \vec{B}_1 + \vec{B}_2 = \boxed{\left(2.80\times10^{-6}\,\hat{i} - 0.800\times10^{-6}\,\hat{j}\right)\ \text{T}}$$
We can also express this in terms of magnitude and direction.	$$B = \sqrt{B_x^2 + B_y^2}$$ $$B = \sqrt{\left(2.80\times10^{-6}\right)^2 + \left(-0.800\times10^{-6}\right)^2}\ \text{T}$$ $B = 2.91\times10^{-6}\ \text{T}$ $$\theta = \tan^{-1}\left(\frac{B_y}{B_x}\right) = -15.9°$$ or $15.9°$ below the $+x$ axis in the xy plane

Chapter 30 – Magnetic Fields and Forces

82. (N) A section of a long, straight wire is twisted into a coplanar loop of radius $R =$ 45.0 cm (Fig. P30.82). If the current in the wire is $I = 5.00$ A, what are the magnitude and direction of the magnetic field at point O, the center of the loop? (The wire is sheathed in insulation.)

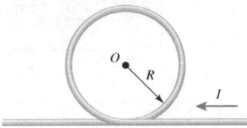

Figure P30.82

INTERPRET and ANTICIPATE

We can view this shape as a loop of current (going counter-clockwise in the loop) plus a long straight wire (with a current going to the left). This makes our lives easier since we can determine the field due to each of these known shapes and add them together.

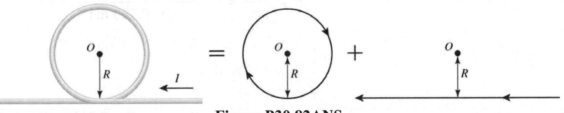

Figure P30.82ANS

SOLVE

First, consider the loop of current. The magnetic field at the center is given by Equation 30.10. R is the radius of the loop and y is the distance along the central axis of the loop from the center of the loop. Since we want to determine the field *at* the center of the loop, $y = 0$. The direction can be found from the simple right hand rule. For instance, consider the left side of the loop: point your thumb up in the direction of the current and wrap your fingers around the wire to find the

$$B_{loop} = \frac{\mu_0 I R^2}{2\left(R^2 + y^2\right)^{3/2}}$$

$$B_{center\,of\,loop} = B(y = 0) = \frac{\mu_0 I R^2}{2\left(R^2\right)^{3/2}}$$

$$B_{center\,of\,loop} = \frac{\mu_0 I}{2R}\,\text{into the page}$$

direction of the magnetic field, into the page to the right side (inside the loop) and out of the page on the left side (outside of the loop). Thus, the field points into the page at the center of the loop.	
Now, consider the line of charge. The field is given by Equation 30.9. The distance from the line of charge to the center of the loop is equal to the radius of the loop. The direction is again found with the simple right hand rule. Point your thumb to the left along the direction of current and wrap your fingers around to find the direction of the field, into the page above the wire (and at the point at the center of the loop) and out of the page below the wire.	$B_{\text{line}} = \dfrac{\mu_0 I}{2\pi r} = \dfrac{\mu_0 I}{2\pi R}$ into the page
Both fields point into the page at the center of the loop, so the total field is simply the sum of these two.	$B_{\text{tot}} = B_{\text{loop}} + B_{\text{line}} = \dfrac{\mu_0 I}{2R} + \dfrac{\mu_0 I}{2\pi R}$ $B = \left(1 + \dfrac{1}{\pi}\right)\dfrac{\mu_0 I}{2R}$ $B = \left(1 + \dfrac{1}{\pi}\right)\dfrac{\left(4\pi \times 10^{-7}\ \text{T·m/A}\right)\left(5.00\ \text{A}\right)}{2\left(0.450\ \text{m}\right)}$ $B = \boxed{9.20 \times 10^{-6}\ \text{T} \quad \text{into the page}}$

CHECK and THINK

While the current seems more complicated than many we've seen, it is essentially just a loop and a wire and we know how to deal with each separately. Since the total magnetic field is the vector sum due to all currents, we can find each and add them to determine the final answer.

87. (A) A charged particle with charge q and velocity $\vec{v} = v_x \hat{i} + v_y \hat{j}$ is moving into a region of space with a uniform magnetic field $\vec{B} = B_x \hat{i} + B_y \hat{j}$. Find the resultant magnetic force on the particle the instant it enters the field.

INTERPRET and ANTICIPATE

The magnetic force can be found using Eq. 30.17. We must be able to find a vector cross product using the component forms of the velocity and magnetic field. A vector cross product in three dimensions can be calculated in the following way:

$$\vec{A} \times \vec{B} = \begin{vmatrix} \hat{i} & \hat{j} & \hat{k} \\ A_x & A_y & A_z \\ B_x & B_y & B_z \end{vmatrix} = \left(A_y B_z - A_z B_y \right)\hat{i} - \left(A_x B_z - A_z B_x \right)\hat{j} + \left(A_x B_y - A_y B_x \right)\hat{k}$$

Given that we only have x and y components for the velocity and magnetic field, we expect that the magnetic force will only have a z component, when we compare the cross product above to Eq. 30.17.

SOLVE

Writing Eq. 30.17, and using the two vectors from the problem statement, we find an expression for the magnetic force on the particle.

$$\vec{F}_B = q\left(\vec{v} \times \vec{B} \right) = q \begin{vmatrix} \hat{i} & \hat{j} & \hat{k} \\ v_x & v_y & 0 \\ B_x & B_y & 0 \end{vmatrix}$$

$$\vec{F}_B = \boxed{q\left(v_x B_y - v_y B_x \right)\hat{k}}$$

CHECK and THINK

As expected, the magnetic force points along the z direction. Note that one of the features of any mathematical vector cross product is that the resultant vector must be perpendicular to the other two. Knowing that the velocity and the magnetic field reside entirely in the xy plane, the resultant force must be along the z direction.

90. (N) A mass spectrometer (Fig. 30.40, page 956) operates with a uniform magnetic field of 20.0 mT and an electric field of 4.00×10^3 V/m in the velocity selector. What is the radius of the semicircular path of a doubly ionized alpha particle ($m_a = 6.64 \times 10^{-27}$ kg)?

INTERPRET and ANTICIPATE

We can look at Example 30.5 to help us think through this problem. By knowing the magnitudes of the electric and magnetic fields, we can determine the velocity that will be selected by the spectrometer using $v = E/B$. We can then use Eq.30.23 to find the radius of the path of the alpha particle. Note that the particle is doubly ionized, so $q = 2e$.

SOLVE	
First, similar to the process used in Example 30.5, find the velocity of the alpha particles that will be selected by the velocity selector.	$v = \dfrac{E}{B} = \dfrac{4.00 \times 10^3 \text{ V/m}}{0.0200 \text{ T}} = 2.00 \times 10^5 \text{ m/s}$
Now, use Eq. 30.23 along with the fact that $q = 2e$.	$r = \dfrac{mv}{qB} = \dfrac{\left(6.64 \times 10^{-27} \text{ kg}\right)\left(2.00 \times 10^5 \text{ m/s}\right)}{2\left(1.60 \times 10^{-19} \text{ C}\right)\left(0.0200 \text{ T}\right)}$ $r = \boxed{0.208 \text{ m}} = 20.8 \text{ cm}$

CHECK and THINK

The strength of the fields could be adjusted to select particles with a different velocity. Then, assuming the mass and charge are the same, the radius of the path will be larger or smaller based on whether or not the velocity is larger or smaller.

95. (N) A proton enters a region with a uniform electric field $\vec{E} = 5.0\,\hat{k}$ V/m and a uniform magnetic field $\vec{B} = 5.0 \times 10^{-4}\,\hat{k}$ T. The proton has initial velocity $\vec{v} = 2.5 \times 10^5\,\hat{i}$ m/s. How far along the z axis does the proton travel after it undergoes three complete revolutions?

INTERPRET and ANTICIPATE

We need to use the Lorentz force law to solve this problem. The proton is initially moving perpendicular to the magnetic field, which will cause it to undergo circular motion. The electric field will then cause it to accelerate along the z direction. Since this is in the same direction as the magnetic field (i.e. this component is parallel to the magnetic field), there will be no effect due to the new z component of velocity on the rotation due to the magnetic force. Therefore, we can ask ourselves: (*i*) how long does it take for the particle to complete three revolutions in this magnetic field and then (*ii*) how far is the proton accelerated by the electric field in this time?

SOLVE	
The force on a particle subjected to both electric and magnetic fields is given by the Lorentz force (Equation 30.21). As described above, there is a magnetic force that causes rotation in the *xy* plane while	$\vec{F}_{\text{L}} = q\left(\vec{E} + \vec{v} \times \vec{B}\right)$

the electric field causes acceleration along z.	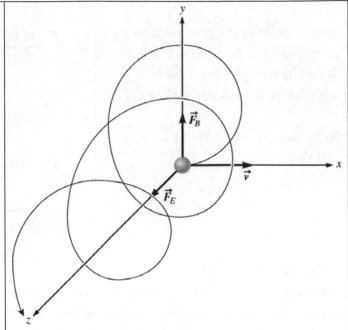
	Figure P30.95ANS
Since the z component of velocity is parallel to the magnetic field, it does not contribute to the magnetic force (the cross product picks out perpendicular components of vectors and is zero for parallel vectors). So, the magnetic force (and the resulting circular motion) depends only on the initial x velocity. The magnetic force acting on the proton provides the centripetal acceleration, holding the proton in the circular path with a radius given by Eq. 30.23.	$qvB = \dfrac{1}{2}mv^2 \quad \rightarrow \quad r = \dfrac{mv}{qB}$
The time that it takes to make one revolution (or the period) is equal to the circumference of one revolution ($2\pi r$) divided by the speed v. (You can also think of it as "the speed equals the circumference divided by the period.") Use the equation for the radius above.	$T = \dfrac{2\pi r}{v} = \dfrac{2\pi m}{qB} \qquad\qquad (1)$

The z component of the force is constant and produces a constant acceleration a_z along the z axis. We can use kinematics to find the z position as a function of time t.	$a_z = \dfrac{F_z}{m} = \dfrac{qE}{m}$ $z = \dfrac{1}{2} a_z t^2 = \left(\dfrac{qE}{2m}\right) t^2$ $\qquad$ (2)
We want to know the position after three revolutions, or $t = 3T$. Combine Equations (1) and (2).	$z = \left(\dfrac{qE_0}{2m}\right)\left(3 \cdot \dfrac{2\pi m}{qB}\right)^2 = \dfrac{18\pi^2 E m}{qB^2}$
Finally, insert numerical values with the proton charge $q = e = 1.6 \times 10^{-19}$ C and mass $m = 1.67 \times 10^{-27}$ kg.	$z = \dfrac{18\pi^2 (5.0 \text{ V/m})(1.67 \times 10^{-27} \text{ kg})}{(1.6 \times 10^{-19} \text{C})(5.0 \times 10^{-4} \text{ T})^2}$ $z = \boxed{37 \text{ m}}$

CHECK and THINK

While this charge experience both a magnetic and electric force, the orientation of the fields relative to the charge mean that the magnetic force leads to rotation in one plane while the electric field causes the proton to accelerate perpendicular to this plane. In fact, helical motion is a common type of motion observed for charges in electric and magnetic fields.

31

Gauss's Law for Magnetism and Ampère's Law

5. (N) In a laboratory exercise, you place a compass 2.00 cm from a wire as shown in Figure P31.5. Using a variable power supply, you slowly increase the current through the wire starting with $I = 0$. Assume the magnitude of the Earth's magnetic field is 5.0×10^{-5} T and that it points to the right in the figure.

a. If you can detect a deflection of only 2°, what is the minimum current you can detect?

b. Very large currents cause the compass needle to essentially align with the field due to the wire. Assuming you can't distinguish currents that lead to more than 88° of deflection, what is the largest current you can detect?

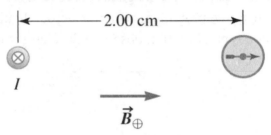

Figure P31.5

INTERPRET and ANTICIPATE
The deflection of the needle depends on the magnetic field created by the wire and the reference field when there is no current through the wire (the magnetic field of the earth). The direction of the field read by the compass corresponds to the resultant vector of these two fields added together.

SOLVE	
a. According to the right hand rule, the magnetic field due to the wire, $\vec{B}$, points downward at the location of the compass and is perpendicular to the earth's magnetic field. This is exactly the situation described by Equation 31.1. The reference field	$\tan\theta = \dfrac{B}{B_\oplus} \quad \rightarrow \quad B = B_\oplus \tan\theta$ $B = \left(0.5\times10^{-4}\ \text{T}\right)\tan\left(2°\right)$ $B = 1.7\times10^{-6}\ \text{T}$

$\vec{B}_\oplus$ is the magnetic field of the earth. Using a deflection angle of 2°, we can determine the magnetic field due to the wire.	 **Figure P31.5ANS**
The magnetic field due to an infinitely long wire was determined in Chapter 30. Solve this for current I.	$B = \dfrac{\mu_0 I}{2\pi R}$ $I = \dfrac{2\pi R B}{\mu_0}$
Using the information provided, we can now determine the current through the wire.	$I = \dfrac{2\pi\left(0.02\ \text{m}\right)\left(1.7\times10^{-6}\ \text{T}\right)}{4\pi\times10^{-7}\ \dfrac{\text{T}\cdot\text{m}}{\text{A}}} = \boxed{0.17\ \text{A}}$
b. Repeat the calculation for a deflection angle of 88°.	$B = B_\oplus \tan\theta$ $B = \left(0.5\times10^{-4}\ \text{T}\right)\tan\left(88°\right)$ $B = 1.4\times10^{-3}\ \text{T}$ $I = \dfrac{2\pi R B}{\mu_0}$ $I = \dfrac{2\pi\left(0.02\ \text{m}\right)\left(1.4\times10^{-3}\ \text{T}\right)}{4\pi\times10^{-7}\ \dfrac{\text{T}\cdot\text{m}}{\text{A}}} = \boxed{1.4\times10^{2}\ \text{A}}$

CHECK and THINK

The deflection angle of the compass needle relative to the reference field (earth's magnetic field) depends on the strength of the magnetic field due to the current. The small deflection angle of 2° corresponds to a current of $0.17 A$, which is the minimum current that could be detected assuming that 2° is the smallest measureable angle. As the current is turned up, the compass deflects more. The large deflection angle of 88° corresponds to a much larger current of 140 A as expected.

7. (A) The magnetic field in some region is given by $\vec{B} = B_x \hat{i} + B_y \hat{j}$. The areas A_x, A_y, A_z, B_x, and B_y are constants. What is the magnetic flux through each of these areas:

a. $\vec{A} = A_x \hat{i}$

b. $\vec{A} = A_y \hat{j}$

c. $\vec{A} = A_z \hat{k}$?

INTERPRET and ANTICIPATE	
Magnetic flux is a measure of the magnetic field through a surface. It depends on the magnetic field, the area of the surface, and their relative orientation.	

SOLVE	
The magnetic flux can be calculated using Equation 31.2. The area vector is perpendicular to the surface with a magnitude equal to the area of the surface. The flux is, roughly speaking, how much magnetic field goes through the surface, which is found as the dot product of the magnetic field and area vector.	$\Phi_B = \int \vec{B} \cdot d\vec{A}$ $\Phi_B = \int \left(B_x dA_x + B_y dA_y + B_z dA_z \right)$ $\Phi_B = \int B_x dA_x + \int B_y dA_y \quad \left(\text{since } B_z = 0 \right)$
a. For this surface, $A_y = 0$.	$\Phi_B = \boxed{B_x A_x}$
b. For this surface, $A_x = 0$.	$\Phi_B = \boxed{B_y A_y}$
c. For this surface, both A_x and $A_y = 0$. Since the magnetic field only has x and y components (in other words, it's in the xy plane), the magnetic flux through a surface with only a z component (i.e. the surface itself is in the xy plane) is zero. This makes sense, since the magnetic field lines go *along* the surface but not *through* it.	$\Phi_B = \boxed{0}$

Chapter 31 – Gauss's Law of Magnetism and Ampère's Law

12. (N) A uniform magnetic field of 0.25 T points in a direction 30.0° from the positive x axis, 60.0° from the negative y axis in a region of space, and 90.0° from the positive z axis. In each of the following cases, sketch the surface and determine the magnetic flux through the surface:

a. $\vec{A} = 0.010\,\hat{j}$ m^2

b. $\vec{A} = 1.0\,\hat{k}$ m^2

INTERPRET and ANTICIPATE
The magnetic flux through a surface depends on the area of the surface and the magnitude and direction of the magnetic field relative to the surface.

SOLVE

a. The magnetic flux through a surface is given by Equation 31.2. The flux depends on the dot product of the magnetic field with the area vector (which points perpendicular to the surface and has a magnitude equal to the area of the surface). Therefore, for a flat surface, the flux depends on the component of the magnetic field perpendicular to the surface. This can also be expressed as the magnitude of the magnetic field times the area times the angle between these two vectors.

$$\Phi_B = \int \vec{B} \cdot \vec{A}$$

$$\Phi_B = BA\cos\theta$$

Sketch the field and the area vector. The area points along the y axis and the magnetic field 60° from the $-y$ direction, so the two vectors are 120° apart.

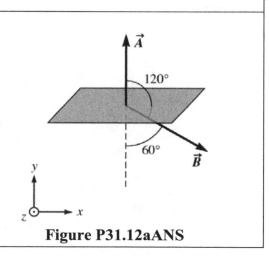

Figure P31.12aANS

209

We now calculate the flux. Since the magnetic field and area vector point in opposite directions, the flux is negative. The unit of magnetic flux is the weber.	$\Phi_B = (0.25 \text{ T})(0.01 \text{ m}^2)\cos(120°)$ $\Phi_B = \boxed{-0.00125 \text{ Wb}}$
b. Sketch the situation described. The area vector points along the *z* direction but the field is in the *xy* plane (90.0° from the positive *z* axis), so they are perpendicular with an angle of 90°.	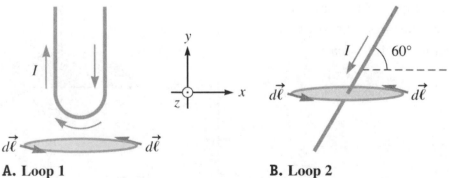 **Figure P31.12bANS**
Since the field points *along* the plane of the area, no field lines actually go *through* the area and the flux is zero. Our calculation confirms this.	$\Phi_B = BA\cos\theta$ $\Phi_B = (0.25 \text{ T})(1.0 \text{ m}^2)\cos(90°) = \boxed{0}$

CHECK and THINK

The flux was found given the magnetic field, the area vector, and their relative orientations. In the first case, the magnetic field points in a direction opposite how we defined the area and the flux is negative. In the second case, the field points along the plane (perpendicular to the area vector), so the flux through the loop is zero.

18. (N) Suppose the current in each wire shown in Figure P31.18 is $I = 0.50$ A. What is the result of the circulation integral around the Ampèrian loop in parts *A* and *B* of the figure?

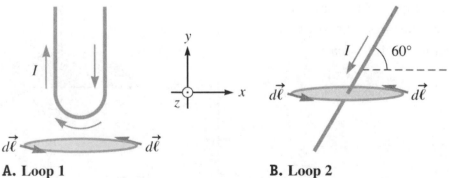

A. Loop 1   **B. Loop 2**

Figure P31.18

INTERPRET and ANTICIPATE The circulation integral is the integral of the magnetic field along a closed path. Ampère's law relates the circulation integral to the current through the loop, which we can easily calculate using the figure.	

SOLVE The circulation integral is the left side of Eq. 31.5. We can easily calculate the right hand side.	$\oint \vec{B} \cdot d\vec{\ell} = \mu_0 I_{thru}$
Loop A: There is no current that actually goes through the loop, so I_{thru} = 0.	$\oint \vec{B} \cdot d\vec{\ell} = \mu_0 (0) = 0$
Loop B: There is one wire going through the loop, so I_{thru} = 0.5 A.	$\oint \vec{B} \cdot d\vec{\ell} = \mu_0 I_{thru} = \left(4\pi \times 10^{-7} \dfrac{\text{T} \cdot \text{m}}{\text{A}} \right)(0.5 \text{ A})$ $\oint \vec{B} \cdot d\vec{\ell} = \boxed{6.3 \times 10^{-7} \text{ T} \cdot \text{m}}$

CHECK and THINK Amperc's law allows us to relate the circulation integral to the current going through the loop.	

23. (A) A steady current I flows through a wire of radius a. The current density in a wire varies with r as $J = kr''$, where k is a constant and r is the distance from the axis of the wire. Find expressions for the magnitudes of the magnetic field inside and outside the wire as a function of r. (Hint: Find the current through an Ampèrian loop of radius r using $I_{thru} = \int \vec{J} \cdot d\vec{A}$.)

INTERPRET and ANTICIPATE Based on the hint, we need to integrate the current density to find the current through an Ampèrian loop. As usual, we'll pick a circular loop centered on the wire based on the symmetry of the wire to relate $B(r)$ around the loop to the current through the loop.	

SOLVE Ampère's law relates the magnetic field around the loop to the current through the loop, which we can find by integrating as suggested by the	$\oint \vec{B} \cdot d\vec{\ell} = \mu_0 I_{thru} = \mu_0 \int \vec{J} \cdot d\vec{A}$

211

hint. We must integrate because the current density is not uniform and depends on r.	
If we imagine a current into the page in wire of radius a, the magnetic field lines form clockwise loops both inside and outside the wire. We imagine an Ampèrian loop at a radius r that could either be smaller or larger than R with a clockwise integration path $d\vec{\ell}$.	 $r < R$ $r > R$ **Figure P31.23ANS**
Outside the wire: For $r > R$, the entire current I goes through the loop. The magnetic field has a constant magnitude around the loop, so the integral is simply the magnetic field magnitude B times the length of the circular loop, $2\pi r$: $\oint \vec{B} \cdot d\vec{\ell} = B \oint d\ell = B(2\pi r)$	$B_{out}(2\pi r) = \mu_0 I$ $\boxed{B_{out} = \dfrac{\mu_0 I}{2\pi r}}$
Inside the wire: For $r < R$, the Ampèrian loop encloses a current that we can calculate by integrating the current density (current per area) over the area of the Ampèrian loop. The area element can be expressed as $dA = 2\pi r' dr'$ (basically adding up the total area of rings of width dr' with radii that range from $r' = 0$ to r).	$I_{thru} = \int \vec{J} \cdot d\vec{A}$ $I_{thru} = \int_0^r (kr'^2) 2\pi r' \, dr'$ $I_{thru} = \pi k \dfrac{r^4}{2}$ (1)
If we integrate the current density for the entire wire (up to a radius a), this must equal the	$I = \int_0^a (kr^2) 2\pi r \, dr = \pi k \dfrac{a^4}{2}$

212

total current in the wire, I. This allows us to relate the constant k to known quantities.	$k = \dfrac{2I}{\pi a^4}$ (2)
Insert Eq. (2) into (1).	$I_{thru} = \pi \left(\dfrac{2I}{\pi a^4} \right) \dfrac{r^4}{2} = I \left(\dfrac{r}{a} \right)^4$
Finally, evaluate Ampère's law.	$B_{in}(2\pi r) = \mu_0 I \left(\dfrac{r}{a} \right)^4$ $\boxed{B_{in} = \dfrac{\mu_0 I r^3}{2\pi a^4}}$

CHECK and THINK

In both cases, we solve Ampère's law to determine the magnetic field due to a current. Inside the wire, we need to integrate to find the total current going through the Ampèrian loop. The field increases at positions closer to the edge of the wire and then fall off as $1/r$ as we expect at points moving away from the wire.

25. (N) A magnetic field of 4.00 μT is measured at a distance of 25.0 cm from a long, straight wire with a current of 5.00 A. What is the distance from the wire at which a field of 0.500 μT will be measured?

INTERPRET and ANTICIPATE	
This is a straightforward application of the magnetic field near a long straight wire.	

SOLVE The magnetic field due to a long straight wire (Equation 30.9) was found using Ampère's law in Example 31.5.	$B = \dfrac{\mu_0 I}{2\pi r}$
One way to calculate the answer is to note that the magnetic field strength is inversely proportional to the distance from the wire. So, 0.500 μT, which is $1/8^{th}$ the field of 4.00 μT, must be measured at a distance	$B \propto \dfrac{1}{r}$ so $Br = \text{constant}$

that is 8 times as far from the wire, or 8(25.0 cm) = 200 cm.	$B_1 r_1 = B_2 r_2$ $r_2 = \dfrac{B_1}{B_2} r_1 = \dfrac{4.00\ \mu T}{0.500\ \mu T} \cdot 25.0\ \text{cm} = 200\ \text{cm}$
A second way to find the answer is to use Eq. 30.9 directly and solve for the radius.	$r = \dfrac{\mu_0 I}{2\pi B}$ $r = \dfrac{\left(4\pi \times 10^{-7}\ \text{T} \cdot \text{m/A}\right)\left(5.00\ \text{A}\right)}{2\pi \left(0.500 \times 10^{-6}\ \text{T}\right)}$ $r = \boxed{2.00\ \text{m}}$

CHECK and THINK

There are a couple ways to approach this problem, but it is a straightforward application of the magnetic field due to a current-carrying wire.

---

29. (N) A coaxial cable (Fig. P31.27) carries a current of 3.25 A. The radii are $r_1 = 0.500$ cm, $r_2 = 1.00$ cm, $r_3 = 1.50$ cm, and $r_4 = 2.00$ cm.

a. Where is the magnetic field strongest?

b. What is the magnitude of the magnetic field at this point?

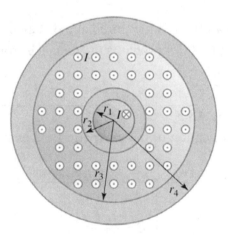

Figure P31.27

INTERPRET and ANTICIPATE

According to Ampère's law, the integral of the magnetic field around a closed loop is related to the enclosed current. Imagine an Ampèrian loop centered on the wire with a radius that increases from zero. As the loop gets to $r = r_1$, the largest net current is enclosed. The current is going in the opposite direction in the outer conductor, so as the loop is increased further, the enclosed current begins to decrease back towards zero. So, we expect the strongest magnetic field to be at $r = r_1$, but we'll calculate this to be sure.

SOLVE Let's first follow the steps carried out in Problems 27 and 28 just to confirm our intuition above. Apply Ampère's law (Eq. 31.5) as we did in Example 31.7.	$\oint \vec{B} \cdot d\vec{\ell} = \mu_0 I$
For $r < r_1$, the enclosed current is the fraction of the total current in the Ampèrian loop. So, it's I times the ratio of the area of the Amperian loop to the total cross sectional area of the wire: $I_{thru} = I \dfrac{\pi r^2}{\pi r_1^2}$	$B(2\pi r) = \mu_0 I \dfrac{\pi r^2}{\pi r_1^2}$ $B_{r<r_1} = \dfrac{\mu_0 I}{2\pi r_1^2} r$
For $r_1 < r < r_2$, the entire current of the inner wire I is enclosed.	$B(2\pi r) = \mu_0 I$ $B_{r_1<r<r_2} = \dfrac{\mu_0 I}{2\pi r}$
For $r_2 < r < r_3$, the current in the outer shell is the current I times the ratio of the area of the annulus from r_2 to r to the total area of the annulus, $\dfrac{r^2 - r_2^2}{r_3^2 - r_2^2}$. This current is in the opposite direction of the current in the inner wire. We can now calculate the net current enclosed, I_{thru}.	$I_{thru} = I - I \dfrac{r^2 - r_2^2}{r_3^2 - r_2^2}$ $B(2\pi r) = \mu_0 I \left[1 - \dfrac{r^2 - r_2^2}{r_3^2 - r_2^2} \right]$ $B_{r_2<r<r_3} = \dfrac{\mu_0 I}{2\pi r} \left[1 - \dfrac{r^2 - r_2^2}{r_3^2 - r_2^2} \right]$
For $r > r_3$, the total current enclosed is zero.	$B(2\pi r) = \mu_0 (0)$ $B_{r>r_3} = 0$

We can plot the magnetic field to confirm that the largest value is at r_1 as we anticipated above.	*B* (T) *r* (m) **Figure P31.28ANS**
Finally, evaluate the magnetic field at $r = r_1$.	$$B = \frac{\mu_0 I}{2\pi r_1}$$ $$B_{r=r_1=5\text{cm}} = \frac{\left(4\pi \times 10^{-7} \dfrac{\text{T}\cdot\text{m}}{\text{A}}\right)(3.25\,\text{A})}{2\pi\left(5.00\times 10^{-3}\ \text{m}\right)} = \boxed{1.30\times 10^{-4}\ \text{T}}$$

CHECK and THINK

Based on Ampère's law, the integral of the magnetic field is related to the enclosed current. An Ampèrian loop with $r = r_1$ is, roughly speaking, the smallest loop enclosing the most net current, so it's the location with the largest magnetic field. Knowing this, we can calculate the specific value with Ampère's law and the values given.

---

31. (N) Figure P31.31 shows a cylindrical conducting shell carrying current I to the right. The nonuniform current density in the conductor is given by $\underset{!}{J} = c\left(r - R_{_"}\right)$ for the region $R_1 < r < R_2$, where c is a constant. What is the magnitude of the magnetic field at a radial distance r_A within the conductor? (Hint: Find the current through an Ampèrian loop of radius r using $I_{\text{thru}} = \int \vec{J}\cdot d\vec{A}$.)

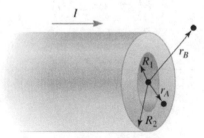

Figure P31.31

216

INTERPRET and ANTICIPATE This is an application of Ampère's law. Use a circular Ampèrian loop centered on the axis of the cylindrical shell. We can then integrate to find the current passing through the loop to relate it to the circulation integral (the integral of the magnetic field around the loop).	

SOLVE Apply Ampère's law (Eq. 31.5) using the hint to express this in terms of current density J.	$\oint \vec{B} \cdot d\vec{\ell} = \mu_0 I$ $\oint \vec{B} \cdot d\vec{\ell} = \mu_0 \int \vec{J} \cdot d\vec{A}$

Imagine that the wire is oriented as shown, with the current going into the page. By the simple right hand rule, the magnetic field will form closed clockwise loops. We also draw an Ampèrian loop of radius r that we will integrate clockwise, such that the integration path always points along the magnetic field direction.	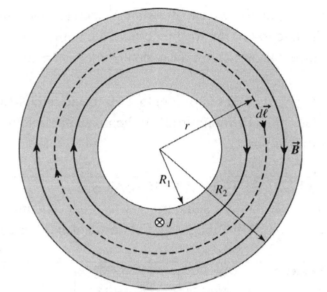 **Figure P31.31ANS**

Before completing the question asked, note that if $r < R_1$, the loop encloses no current, so according to Ampère's law, the field is zero. If $r > R_2$, it encloses the entire current I, so the field just looks like that due to a long straight wire.	$B_{r<R_1} = 0$ $B_{r>R_2} = \dfrac{\mu_0 I}{2\pi r}$

For $R_1 < r < R_2$, we need to perform the integral above. Since the field and integration length are always along the same direction, $\oint \vec{B} \cdot d\vec{\ell} = B \oint d\ell = B(2\pi r_A)$, where	

$2\pi r_A$ is the circumference of the Ampèrian loop. For the right hand side, the area can be integrated by adding up the area of a bunch of rings with radius r and thickness dr. That is, $dA = 2\pi r\, dr$ and we integrate from R_1 to r_A.	$B(2\pi r_A) = \mu_0 \int_{R_1}^{r_A} c(r - R_1)(2\pi r\, dr)$ $B(2\pi r_A) = 2\pi\,\mu_0 c\left[\dfrac{r^3}{3} - \dfrac{r^2}{2}R_1\right]_{R_1}^{r_A}$ $B = \dfrac{\mu_0 c}{r_A}\left(\left[\dfrac{r_A^3}{3} - \dfrac{r_A^2}{2}R_1\right] - \left[\dfrac{R_1^3}{3} - \dfrac{R_1^3}{2}\right]\right)$ $\boxed{B_{R_1 < r < R_2} = \mu_0 c\left[\dfrac{r_A^2}{3} - \dfrac{R_1 r_A}{2} + \dfrac{R_1^3}{6r_A}\right]}$

CHECK and THINK

Conceptually, we can draw an Ampèrian loop and relate the current through the loop to the integral of the magnetic field along the loop. In this case, we need to integrate the current density in order to determine the enclosed current.

---

34. (N) A solenoid of length 2.00 m and radius 1.00 cm carries a current of 0.100 A. Determine the magnitude of the magnetic field inside if the solenoid consists of 2000 turns of wire.

INTERPRET and ANTICIPATE
The solenoid is a standard way to produce a uniform magnetic field. We can apply the equation for the magnetic field in a solenoid directly.

SOLVE The magnetic field in a solenoid is given by Eq. 31.6 where $n = N/L$ is the number of turns divided by the length of the solenoid.	$B = \mu_0 n I = \mu_0 \dfrac{N}{L} I$
Insert numerical values.	$B = \left(4\pi \times 10^{-7}\ \text{T}\cdot\text{m/A}\right)\left(\dfrac{2000}{2.00\ \text{m}}\right)(0.100\,\text{A})$ $B = \boxed{1.26 \times 10^{-4}\ \text{T}}$

CHECK and THINK
This field is around the same order of magnitude as the Earth's magnetic field. While we typically have no need for a solenoid that is two meters long, producing a solenoid with 1000 turns per meter and a current of 0.1 A is well within the limits of what we can do in the lab.

39. (N) The magnetic field inside a solenoid 23.7 cm long is 0.235 T. If the current is 4.53 A and the diameter of the solenoid is 2.37 cm, what are the total number of turns and the length of the wire?

INTERPRET and ANTICIPATE	
We can apply the equation for the magnetic field inside a solenoid. We are given all of the pieces of information we need to determine the number of turns in the solenoid and can then calculate the length of wire needed to create this solenoid.	
SOLVE The magnetic field in a solenoid is given by Eq. 31.6 with $n = N/L$ the number of turns in the solenoid divided by its length.	$B = \mu_0 n I = \mu_0 \dfrac{N}{L} I$
Rearrange this expression to solve for the number of turns and plug in values.	$N = \dfrac{BL}{\mu_0 I}$ $N = \dfrac{(0.235 \text{ T})(0.237 \text{ m})}{(4\pi \times 10^{-7} \text{ T} \cdot \text{m/A})(4.53 \text{ A})} = \boxed{9.78 \times 10^3}$
The length of the wire must equal N loops times the circumference $2\pi r$ of each loop. Note that the diameter is given and we need the radius, (2.37 cm)/2.	$\ell = N 2\pi r$ $\ell = (9790) 2\pi \left(\dfrac{0.0237 \text{ m}}{2}\right) = 728 \text{ m} = \boxed{7.28 \times 10^2 \text{ m}}$
CHECK and THINK The solenoid is a common way to create a magnetic field. Given the current, length, and field created, we can determine how many loops and how much wire we need—in this case, nearly 10,000 loops and approaching a kilometer long wire!	

40. (N) A square conducting loop with side length $a = 1.25$ cm is placed at the center of a solenoid 40.0 cm long with 300 turns and aligned so that the plane of the loop is perpendicular to the long axis of the solenoid. According to an observer, the current in the single turn of the loop is 0.800 A in the counterclockwise direction, and the current in the windings of the solenoid is 8.00 A in the counterclockwise direction. **a.** What is the magnetic force on each side of the square loop? **b.** What is the net torque acting on the square loop?

INTERPRET and ANTICIPATE

A solenoid creates a uniform magnetic field in its interior. A current in a magnetic field experiences a force, so we can determine the force on each segment of the square loop and then the torque.

SOLVE

a. The magnetic field produced by the solenoid in its interior is given by Eq. 31.6, where we choose the magnetic field to point in the positive x direction as shown.

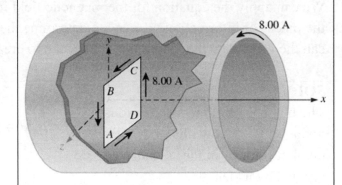

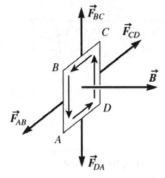

Figure P31.40ANS

$$\vec{B} = \mu_0 n I\, \hat{i}$$

We can plug in values to determine the field strength.	$\vec{B} = \left(4\pi \times 10^{-7} \text{ T·m/A}\right)\left(\dfrac{300}{0.400 \text{ m}}\right)\left(8.00 \text{ A}\right)\hat{i}$ $\vec{B} = 7.54 \times 10^{-3}\hat{i} \text{ T}$
The current through the square loop is counter-clockwise also. Consider side AB of the square current loop first. The force on side AB can be calculated as in Chapter 30.	$\left(\vec{F}_B\right)_{AB} = I\vec{\ell} \times \vec{B}$

Plug in values. To get the direction, use the right hand rule: If you point your fingers in the negative y direction $\left(\vec{\ell}\right)$ and wrap your fingers towards the positive x direction $\left(\vec{B}\right)$, your thumb points in the positive z direction $\left(\left(\vec{F}_B\right)_{AB}\right)$. (It's also the case mathematically

that $-\hat{j}\times\hat{i}=\hat{k}$.) $(\vec{F}_B)_{AB}=(0.800\ \text{A})\left[(1.25\times10^{-2}\text{m})(-\hat{j})\times(7.54\times10^{-3}\ \text{T})\hat{i}\right]$ $(\vec{F}_B)_{AB}=7.54\times10^{-5}\hat{k}\ \text{N}$	

Four each of the four segments of the square loop, the current, magnetic field, and length are all the same, so the magnitude of all four forces are the same. To find the directions, use the right hand rule four the other three segments: *BC*: Current in negative *z* direction, field in positive *x* direction, so force points in positive *y* direction. *CD*: Current in positive *y* direction, field in positive *x* direction, so force points in negative *z* direction. *DA*: Current in negative *z* direction, field in positive *x* direction, so force points in negative *y* direction.	$(\vec{F}_B)_{AB}=7.54\times10^{-5}\hat{k}\ \text{N}$ $(\vec{F}_B)_{BC}=7.54\times10^{-5}\hat{j}\ \text{N}$ $(\vec{F}_B)_{CD}=-7.54\times10^{-5}\hat{k}\ \text{N}$ $(\vec{F}_B)_{DA}=-7.54\times10^{-5}\hat{j}\ \text{N}$ The force on each segment is 75.4 μN directed away from the center of the loop.
b. Since the force on each segment of the loop is radially outward, along the direction of $\vec{r}$, the radial vector from the center of the loop, the torque due to each segment is zero and therefore the net torque is zero. (For instance, the radial direction from the center of the loop to segment *AB* is in the positive *z* direction, the same direction as the force, so the cross product $\vec{r}\times\vec{F}$ is zero.)	$\vec{\tau}=\vec{r}\times\vec{F}$

CHECK and THINK
Since each segment of the square has the same length and current and is in a uniform field, the force on each side is the same. The direction of each is toward the outside of the loop, so there are no forces *around* the center causing a torque. This makes sense from what we learned in Chapter 30 as well—the magnetic moment of the loop (using the right hand rule) is in the positive x direction and the torque due to a uniform magnetic field tends to align the magnetic moment to the field… since they are *already* aligned, there is no torque on the loop!

44. (N) You have a spool of copper wire 5.00 mm in diameter and a power supply. You decide to wrap the wire tightly around a soda can that is 12.0 cm long and has a diameter of 6.50 cm, forming a coil, and then slide the wire coil off the can to form a solenoid. If the power supply can produce a maximum current of 125 A in the coil, what is the maximum magnetic field you would expect to produce in this solenoid? Assume the resistivity of copper is 1.68×10^{-8} Ω·m.

INTERPRET and ANTICIPATE
Using the information given, we can determine the spacing of the loops and the magnetic field produced in the solenoid, one of the typical geometries we've seen.

SOLVE The magnetic field produced in a solenoid is given by Equation 31.6.	$B = \mu_0 n I$
The quantities are all given except for the number of loops per meter, n. Given the diameter of the wire, we can only put them as close as 1 loop per 5.00 mm, which allows 200 loops per meter.	$n = \dfrac{1 \text{ loop}}{5 \text{ mm}} = \dfrac{1}{0.005 \text{ m}} = 200 \text{ m}^{-1}$
Filling in these values, we can determine the magnetic field.	$B = \left(4\pi \times 10^{-7} \dfrac{\text{T} \cdot \text{m}}{\text{A}} \right) \left(200 \text{ m}^{-1} \right) \left(125 \text{ A} \right)$ $B = \boxed{0.0315 \text{ T}}$

CHECK and THINK
The magnetic field was found using the equation for the solenoid and the values given. Despite the large current the wires can carry, their thickness limits how closely they will pack. The resulting magnetic field is significantly larger than that due to the Earth's field $\left(\approx 5 \times 10^{-4} \text{ T} \right)$, though not huge.

Chapter 31 – Gauss's Law of Magnetism and Ampère's Law

47. (N) A toroid with an inner radius of 50.0 cm and outer radius of 75.0 cm is wound with 760.0 turns of wire carrying a current of 955 A. What is the magnitude of the magnetic field along
a. the inner radius of the toroid and
b. the outer radius of the toroid?

INTERPRET and ANTICIPATE	
This is a direct application of the magnetic field in a toroid. The magnetic field depends on the number of loops and current, similar to the solenoid, but the field decreases with distance from the center of the toroid.	

SOLVE	
The magnetic field of a toroid is given by Eq. 31.7.	$B = \dfrac{\mu_0 NI}{2\pi r}$
a. Insert values for the inner radius.	$B_{inner} = \dfrac{\left(4\pi\times10^{-7}\ \text{T}\cdot\text{m/A}\right)(760)(955\ \text{A})}{2\pi(0.500\ \text{m})} = \boxed{0.290\ \text{T}}$
b. Insert values for the outer radius.	$B_{outer} = \dfrac{\left(4\pi\times10^{-7}\ \text{T}\cdot\text{m/A}\right)(760)(955\ \text{A})}{2\pi(0.750\ \text{m})} = \boxed{0.194\ \text{T}}$

CHECK and THINK	
As expected, the magnetic field decreases with distance from the center of the toroid.	

51. (N) A toroid has an outer radius of 5.75 cm. The strongest magnetic field produced by the torus is 0.632 T and the weakest magnetic field is 0.316 T. What is the toroid's inner radius?

INTERPRET and ANTICIPATE	
The magnetic field in a toroid depends on the number of loops and current, similar to the solenoid, but the field is not uniform inside the toroid. Instead, the field decreases with distance from the center of the toroid.	

SOLVE	
The magnetic field of a toroid is given by Eq. 31.7.	$B = \dfrac{\mu_0 NI}{2\pi r}$

We can express the fields at the minimum and maximum radii (i.e. the inner and outer edges of the toroid).	$B_{max} = \dfrac{\mu_0 NI}{2\pi r_{min}}$ and $B_{min} = \dfrac{\mu_0 NI}{2\pi r_{max}}$
Note that the quantity $\mu_0 NI/2\pi$ is the same for both, which allows us to write $B_{max}r_{min} = B_{min}r_{max}$.	$B_{max}r_{min} = \dfrac{\mu_0 NI}{2\pi} = B_{min}r_{max}$
We can now solve for the minimum radius. The ratio of the radii turns out to equal the ratio of the magnetic fields.	$r_{min} = \dfrac{B_{min}}{B_{max}} r_{max}$ $r_{min} = \left(\dfrac{0.316 \text{ T}}{0.632 \text{ T}} \right) 5.75 \text{ cm} = \boxed{2.88 \text{ cm}}$

CHECK and THINK

Using the formula for the magnetic field in a toroid, we're able to relate the magnetic field at the inner and outer boundary to their radii. In particular, the ratio of the fields at the inner and outer boundary is equal to the ratio of their radii.

54. (N) A parallel-plate capacitor is made using circular plates of radius 10.0 cm separated by a distance of 0.50 cm. It is connected to a battery, and the voltage between the capacitor plates increases at a rate of 125,000 V/s. What is the displacement current between the capacitor plates?

INTERPRET and ANTICIPATE	
The displacement current depends on the rate of change of the electric flux. The electric field, and therefore the flux, varies with the changing voltage.	

SOLVE The displacement current is given by Equation 31.9.	$I_d = \varepsilon_0 \dfrac{d\Phi_E}{dt}$
The electric flux depends on the electric field and the area of the plates. The electric field is equal to the voltage between the plates divided by the distance between them. The area of the circular plate is πr^2. Since the area vector (perpendicular to the area between the plates) points along the field direction,	$\Phi_E = \vec{E} \cdot \vec{A}$ $\Phi_E = \dfrac{V}{d} \cdot \pi R^2$

$\vec{E} \cdot \vec{A} = EA\cos(0°) = EA$. The rate of change of the flux then depends on the rate of change of the voltage, which is given.	
We can now determine the rate of change of the flux.	$\dfrac{d\Phi_E}{dt} = \dfrac{dV}{dt} \cdot \dfrac{\pi R^2}{d}$ $\dfrac{d\Phi_E}{dt} = \left(125{,}000\,\dfrac{V}{s}\right)\left(\dfrac{\pi(0.10\text{ m})^2}{0.005\text{ m}}\right)$ $\dfrac{d\Phi_E}{dt} = 790{,}000\,\dfrac{V\cdot m}{s}$
Finally, we can calculate the displacement current.	$I_d - \varepsilon_0 \dfrac{d\Phi_E}{dt}$ $I_d = \left(8.85\times10^{-12}\,\dfrac{C}{V\cdot m}\right)\left(790{,}000\,\dfrac{V\cdot m}{s}\right)\text{A}$ $I_d = \boxed{7.0\times10^{-6}\text{ A}}$

CHECK and THINK

The displacement current depends on the rate of change of the electric flux, which depends on the rate of change of the voltage. The displacement current in this case is quite small.

59. (N) A uniform magnetic field $\vec{B} = 5.44\times10^4\,\hat{i}$ T passes through a closed surface with a slanted top as shown in Figure P31.59.

a. Given the dimensions and orientation of the closed surface shown, what is the magnetic flux through the slanted top of the surface?

b. What is the net magnetic flux through the entire closed surface?

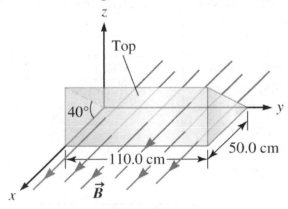

Figure P31.59

Chapter 31 – Gauss's Law of Magnetism and Ampère's Law

INTERPRET and ANTICIPATE

In order to find the flux through the top surface, we need to express the direction of the area vector of that surface. Preferably, we would like to know the angle between the area vector of that surface and the direction of the magnetic field. Recall that the area vector of a surface is always perpendicular to the surface. Since the surface itself is 40° with respect to the field, the area vector must be 130° from the positive *x*-direction, as seen in the sketched side-view shown in Figure P31.59ANS. There, we preferentially drew the area vector perpendicular to the slated top such that it happens to hit the bottom corner of the closed surface, for ease of interpreting the angle it makes with the horizontal. We can then use Eq. 31.2 to express the magnetic flux through the top surface. We expect he magnetic flux through the entire closed surface (part (b)) to be 0 since there are no magnetic monopoles (Eq. 31.4).

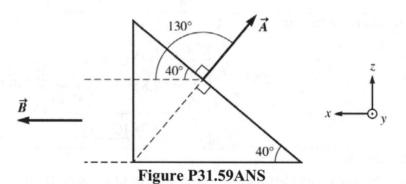

Figure P31.59ANS

SOLVE

a. Because the angle between the slanted surface and the magnetic field is never changing, we can replace the integral of the dot product in Eq. 31.2 with just the dot product between the field and the area vector. We can then write the magnitude of the dot product between two vectors as $BA\cos\theta$, by definition.

$$\Phi_B = \int \vec{B} \cdot d\vec{A} = \vec{B} \cdot \vec{A} = BA\cos\theta$$

In order to find this magnetic flux, we need the area of the slanted top. The length of one side is 110.0 cm = 1.100 m. The width, *w*, of the other side can be found through some geometry from our side view in Figure P31.59ANS.

$$w = \frac{0.500 \text{ m}}{\cos(40°)}$$

226

Thus, the area of the slanted top is given by the length times the width.	$A = (1.100 \text{ m}) \left(\dfrac{0.500 \text{ m}}{\cos(40°)} \right)$
Then, because the angle between the area vector of the slanted top and the magnetic field is 130°, we can find the magnetic flux through the top surface.	$\Phi_B = BA\cos\theta$ $\Phi_B = (5.44 \times 10^4 \text{ T})(1.100 \text{ m}) \left(\dfrac{0.500 \text{ m}}{\cos(40°)} \right) \cos(130°)$ $\Phi_B = \boxed{-2.51 \times 10^4 \text{ Wb}}$
b. The net magnetic flux through the closed surface must be $\boxed{0}$ since there are no known magnetic monopoles, such as expressed in Eq. 31.4.	$\Phi_B = \oint \vec{B} \cdot d\vec{A} = \boxed{0}$

CHECK and THINK

The net magnetic flux through any closed surface must be 0, however, each individual surface may have a magnetic flux through it alone. In this example, we can reason that the magnetic flux through the bottom of the closed surface must be 0, because its area vector would be perpendicular to the field, and that the magnetic flux through the other side of the closed surface must be 2.51×10^4 Wb, so that the net magnetic flux through the entire closed surface is 0.

---

65. (N) A coaxial cable is constructed from a central cylindrical conductor of radius $r_A = 1.00$ cm carrying current $I_A = 8.00$ A in the positive x direction and a concentric conducting cylindrical shell with inner radius $r_B = 14.0$ cm and outer radius $r_C = 15.0$ cm with a current of 20.0 A in the negative x direction (Fig. P31.65). What are the magnitude and direction of the magnetic field

a. at point O, a distance of 10.0 cm from the center of the coaxial cable along the y axis, and

b. at point P, a distance of 20.0 cm from the center of the coaxial cable along the y axis?

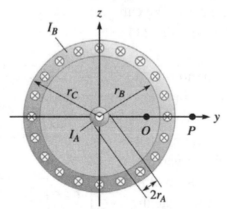

Figure P31.65

Chapter 31 – Gauss's Law of Magnetism and Ampère's Law

INTERPRET and ANTICIPATE

Ampère's law relates the integral of the magnetic field around a closed loop to the enclosed current. We can evaluate this using an Ampèrian loop with each of the radii indicated and solve for the field strength.

SOLVE **a.** Evaluate Ampère's law (Eq. 31.5) with a radius $r = r_O$. In this case, the loop is between the two conductors, so it encloses the wire at the center and the current through the loop is $I_{thru} = I_A = 8.00$ A.	$\oint \vec{B} \cdot d\vec{\ell} = \mu_0 I_{thru}$ $B(2\pi r_O) = \mu_0 I_A$ $B_O = \dfrac{\mu_0 I_A}{2\pi r_O}$ (1)
Insert numbers to calculate the value. We can determine the direction using the simple right hand rule. Point your thumb in the direction of the current (out of the page) and your fingers wrap counter-clockwise, indicating the direction of the magnetic field, which forms closed circular loops. At the location of point O, your fingers point upwards, or in the positive z direction according to the axes in the figure.	$B_O = \dfrac{\left(4\pi \times 10^{-7} \text{ T} \cdot \text{m/A}\right)\left(8.00 \text{ A}\right)}{2\pi\left(1.00 \times 10^{-1} \text{ m}\right)}$ $B_O = 1.60 \times 10^{-5}$ T in the $+z$ direction or $\vec{B}_O = \boxed{1.60 \times 10^{-5}\, \hat{k} \text{ T}}$
b. We perform the same calculation at point P, which is outside the coaxial cable. In this case, a loop of radius $r = r_P$ encloses a net current that includes both the 8.00 A current out of the page and the 20.0 A current into the page. The net current through the loop is therefore $I_P = 12.0$ A into the page. We can now write an expression like Equation (1) above.	$B_P = \dfrac{\mu_0 I_P}{2\pi r_P}$
Again using the simple right hand rule, point your thumb into the page in the direction of the net current and your fingers wrap around the loop clockwise indicating the magnetic field direction. The field is downward (negative z direction) at the location of point P. We can insert numbers to calculate the magnitude.	$B_P = \dfrac{\left(4\pi \times 10^{-7} \text{ T} \cdot \text{m/A}\right)\left(12.0 \text{ A}\right)}{2\pi\left(2.00 \times 10^{-1} \text{ m}\right)}$ $B_P = 1.20 \times 10^{-5}$ T in the $-z$ direction or $\vec{B}_P = \boxed{-1.20 \times 10^{-5}\, \hat{k} \text{ T}}$

CHECK and THINK
Using Ampère's law, we are able to calculate the field at both locations. Note that the field is related to the total (net) current enclosed by the loop with the magnitude found using Ampère's law and the direction from the simple right hand rule.

66. (N) Review In a laboratory vacuum chamber, a long, straight, horizontal wire with a current of 55.0 μA in the positive x direction is used to "levitate" a doubly ionized helium nucleus moving parallel to the wire with a speed of 5.10×10^5 m/s in the negative x direction. What is the distance above the wire where the magnetic force of the wire balances the gravitational force on the helium nucleus?

INTERPRET and ANTICIPATE
Picture the wire with a current to the right. Using the simple right hand rule, the magnetic field due to the current is coming out of the page above the wire and into the page below the wire. The Helium ion, with a positive charge (as it's lost two electrons), is moving to the left above the wire, so the magnetic force points upwards. This makes sense, since it would balance the downward force of gravity.

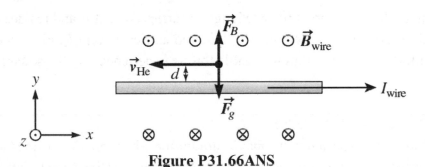

Figure P31.66ANS

SOLVE	
Let d signify the distance above the wire where the helium nucleus levitates. At the location of the helium nucleus, the current creates a field $B = \dfrac{\mu_0 I}{2\pi d}$ in the positive z direction.	$\vec{B}_{wire} = \dfrac{\mu_0 I}{2\pi d}\hat{k}$
The magnetic force is then $\vec{F}_B = q\vec{v} \times \vec{B}$. The resulting force, according to the right hand rule is upwards ($+y$).	$\vec{F}_B = qv\left(-\hat{i}\right) \times \dfrac{\mu_0 I}{2\pi d}\left(\hat{k}\right)$ $\vec{F}_B = \dfrac{qv\mu_0 I}{2\pi d}\hat{j}$

The gravitational force is downward.	$\vec{F}_g = -mg\,\hat{j}$
We now require that the net force is zero $\left(\vec{F}_B + \vec{F}_g = 0\right)$, which means that the magnitudes of the two forces must be equal.	$mg = \dfrac{qv\mu_0 I}{2\pi d}$
Solve for the distance at which the forces are balanced.	$d = \dfrac{qv\mu_0 I}{2\pi mg}$

Finally, insert numerical values:

$$d = \frac{2\left(1.60\times10^{-19}\text{ C}\right)\left(5.10\times10^5\text{ m/s}\right)\left(4\pi\times10^{-7}\text{ T·m/A}\right)\left(0.550\times10^{-6}\text{ A}\right)}{2\pi\left(6.64\times10^{-27}\text{ kg}\right)\left(9.80\text{ m/s}^2\right)} = \boxed{0.276\text{ m}}$$

CHECK and THINK

In this geometry, we find the magnetic force is upwards and can therefore balance the gravitational force at this instant in time.

---

71. (A) A square loop of wire with side length s carries current I_1 and resides inside a solenoid with n turns per unit length. The solenoid carries current I_2. Find an expression for the magnitude of the maximum possible torque experienced by the square loop of wire.

INTERPRET and ANTICIPATE

The square loop will experience the maximum torque when the loop resides in a plane perpendicular the cross-sectional area of the solenoid. Under this orientation, two sides of the loop will have currents running in opposite directions and perpendicular to the field. So, the magnetic force on each of those two sides is a maximum and in opposite directions, causing the loop to rotate in the field. Under any other orientation, the amount of magnetic force will not change, but the torque each force causes will be less. The maximum value of the torque can be expressed using Eq. 30.45, as $\tau = IAB$. The maximum torque should depend on the amount of both the current in the loop and the current in the solenoid.

SOLVE

In Eq. 30.45, the current I refers to the current in the loop and the area A is the area of the loop. Using the information in the problem statement, we express The maximum torque.	$\tau = IAB = I_1 s^2 B$

The magnitude of the magnetic field is that caused by the solenoid, Eq. 31.6, where the current in the solenoid is I_2.	$B = \mu_0 n I_2$
We can then write the final expression for the maximum torque based on the magnetic field due to the solenoid.	$\tau = I_1 s^2 \mu_0 n I_2 = \boxed{\mu_0 n I_1 I_2 s^2}$

CHECK and THINK

The strength of the magnetic field due the solenoid affects the magnetic force that could be exerted on a current-carrying wire within the solenoid. That force is also affected by the amount of current running through it, and the torque is directly related to the amount of the force exerted on each side of the loop.

74. (A) Two long coaxial solenoids each carry current I but in opposite directions as shown in Figure P31.74. The inner solenoid of radius R_1 has n_1 turns per unit length, and the outer solenoid with radius R_2 has n_2 turns per unit length. Find expressions for the magnetic fields inside the inner solenoid, between the two solenoids, and outside both solenoids.

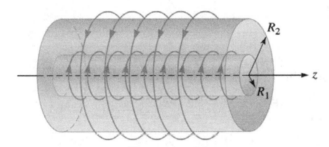

Figure P31.74

INTERPRET and ANTICIPATE
The solenoid is a standard geometry that produces a uniform magnetic field in its interior and we can calculate the field due to each given the current and number of turns per length. The total magnetic field in each region is the sum of the fields due to each coil.

SOLVE Each ideal solenoid creates a uniform field inside the coils according to Eq. 31.6 and no field outside of the coil. The direction of the field can be found using the simple right hand rule. For instance,	$B = \mu_0 n I$

for the inner loop, the current goes up on the front side of the loop— point your thumb up along the current and your fingers wrap in the direction of the field, to the left ($-z$) inside the solenoid and to the right ($+z$) outside the solenoid. For the larger coil, the current is in the other direction, so we find that the field is to the right ($+z$) inside the solenoid and to the left ($-z$) outside the solenoid.	
Now consider each region separately: $r < R_1$ (inside the inner coil): The field due to the smaller and larger coils point in the $-z$ and $+z$ directions respectively.	$\vec{B}_{r<R_1} = B_2\hat{k} - B_1\hat{k}$ $\vec{B}_{r<R_1} = \boxed{\mu_0 I (n_2 - n_1)\hat{k}}$
$R_1 < r < R_2$ (between inner and outer): The field outside the smaller coil is zero but inside the larger coil points in the $+z$ direction.	$\vec{B}_{R_1<r<R_2} = B_2\,\hat{k}$ $\vec{B}_{R_1<r<R_2} = \boxed{\mu_0 n_2 I\,\hat{k}}$
$r > R_2$ (outside both coils): The field is zero outside the coils.	$\vec{B}_{r>R_2} = \boxed{0}$

CHECK and THINK

The net field in each case is the sum of the fields due to each solenoid.

32

Faraday's Law of Induction

1. A constant magnetic field of 0.275 T points through a circular loop of wire with radius 3.50 cm as shown in Figure P32.1.
a. (N) What is the magnetic flux through the loop?
b. (C) Is a current induced in the loop? Explain.

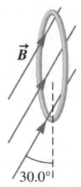

30.0°

Figure P32.1

INTERPRET and ANTICIPATE	
The magnetic flux through the loop depends on the magnetic field, the area of the loop, and their relative orientation. An emf and current are induced if there is a *change* in flux.	

SOLVE	
a. The flux is given by Equation 32.3. The magnetic field is shown in the figure and the area vector points perpendicular to the surface of the loop. In this case, the angle between these vectors is 60 degrees. The area of the loop is the area of a circle, πr^2.	$\Phi_B = \int \vec{B} \cdot d\vec{A} = BA\cos\theta$ $\Phi_B = B(\pi r^2)\cos\theta$
We can now evaluate this expression.	$\Phi_B = (0.275 \text{ T})\pi(0.0350 \text{ m})^2 \cos 60°$ $\Phi_B = \boxed{5.29 \times 10^{-4} \text{ Wb}}$

233

b. The magnetic field, the loop, and their relative orientation are constant in time, so the flux remains unchanged. While the flux is not zero, a current is induced when there is a *change* in flux. Therefore, there is no current induced.

CHECK and THINK

The magnetic flux was calculated using the magnetic field, the area of the loop, and the angle between the field and the area vector. Since the orientation of the loop and field are not changing, there is no induced emf or induced current.

5. (N) The intensity of the Earth's magnetic field near the equator is 35.0 μT. A circular coil with 40 turns and a radius of 75.0 cm is placed so its axis points along the direction of the Earth's magnetic field. The coil is then rotated through an angle of 225° in 50.0 ms. What is the magnitude of the average emf generated in the circular coil?

INTERPRET and ANTICIPATE

We can determine the magnetic flux through the loop before and after and then determine the induced emf based on the average rate of change of the flux.

SOLVE	
An induced emf is caused by a *change* in flux. The magnetic flux through one loop is given by Eq. 32.3.	$\Phi_B = BA\cos\varphi$
Initially, the area vector (along the axis of the loop) is in the same direction as the magnetic field, so $\varphi_i = 0°$.	$\Phi_{B,i} = BA\cos(0°) = BA$
After the rotation, the angle is 225 degrees.	$\Phi_{B,f} = BA\cos(225°)$
For N loops, the induced emf is multiplied by this factor, as seen in Eq. 32.4. We take the absolute value since we are only asked for the magnitude.	$\|\mathcal{E}\| = N\left\|\dfrac{d\Phi_B}{dt}\right\| = N\left\|\dfrac{\Phi_{B,f}-\Phi_{B,i}}{\Delta t}\right\| = NBA\left\|\dfrac{\cos(225°)-1}{\Delta t}\right\|$

Finally, we can insert numerical values.	$$\|\mathcal{E}\| = (40)(35.0 \times 10^{-6}\ \text{T})\pi(0.750\ \text{m})^2 \left\| \frac{\cos(225°) - 1}{50.0 \times 10^{-3}\ \text{s}} \right\|$$ $$\|\mathcal{E}\| = \boxed{8.45 \times 10^{-2}\ \text{V}}$$

CHECK and THINK

As the coil is rotated, the magnetic flux through it changes, producing an induced emf. Given the field, geometry of the coil, and rotation angle over time, we are able to determine this emf.

10. (N) A wire is formed into a square loop with sides $d = 18.0$ cm and positioned in a spatially uniform magnetic field with its plane perpendicular to the direction of the field. What is the magnitude of the average emf induced in the loop if the magnitude of the magnetic field is increased by 30.0 mT per second?

INTERPRET and ANTICIPATE

Since the magnitude of the magnetic field is increasing in time, so is the magnetic flux. The induced emf depends on the rate of change of the flux.

SOLVE The induced emf is given by Eq. 32.2.	$$\|\mathcal{E}\| = \left\| \frac{\Delta \Phi_B}{\Delta t} \right\|$$
The magnetic flux can be found using Eq. 32.3.	$$\Phi_B = BA\cos\varphi$$
In this case, the magnetic field is perpendicular to the plane of the loop. In other words, the area vector (which is perpendicular to the plane) is parallel to the field, so $\varphi = 0°$. Since only the field strength is changing in time, the rate of change of the flux depends only on the rate of change of the field, $\Delta B / \Delta t$, which is given. The area is the area of the square with side d.	$$\|\mathcal{E}\| = \left\| \frac{\Delta \Phi_B}{\Delta t} \right\| = \frac{\Delta B}{\Delta t} A = \frac{\Delta B}{\Delta t} d^2$$
We can now calculate the value.	$$\mathcal{E} = (30.0 \times 10^{-3}\ \text{T/s})(0.180\ \text{m})^2$$ $$\mathcal{E} = \boxed{9.72 \times 10^{-4}\ \text{V}}$$

Chapter 32 – Faraday's Law of Induction

12. Suppose a uniform magnetic field is perpendicular to the 8-1/2 × 11-in. page of your homework and a rectangular metal loop is perpendicular to the page such that one of its sides bisects the page into two long strips. The loop has the same dimensions as the page. The magnetic field is changing with time as described by $B = 3.75 \times 10^{-3} \, t^{-2}$, where B is in teslas and t is in seconds.

a. (C) Is the magnetic field increasing or decreasing?

We sketch the situation described and plot the function given. The magnetic field is falling off as $\frac{1}{t^2}$. That is, it is positive and decreasing towards zero.

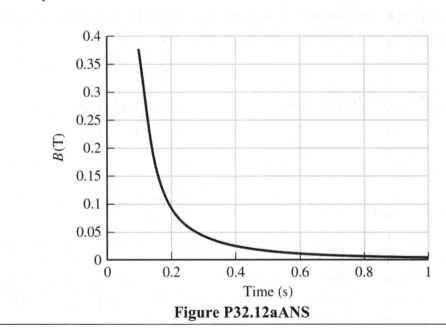

Figure P32.12aANS

b. (N) Find the emf induced in the loop.

The field is parallel to the plane of the loop. (Or, in other words, the area vector of the loop is perpendicular to the magnetic field.) Therefore, while the magnetic field is changing, there is no flux through the loop. Since there is no flux (and, more importantly, no change in flux) there is no induced emf according to Faraday's law (Equation 32.6):

$$\mathcal{E} = -N\frac{d\Phi_B}{dt} \text{ where } \frac{d\Phi_B}{dt} = 0.$$

Figure P32.12bANS

15. (A) The magnetic field in a region of space is given by $\vec{B} = B_x\hat{i} + B_y\hat{j}$. A coil of N turns is oriented so that its cross-sectional area is in the x direction: $\vec{A} = A_x\hat{i}$. A small bulb is connected across the ends of the coil. The total resistance of the coil and the bulb is R. Find an expression for the current through the bulb if $B_x(t) = B_0$ and $B_y = B_0\left(t/t_0\right)^2$, where B_0 and t_0 are constants.

INTERPRET and ANTICIPATE
Given the magnetic field and area, we can determine the magnetic flux. Faraday's law allows us to determine the induced emf from the rate of change of the flux. We then find the current using Ohm's law.

SOLVE	
First, we determine the magnetic flux as the dot product of the magnetic field and area vector, keeping in mind that $A_y = 0$. Since the area vector (which is perpendicular to the plane of the are) is in the x direction, the x component of the magnetic field points through the coil.	$\Phi_B = \vec{B} \cdot \vec{A} = B_x A_x + B_y A_y$ $\Phi_B = B_x A_x$
Now apply Faraday's law (Eq. 32.6). Note thought that while B_y varies in tume, B_x and A_x are constant. Therefore, the derivative with respect to time is zero.	$\mathcal{E} = -N\frac{d\Phi_B}{dt}$ $\mathcal{E} = 0$

Since the induced emf is zero, the current is zero.	$I = \mathcal{E}/R = \boxed{0}$

CHECK and THINK

While there is a changing magnetic field, the flux is not changing in time. Therefore, the induced emf and current are zero. Another way to say this is that the area vector has only an x component. Only the x component of the magnetic field points through the area, and this component is constant. The y component, which varies, points along the surface and does not contribute to the flux.

21. (N) Figure P32.21 shows a circular conducting loop with a 5.00-cm radius and a total resistance of 1.30 Ω placed within a uniform magnetic field pointing into the page.
a. What is the rate at which the magnetic field is changing if a counterclockwise current $I = 4.60 \times 10^{-2}$ A is induced in the loop?
b. Is the induced current caused by an increase or a decrease in the magnetic field with time?

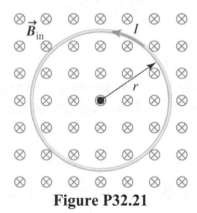

Figure P32.21

INTERPRET and ANTICIPATE

Faraday's law relates the rate of change of the magnetic flux to the induced emf, which then determines the current. The direction of the current is determined by whether the field is increasing or decreasing and Lenz's law, with the induced field opposing the change in flux.

SOLVE **a.** The normal to the enclosed area can be taken to be parallel to the magnetic field, so using Eq. 32.3, the flux through the loop is: $\Phi_B = BA \cos(0°) = BA$.	$\Phi_B = BA$

Since the position and size of the loop is constant, the rate of change of the flux depends on the rate of change of the magnetic field.	$$\frac{d\Phi_B}{dt} = \frac{d}{dt}(BA) = A\frac{dB}{dt}$$
The induced emf can now be found using Faraday's law (Eq. 32.6) and related to the current using Ohm's law.	$$\left|\mathcal{E}\right| = \left|-\frac{d\Phi_B}{dt}\right| \quad \text{and} \quad \mathcal{E} = IR$$
Combine these, including the area of the circular loop, πr^2.	$$IR = A\frac{dB}{dt} = \pi r^2\frac{dB}{dt}$$
Solve for the rate of change of the magnetic field.	$$\frac{dB}{dt} = \frac{IR}{\pi r^2}$$
Insert numerical values.	$$\frac{dB}{dt} = \frac{\left(4.60\times10^{-2}\ \text{A}\right)\left(1.30\ \Omega\right)}{\pi\left(5.00\times10^{-2}\ \text{m}\right)^2} = \boxed{7.61\ \text{T/s}}$$

b. If the magnetic field was increasing, the flux would be increasing into the page, so the induced current would tend to oppose the increase by generating a field out of the page. Using the simple right hand rule, the direction of a current needed to produce a field out of the pate would be counter-clockwise. This is the case here, so the field is $\boxed{\text{increasing}}$.

CHECK and THINK
We are able to use Faraday's law and Lenz's law to determine the change in field and whether it's increasing or decreasing.

25. (N) A coil with cross-sectional area $1.50 \times 10^{-2}\ \text{m}^2$ and 225 turns is positioned in a uniform magnetic field so that the plane of the coil is perpendicular to the direction of the field. The time-varying magnitude of the magnetic field is given by $B = 4.00 - 0.0200\ t - 0.00500\ t^2$, with B in teslas and t in seconds. What is the emf induced in the coil when $t = 2.00\ \text{s}$?

INTERPRET and ANTICIPATE
As the field changes, the magnetic flux through the coil changes, producing an induced emf. We can express the magnetic flux through the coil and take a derivative to determine its rate of change at the given time.

SOLVE The induced emf depends on the change in magnetic flux according to Faraday's law, Eq. 32.6.	$$\mathcal{E} = -N\frac{d\Phi_B}{dt}$$
The magnetic flux depends on the field, cross-sectional area of the coil, and their relative orientation. Since the area vector is perpendicular to the plane of the coil, the magnetic field and area vector are parallel and $\varphi = 0°$.	$$\Phi_B = BA\cos 0° = BA$$
Now combine these, using the fact that only the magnetic field is changing in time.	$$\mathcal{E} = -N\left(\frac{dB}{dt}\right)A$$ $$\mathcal{E} = -N\left[\frac{d}{dt}\left(4.00 - 0.0200t - 0.00500t^2\right)\right]A$$ $$\mathcal{E} = -N\left(-0.02 - 0.01t\right)A$$ $$\mathcal{E} = N\left(0.02 + 0.01t\right)A$$
Evaluate the expression at 2.00 seconds.	$$\mathcal{E} = 225\left(0.02 + 0.01(2.00)\right)\left(1.50\times 10^{-2}\ \text{m}^2\right)$$ $$\mathcal{E} = \boxed{0.135\ \text{V}}$$

CHECK and THINK
The changing magnetic field produces a changing magnetic flux and therefore an induced emf. The positive sign indicates that the induced current is in such a way that it produces an induced field in the positive direction based on the original field. Since the magnetic field is decreasing (becoming more negative), the induced field points in the positive direction to oppose this change.

28. (A) A solenoid of area A_{sol} produces a uniform magnetic field (Fig. P32.28; shown in cross section). The solenoid's magnetic field points out of the page and is decreasing according to $B = B_0\left(t_0/t\right)^2$. A single conducting loop of area A_{loop} and resistance R is coaxial with the solenoid, with $A_{\text{loop}} < A_{\text{sol}}$. Find an expression for the current in the loop. What does the sign of your answer mean?

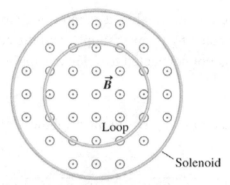

Figure P32.28

INTERPRET and ANTICIPATE The solenoid creates a magnetic field that goes through the loop. The induced emf in the loop depends on the rate of change of the magnetic flux (Faraday's law).	
SOLVE Apply Faraday's law (Eq. 32.6) to determine the induced emf.	$\mathcal{E} = -N\dfrac{d\Phi_B}{dt}$
Since the field is perpendicular to the plane of the loop, $\varphi = 0$, and according to Eq. 32.3, $\Phi_B = BA$, where the area is that of the loop. The area and orientation of the loop are constant, so the change in flux depends on the change in the magnetic field.	$\Phi_B = BA_{\text{loop}}$ $\dfrac{d\Phi_B}{dt} = A_{\text{loop}}\dfrac{dB}{dt}$
Take the derivative of the magnetic field that is given and insert this into Faraday's law.	$\mathcal{E} = -NA_{\text{loop}}\dfrac{dB}{dt} = -NA_{\text{loop}}\left[-2B_0\dfrac{t_0^2}{t^3}\right]$ $\mathcal{E} = \dfrac{2NA_{\text{loop}}B_0 t_0^2}{t^3}$
Use Ohm's law to find the current. The sign indicates the direction. The positive sign means that the current creates a magnetic field in the same (positive) direction as the original field. This makes sense because the induced field opposes the *change* in the original field, which is decreasing field strength, and therefore points in the same direction to reinforce it.	$I = \dfrac{\mathcal{E}}{R} = \boxed{\dfrac{2NA_{\text{loop}}B_0 t_0^2}{Rt^3}}$
CHECK and THINK This is an application of Faraday's law. Since the field strength is decreasing, an induced emf is created, leading to a current that tends to reinforce the decreasing field.	

Chapter 32 – Faraday's Law of Induction

30. (C) Two circular conducting loops labeled A and B are close together and their axes are parallel. Loop A is connected to a power supply, and the current in A is increasing linearly with time. Loop B is not connected to a power supply, but the current in B runs through a bulb.

a. Describe the current induced in loop B and the brightness of the bulb as a function of time.

> The magnetic field created by loop A depends on the current through the loop. Since the current through A is increasing linearly, the magnetic field it produces increases linearly. The magnetic flux through loop B, which depends on the field according to Eq. 32.3, also increase linearly. The induced emf depends on the *derivative* of the flux (Eq. 32.6, Faraday's Law), so the induced emf is *constant*. From Ohm's law, the current induced in B is also constant and therefore so is the brightness of the bulb.

b. Explain how your answer is consistent with the principle of conservation of energy.

> The energy supplied by the power supply is the source of the energy that ultimately drives the current in loop B and lights the bulb. So, energy is put in by the power supply, transmitted by the changing electromagnetic field, and dissipated in loop B. This energy is converted from one form to another in this process.

c. If the current in loop A is decreasing instead of increasing, how do your answers change?

> The answers remain unchanged. A decreasing current in loop A still generates a changing flux, which induces a constant current in B and lights the bulb. In this case, the current induced in loop B will be in the opposite direction, but the bulb still lights and the energy source is still the power supply.

37. (N) The slide generator in Figure 32.14 (page 1020) is in a uniform magnetic field of magnitude 0.0500 T. The bar of length 0.365 m is pulled at a constant speed of 0.500 m/s. The U-shaped conductor and the bar have a resistivity of 2.75×10^{-8} $\Omega \cdot m$ and a cross-sectional area of 8.75×10^{-4} m^2. Find the current in the generator when $x = 0.650$ m.

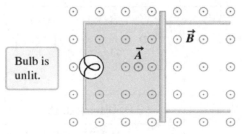

Bulb is unlit.

A.

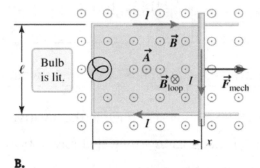

Bulb is lit.

B.

Figure P32.14

INTERPRET and ANTICIPATE
The slide generator generates a motional emf. We can determine the current by determining the emf and resistance of the loop. We'll first derive an expression, which is actually the goal of Problem 36, and then insert values to determine the answer for this problem.

SOLVE	
The voltage driving the current is the motional emf (Eq. 32.7).	$\mathcal{E} = B\ell v$
To find the current, use Ohm's law.	$I = \dfrac{\mathcal{E}}{R}$
We don't know the resistance directly, but we can express it in terms of the resistivity, length, and cross-sectional area of the conductor. The total length is the perimeter of the rectangular loop, $L = 2\ell + 2x$.	$R = \rho\dfrac{L}{A} = \rho\dfrac{2\ell + 2x}{A}$

Combine these formulas to write an expression for current.	$I = \dfrac{\mathcal{E}}{R} = \dfrac{B\ell v}{\rho\dfrac{L}{A}} = \dfrac{B\ell vA}{2\rho(\ell+x)}$
Finally, insert numerical values.	$I = \dfrac{(0.0500\ \text{T})(0.365\ \text{m})(0.500\ \text{m/s})(8.75\times10^{-4}\ \text{m}^2)}{2(2.75\times10^{-8}\ \Omega\cdot\text{m})(0.365\ \text{m}+0.650\ \text{m})}$ $I = \boxed{143\ \text{A}}$

CHECK and THINK
The slide generator generates 143 A, which is a very significant current!

---

39. A thin conducting bar (60.0 cm long) aligned in the positive y direction is moving with velocity $\vec{v} = (1.25\ \text{m/s})\hat{i}$ in a region with a spatially uniform 0.400-T magnetic field directed at an angle of 36.0° above the xy plane.
a. (N) What is the magnitude of the emf induced along the length of the moving bar?
b. (C) Which end of the bar is positively charged?

INTERPRET and ANTICIPATE
The induced emf in this case is a motional emf. We draw a sketch first to be clear about the geometry.

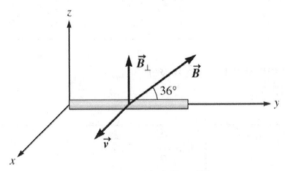

Figure P32.39ANS

SOLVE
a. The motional emf depends on the velocity, length of the bar, and the magnetic field according to Eq. 32.7. Note that the magnetic field here is the component that is *perpendicular* to the bar, as in the slide generator in Section 30.5. To understand why, note that the force due to the perpendicular component of the field,

$\mathcal{E} = B_\perp \ell v$

$B_\perp = B\sin(36.0°)$

$q\vec{v} \times \vec{B}_{\perp}$, points along the bar (in the negative y direction for positive charges and positive y for negative charges), leading to charge separation and an induced voltage. For the component of the field along the bar, the force $q\vec{v} \times \vec{B}_{\parallel}$ points either up or down, which does not lead to charges separating across the length of the bar.	
Insert numerical values.	$\mathcal{E} = \left[(0.400 \text{ T})\sin 36.0°\right](0.600)(1.25 \text{ m/s})$ $\mathcal{E} = \boxed{0.176 \text{ V}}$

b. Apply the right hand rule to determine the force on charges in the bar $\left(q\vec{v} \times \vec{B}_{\perp}\right)$ as described in part (a). Imagine holding your right hand in the x direction, parallel to the bar's velocity, and curling your fingers upward, in the direction of $B_{\perp}$. Your thumb will then be pointing in the negative y direction. By the right-hand rule, the magnetic force on charges in the wire would tend to move positive charges in the negative y direction, such that the end of the bar with the lowest y value is positively charged.

CHECK and THINK
We can use the formula for motional emf along with the right hand rule to determine the polarity of the bar after the charges separate.

45. (N) The magnetic flux through each turn of a 140-turn coil is given by $\Phi_B = 8.75 \times 10^{-3} \sin \omega t$, where ω is the angular speed of the coil and Φ_B is in webers. At one instant, the coil is observed to be rotating at a rate of 8.90×10^2 rev/min.
a. What is the induced emf in the coil as a function of time for this angular speed?
b. What is the maximum induced emf in the coil for this angular speed?

INTERPRET and ANTICIPATE
The induced emf depends on the rate of the change of the magnetic flux, which is given. We can determine the function and the maximum of the function given an expression for the flux.

SOLVE **a.** Faraday's law (Eq. 32.6) allows us to determine the induced emf as	

the rate of change of the flux.	$\mathcal{E} = -N\dfrac{d\Phi_B}{dt}$ $\mathcal{E} = -N\left(8.75\times10^{-3}\right)\omega\cos\omega t$
Convert the angular frequency from revolutions per minute to radians per second.	$\omega = 890\dfrac{\text{rev}}{\text{min}}\left(\dfrac{2\pi\text{ rad}}{\text{rev}}\right)\left(\dfrac{1\text{ min}}{60\text{ s}}\right) = 93.2\dfrac{\text{rad}}{\text{s}}$
Insert numbers into the expression above. The resulting expression gives an answer in volts with the time is in seconds.	$\mathcal{E} = -140\left(8.75\times10^{-3}\right)(93.2)\cos(93.2t)$ $\mathcal{E} = \boxed{-114\cos(93.2t)}$
b. Since cosine can have values between −1 and +1, the peak emf is simply the pre-factor in the expression found in part (a).	$\mathcal{E}_{max} = \boxed{114\text{ V}}$

CHECK and THINK

In this case, since the magnetic flux is given, we simply need to take the derivative to evaluate Faraday's law.

48. (N) A 44-turn rectangular coil with length ℓ = 15.0 cm and width w = 8.50 cm is in a region with its axis initially aligned to a horizontally directed uniform magnetic field of 745 mT and set to rotate about a vertical axis with an angular speed of 64.0 rad/s.

a. What is the maximum induced emf in the rotating coil?

b. What is the induced emf in the rotating coil at t = 1.00 s?

c. What is the maximum rate of change of the magnetic flux through the rotating coil?

INTERPRET and ANTICIPATE

This is an ac generator, which creates an ac voltage when a coil is rotated in a magnetic field, so we can apply the equation for the voltage created by a generator. The voltage oscillates in time and depends on the magnetic field, area, number of loops, and the rotation rate.

SOLVE

a. The voltage produced by an ac generator is given by Eq. 32.13.	$\mathcal{E} = NBA\omega\sin\omega t = \mathcal{E}_{max}\sin\omega t$

The maximum voltage is the prefactor.	$\mathcal{E}_{max} = NBA\omega$ $\mathcal{E}_{max} = 44(0.745 \text{ T})(0.150 \text{ m})(0.0850 \text{ m})\left(64.0 \ \frac{rad}{s}\right)$ $\mathcal{E}_{max} = \boxed{26.7 \text{ V}}$
b. Inserting this value back into Eq. 32.13, we can evaluate the expression at 1.00 s. At this time, $\omega t = 64.0$ rad. Note that it is important when calculating the sine function to make sure your calculator is set to radians instead of degrees. The resulting voltage is less than the maximum value as we would expect.	$\mathcal{E}_{t=1.00 \text{ s}} = (26.7 \text{ V})\sin(64.0) = \boxed{24.6 \text{ V}}$
c. Rearrange Faraday's law (Eq. 32.6) to determine the maximum rate of change of the flux.	$\left\vert\mathcal{E}\right\vert = \left\vert N\dfrac{d\Phi_B}{dt}\right\vert \quad \rightarrow \quad \left\vert\dfrac{d\Phi_B}{dt}\right\vert_{max} = \dfrac{\mathcal{E}_{max}}{N}$
Insert values.	$\left\vert\dfrac{d\Phi_B}{dt}\right\vert_{max} = \dfrac{26.7 \text{ V}}{44} = \boxed{0.607 \text{ Wb/s}}$

CHECK and THINK

The problem concerns an ac generator. With the equation for the generator, we are able to calculate the desired quantities. This particular generator creates a voltage with a peak value of 26.7 V.

56. (N) You have a 750-W hair dryer that is designed to work on the American power grid ($\mathcal{E}_{rms}$ = 120.0 V, 60.0 Hz).

a. What is the rms current in the dryer?

b. If you connected the dryer to the European power grid ($\mathcal{E}_{rms}$ = 240.0 V, 50.0 Hz), what would be the current in the dryer? What would happen to the dryer and why?

c. If you used a transformer to connect the dryer to the European grid, what ratio of N_S/N_P would be required?

INTERPRET and ANTICIPATE The power used by an electric device depends on the rms current and voltage. Since the European power grid is at higher voltage, we might guess that the hair dryer will draw more current and that we need a step-down transformer to use the hair dryer.	

SOLVE **a.** The rms current can be found from the power and rms voltage using Eq. 32.24.	$P_{avg} = I_{rms}\mathcal{E}_{rms} \quad \rightarrow \quad I_{rms} = \dfrac{P_{avg}}{\mathcal{E}_{rms}}$
Insert values.	$I_{rms} = \dfrac{750\text{ W}}{120\text{ V}} = \boxed{6.25\text{ A}}$
b. The key is to note that it is actually the resistance of the hair dryer that stays the same as we bring it to Europe. The resistance is a physical property of the hair dryer which is the same regardless of the voltage source, as opposed to the current or power drawn, which depend on the voltage. So, first we determine the resistance using Ohm's law and the result of part (a).	$R = \dfrac{\mathcal{E}_{rms}}{I_{rms}} = \dfrac{120\text{ V}}{6.25\text{ A}} = 19.2\ \Omega$
Now, calculate the rms current drawn using the rms European voltage and the resistance. The current is much higher, so there is a danger that it would burn out. In fact, since the voltage is double, we see that the current is doubled.	$I_{rms} = \dfrac{\mathcal{E}_{rms}}{R} = \dfrac{240\text{ V}}{19.2\ \Omega} = \boxed{12.5\text{ A}}$
c. Use Eq. 32.27. The primary voltage is the input (European) and the secondary voltage is the desired output (120 V since our hair dryer is designed to work with American voltages). There are fewer turns in the secondary coil, so this is a step-down transformer that reduces the input voltage.	$\dfrac{N_P}{N_S} = \dfrac{V_P}{V_S} = \dfrac{240\text{ V}}{120\text{ V}} = \boxed{2}$

CHECK and THINK

We see from this example that when a device designed to work with lower voltage American outlets is connected to higher voltage European outlets it might burn out. This is why electrical adapters exist for travelers—these are just transformers, but necessary for appliances to operate as designed!

57. (N) An electric toothbrush charger is one example of a small transformer used in your home. The secondary coil in the toothbrush gets an induced emf from the primary coil in the charger when they are aligned. (The toothbrush coil is placed within the primary coil when it is plugged in.)
a. If the maximum value of the emf in the primary coil ($N_P = 1000$) is 112 V, what is the maximum induced emf in the secondary coil ($N_S = 150$)?
b. A maximum current of 0.050 A flows through the primary coil. What is the average power delivered to the toothbrush?

INTERPRET and ANTICIPATE

We can use Eq. 32.27 to find the maximum induced emf in the secondary coil, as well as the maximum current in the secondary coil. This will allow us to use Eq. 32.23 to find the average power that is delivered to the secondary coil. Since the ratio of the emfs is the same as the ratio of the turns in each coil, we expect the secondary emf to be less than that of the primary because the secondary coil has fewer turns. Likewise, the current in the secondary should be higher than that in the primary because the ratio of the currents is related to the inverse ratio of the turns in each coil.

SOLVE	
a. Use Eq. 32.27 to find the maximum induced emf in the secondary coil. Substitute the known quantities and solve.	$\dfrac{N_P}{N_S} = \dfrac{V_P}{V_S}$ $V_S = V_P \dfrac{N_S}{N_P} = (112 \text{ V})\dfrac{150}{1000} = \boxed{16.8 \text{ V}}$
b. We must first find the maximum current that manifests in the secondary coil by using Eq. 32.27	$\dfrac{I_S}{I_P} = \dfrac{N_P}{N_S}$ $I_S = I_P \dfrac{N_P}{N_S} = (0.050 \text{ A})\dfrac{1000}{150}$
Knowing the maximum current in the secondary coil, we can find the power delivered to the toothbrush by using Eq. 32.23.	$P_{avg} = \dfrac{1}{2}I_{max}\mathcal{E}_{max} = \dfrac{1}{2}\left((0.050 \text{ A})\dfrac{1000}{150}\right)(16.8 \text{ V}) = \boxed{2.8 \text{ W}}$

61. (N) A circular loop with a radius of 0.25 m is rotated by 90.0° over 0.200 s in a uniform magnetic field with $B = 1.50$ T. The plane of the loop is initially perpendicular to the field and is parallel to the field after the rotation.
a. What is the average induced emf in the loop?
b. If the rotation is then reversed, what is the average induced emf in the loop?

INTERPRET and ANTICIPATE
The average emf induced in the loop can be found using Eq. 32.6. The magnetic flux through the loop is changing as the loop rotates in the field. In the initial state, the angle between the area vector of the loop and the field is 0° since the plane of the loop is initially perpendicular to the field. This means the flux is initially at a maximum. The magnetic flux is 0 after the first rotation because the final angle between the area vector of the loop and the field will be 90°. We expect then, in the first rotation, that the emf will be positive because the change in magnetic flux will be negative. In the reverse rotation, we expect to find the same magnitude of emf, but negative because the magnetic flux will be increasing.

SOLVE
a. We use Eq. 32.6 to find the average emf induced in the loop. The field strength is not changing, nor is the area of the loop. These quantities can be expressed using the given info, and the angles can be expressed following the discussion in the Interpret and Anticipate section.

$$\mathcal{E} = -N\frac{d\Phi_B}{dt} = -N\frac{\Delta\Phi_B}{\Delta t}$$

$$\mathcal{E} = -N\frac{BA\cos(\theta_f) - BA\cos(\theta_i)}{\Delta t}$$

$$\mathcal{E} = -N\frac{BA(\cos(\theta_f) - \cos(\theta_i))}{\Delta t}$$

$$\mathcal{E} = -1\frac{(1.50 \text{ T})\pi(0.25 \text{ m})^2(\cos(90°) - \cos(0°))}{0.200 \text{ s}}$$

$$\mathcal{E} = \boxed{1.5 \text{ V}}$$

b. If the rotation is reversed, we can again use Eq. 32.6. The values are the same except the initial and final angles are flipped.	$\mathcal{E} = -N\dfrac{d\Phi_B}{dt} = -N\dfrac{\Delta\Phi_B}{\Delta t}$ $\mathcal{E} = -N\dfrac{BA\cos(\theta_f) - BA\cos(\theta_i)}{\Delta t}$ $\mathcal{E} = -N\dfrac{BA\big(\cos(\theta_f) - \cos(\theta_i)\big)}{\Delta t}$ $\mathcal{E} = -1\dfrac{(1.50\text{ T})\pi(0.25\text{ m})^2\big(\cos(0°) - \cos(90°)\big)}{0.200\text{ s}}$ $\mathcal{E} = \boxed{-1.5\text{ V}}$

CHECK and THINK

Note that the sign of the emf from Eq. 32.6 (Faraday-Lenz's Law) tells us about the direction of the current (clockwise or counterclockwise) as viewed by an observer looking in the direction of the field, and rotating with the loop. The current is clockwise when the emf is positive and counterclockwise when it is negative.

66. (N) A helicopter has blades 4.0 m long (beginning at the rotor and extending outward) rotating at 300.0 rpm (revolutions per minute). Determine the maximum voltage that might develop between the two ends of the blade while they are rotating in the Earth's magnetic field (approximately 10^{-4} T). Assume the speed of the blade is determined by the linear speed of its center of mass and its mass is uniformly distributed.

INTERPRET and ANTICIPATE

The voltage induced between the two ends is a motional emf. Given the magnetic field and properties of the moving rotor, we can determine the induced emf.

SOLVE Motional emf is determined using Equation 32.7. We'll use $B = 10^{-4}$ T and $\ell = 4.0$ m, the length of the rotor.	$\mathcal{E} = B\ell v$
The linear speed of the center of the rotor depends on its angular velocity and distance from the rotation axis as $v = \omega r$. The angular velocity of 300 rpm can be converted to radians per second.	$\omega = 300$ rpm $\omega = \dfrac{300\text{ rev}}{\text{min}}\left(\dfrac{1\text{ min}}{60\text{ s}}\right)\left(\dfrac{2\pi\text{ rad}}{1\text{ rev}}\right) = 31.4\dfrac{\text{rad}}{\text{s}}$

Now calculate the linear speed of the center of the rotor, which is 2 meters from the rotation axis.	$v = \omega R = \left(31.4\dfrac{\text{rad}}{\text{s}}\right)(2.0\text{ m})$ $v = 62.8\dfrac{\text{m}}{\text{s}}$
Now, we can calculate the motional emf induced.	$\mathcal{E} = \left(10^{-4}\text{ T}\right)(4.0\text{ m})\left(62.8\ \dfrac{\text{m}}{\text{s}}\right) = \boxed{0.025\text{ V}}$

CHECK and THINK

The induced voltage is around a couple hundredths of a volt. It's not zero, though it is quite small and unlikely to produce any noticeable effect.

67. (N) A circular coil with 75 turns and radius 12.0 cm is placed around an electromagnet that produces a uniform magnetic field through the coil and perpendicular to the plane of the coil. As the electromagnet powers up, the field it produces increases linearly from 0 to a maximum of 3.50 T in 0.110 s. If the total resistance of the coil is 5.00 Ω, what is the magnitude of the average current induced in the circular coil as the electromagnet powers up?

INTERPRET and ANTICIPATE

As the field increases, the magnetic flux through the coil increases, producing an induced emf. This emf can drive a current that we will calculate.

SOLVE The induced emf depends on the rate of change of the magnetic flux according to Faraday's law (Eq. 32.6). We are only asked for the magnitude of the current, so we only concern ourselves with the magnitude of the emf.	$\left\|\mathcal{E}\right\| = \left\|N\dfrac{d\Phi_B}{dt}\right\|$
The magnetic flux depends on the magnetic field and the cross-sectional area of the coil (Eq. 32.3). The field is perpendicular to the plane of the coil (that is, the area vector is parallel to the field), so the angle $\varphi = 0°$.	$\Phi_B = BA\cos 0° = B\pi r^2$

Only the field changes in time.	$$\left|\mathcal{E}\right| = \left| N \frac{\Delta B}{\Delta t} \pi r^2 \right|$$
Use Ohm's law to calculate the current.	$$I = \frac{\mathcal{E}}{R} = \frac{N \dfrac{\Delta B}{\Delta t} \pi r^2}{R}$$
Finally, insert numerical values to determine the current.	$$I = \frac{(75)\left(\dfrac{3.50 \text{ T}}{0.110 \text{ s}}\right)\pi\left(0.120 \text{ m}\right)^2}{5.00 \ \Omega}$$ $$I = \boxed{21.6 \text{ A}}$$

CHECK and THINK

Since the field changes, the flux through the coil changes, inducing an emf of about 100 V. This induced emf drives a pretty substantial current of over 21 amps as the field is increasing.

75. (N) A rectangular conducting loop with dimensions $w = 32.0$ cm and $h = 78.0$ cm is placed a distance $a = 5.00$ cm from a long, straight wire carrying current $I = 7.00$ A in the downward direction (Fig. P32.75).

a. What is the magnitude of the magnetic flux through the loop?

b. If the current in the wire is increased linearly from 7.00 A to 15.0 A in 0.230 s, what is the magnitude of the induced emf in the loop?

c. What is the direction of the current that is induced in the loop during this time interval?

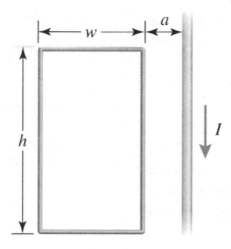

Figure P32.75

INTERPRET and ANTICIPATE

A long straight wire creates a magnetic field that wraps around it according to the simple right hand rule and points into the loop. Since the field decreases with distance from the wire, we will need to integrate to find the total magnetic flux. As the current increases, the field and flux into the loop increase. According to Lenz's law, there is an induced field out of the page to oppose that increase, corresponding to an induced current counter-clockwise in the loop. The rate of the change of the flux allows us to calculate the induced emf using Faraday's law and then the current.

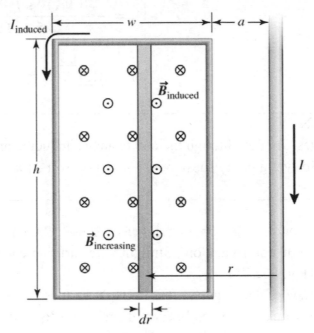

Figure P32.75ANS

SOLVE

a. First, recall the magnetic field created at a distance r from a long straight wire.	$B = \dfrac{\mu_0 I}{2\pi r}$
Now write an expression for the flux through a small rectangular element of height h and width dr within the rectangular loop by applying Eq. 32.3. The field is everywhere perpendicular to the plane of the loop, or in other words, the field is parallel to the area vector and $\varphi = 0°$.	$\Phi_B = BA\cos\varphi = BA$ $d\Phi_B = BdA = \dfrac{\mu_0 I}{2\pi x} h\,dx$

Integrate from $r = a$ to $a + w$.	$$\Phi_B = \int\limits_{a}^{a+w} \frac{\mu_0 I h}{2\pi} \frac{dr}{r} = \frac{\mu_0 I h}{2\pi} \ln\left(\frac{a+w}{a}\right) \quad (1)$$
Insert values to calculate the flux in webers. $$\Phi_B = \frac{\left(4\pi \times 10^{-7} \text{ T·m/A}\right)(7.00 \text{ A})(0.780 \text{ m})}{2\pi} \ln\left(\frac{0.050 + 0.320}{0.050}\right)$$ $$\Phi_B = \boxed{2.19 \times 10^{-6} \text{ Wb}}$$	

b. To determine the induced emf, use Faraday's law (Eq. 32.6). Insert Eq. (1) above, noting that only the current changes in time.	$$\mathcal{E} = -\frac{d\Phi_B}{dt}$$ $$\mathcal{E} = -\frac{\mu_0 h}{2\pi} \ln\left(\frac{a+w}{a}\right) \frac{dI}{dt}$$

Insert numerical values. The negative sign means simply that the induced emf is such that the induced current opposes the change in flux, but we are asked for the magnitude and therefore report the absolute value. The rate of change of the current, $\dfrac{dI}{dt} = \dfrac{\Delta I}{\Delta t}$, is just the change in current over time.

$$\mathcal{E} = -\left[\frac{\left(4\pi \times 10^{-7} \text{ T·m/A}\right)(0.780 \text{ m})}{2\pi} \ln\left(\frac{0.050 + 0.320}{0.050}\right)\right] \frac{(15.0 - 7.00) \text{ A}}{0.230 \text{ s}}$$

$$|\mathcal{E}| = \boxed{1.09 \times 10^{-5} \text{ V}}$$

c. The long, straight wire produces magnetic flux into the page through the rectangle, shown in the figure above. As the magnetic flux increases, the current through the rectangular loop produces its own induced magnetic field out of the page to oppose the increase in flux into the page. The induced current creates this opposing field by traveling $\boxed{\text{counterclockwise}}$ around the loop.

CHECK and THINK

Ultimately, this is an example of Faraday's and Lenz's laws. Since the field is not uniform, we must integrate to find the total flux, but then the rate of change of the flux leads to the induced emf with Lenz's law allowing us to determine the direction of the induced current need to oppose the *change* in flux.

78. (A) A circular loop of radius a and resistance R is placed in a changing magnetic field so that the field is perpendicular to the plane of the loop. The magnetic field varies with time as $B(t) = B_0 e^{-t}$, where B_0 is a constant. Determine the electrical power in the circuit when $t = 0$.

INTERPRET and ANTICIPATE This is an application of Faraday's law. Since the magnetic field is changing, the magnetic flux changes, and therefore there's an induced emf in the loop. The power depends on this induced emf.	

SOLVE Power through a loop with resistance R can be calculated using the voltage across the resistance, which is the induced emf in this case.	$$P = \frac{\Delta V^2}{R} = \frac{\mathcal{E}^2}{R}$$
We can use Faraday's law (Eq. 32.6).	$$\mathcal{E} = -\frac{d\Phi_B}{dt} = -\frac{d(BA)}{dt} = -A\frac{dB}{dt}$$
The magnetic field is given, so we can evaluate the derivative, and the area of the loop is $A = \pi a^2$.	$$\mathcal{E} = -\pi a^2 \frac{d}{dt}\left(B_0 e^{-t}\right)$$ $$\mathcal{E} = \pi a^2 B_0 e^{-t}$$
Determine the emf $\mathcal{E}_0$ at $t = 0$, when the switch is closed using $e^0 = 1$.	$$\mathcal{E}_0 = \mathcal{E}(t = 0) = \pi a^2 B_0$$
Finally we find the electrical power developed in the resistor at the instant the switch is closed.	$$P_0 = \frac{\mathcal{E}_0^2}{R} = \frac{\left(\pi a^2 B_0\right)^2}{R}$$ $$P_0 = \boxed{\frac{\pi^2 a^4 B_0^2}{R}}$$

CHECK and THINK The resulting expression is not obvious, but we see that if the loop has a larger radius or if the magnetic field strength is larger, the power would be higher. This sounds right.	

33

Inductors and AC Circuits

5. A 15.0-mH inductor is connected to a DC power supply. The power supply's emf is fixed at 5.00 V.

a. (N) If you turn down the power supply's current at a constant rate, $dI/dt = -50.0$ A/s, what is the magnitude of the back emf?

b. (N) At what rate would you need to turn down the current so that the back emf is 5.00 V?

c. (C) If you had a lightbulb in this circuit, how would it glow in the two cases? (Ignore the resistance of the bulb.)

INTERPRET and ANTICIPATE	
The back emf is the voltage across the inductor and it depends on the inductance of the circuit and the rate at which the current changes. A larger back emf is associated with a larger rate of change of the current.	

SOLVE **a.** The back emf (the voltage across the inductor) is given by Eq. 33.8.	$$\Delta V_L = -L\frac{dI}{dt}$$
Insert values.	$$\Delta V_L = -\left(15.0 \times 10^{-3}\text{ H}\right)\left(-50.0\text{ A/s}\right)$$ $$\Delta V_L = \boxed{0.750\text{ V}}$$
b. Rearrange Eq. 33.8 to determine the rate the current changes as a function of the inductor voltage and inductance.	$$\frac{dI}{dt} = -\frac{\Delta V_L}{L}$$
Insert values. The rate of 333 A/s is larger than that from part (a), which makes sense since the goal is to create a larger back emf than in part (a).	$$\frac{dI}{dt} = -\frac{5.00\text{ V}}{15.0 \times 10^{-3}\text{ H}}$$ $$\frac{dI}{dt} = \boxed{-3.33 \times 10^2\text{ A/s}}$$

c. The brightness depends roughly on the power dissipated by the bulb, $P = I^2R$, and therefore on the *current* through the bulb. So, in both cases the bulb would become dimmer as the current decreases and eventually stop glowing. In part (a), the current, and therefore the brightness, decreases more slowly compared to part (b).

CHECK and THINK
The back emf depends on the inductance and the rate at which the current changes. A larger back emf is created by a larger change in current.

---

8. (N) A 45.0-cm-long solenoid is 8.00 cm in diameter and has 690 turns.
a. What is the inductance of the solenoid?
b. What is the rate of change of the current dI/dt required to produce an emf of 44.0 μV in the solenoid?

INTERPRET and ANTICIPATE
The inductance of a solenoid depends only on its geometry. We are given the length, diameter, and number of turns, so we have everything we need. The inductance and rate of change of the current then determine the voltage across the inductor. So, with the inductance we calculate in part (a) and the emf produced, we can determine the rate at which current is changing.

SOLVE

a. The inductance depends only on geometry, according to Equation 33.5. This is analogous to capacitance, which depends only on geometrical quantities such as the area and separation of plates.

$$L = \frac{\mu_0 N^2 A}{\ell}$$

We have all the quantities we need to calculate the inductance.

$$L = \frac{\left(4\pi \times 10^{-7}\,\frac{\text{T}\cdot\text{m}}{\text{A}}\right)(690)^2\left[\dfrac{\pi\left(8.00\times 10^{-2}\,\text{m}\right)^2}{4}\right]}{0.450\,\text{m}}$$

$$L = \boxed{6.68 \times 10^{-3}\,\text{H}} = 6.68\,\text{mH}$$

b. We can now apply Eq. 33.8, which says that the voltage across the inductor depends on the inductance and rate of change of the current. (The inductor essentially opposes changes in current.) Since we only care about the magnitude, we can take the absolute value and solve for the rate of change of the current.	$\|\Delta V\| = L\left\|\dfrac{dI}{dt}\right\|$ $\dfrac{dI}{dt} = \dfrac{\|\Delta V\|}{L}$
Insert numerical values and solve.	$\dfrac{dI}{dt} = \dfrac{44.0 \times 10^{-6}\ \text{V}}{6.68 \times 10^{-3}\ \text{H}}$ $\dfrac{dI}{dt} = \boxed{6.58 \times 10^{-3}\ \text{A/s}} = 6.58\ \text{mA/s}$

CHECK and THINK

The inductance depends only on the geometry of the inductor. A value on the order of mH sounds plausible. Given this value, we are able to relate the voltage across the inductor to the rate of change of the current.

---

13. (N) The time constants for a series RC circuit with a capacitance of 5.00 μF and a series RL circuit with an inductance of 2.00 H are identical.
a. What is the resistance R in the two circuits?
b. What is the common time constant for the two circuits?

INTERPRET and ANTICIPATE
The time constant of an RC circuit is $\tau = RC$ (Eq. 29.13) and for an RL circuit is $\tau = \dfrac{L}{R}$ (Eq. 33.13). They indicate the characteristic time for the capacitor to charge or discharge in the first case or for the current to turn on or off in the second case. Given the capacitance and inductance, we can solve for the resistance and the time constant.

SOLVE **a.** Equate the time constants (Eq. 29.13 and 33.13) for the RC and RL circuits.	$\tau = RC = \dfrac{L}{R}$

Solve for the resistance.	$R = \sqrt{\dfrac{L}{C}}$
Insert values. As long as we use metric base units (henries and farads), we will get metric base units in the answer (ohms).	$R = \sqrt{\dfrac{2.00 \text{ H}}{5.00 \times 10^{-6} \text{ F}}} = \boxed{632 \ \Omega}$
b. Now use either Eq. 29.13 or 33.13 to solve for the time constant.	$\tau = RC$ $\tau = \left(632 \ \Omega\right)\left(5.00 \times 10^{-6} \text{ F}\right)$ $\tau = \boxed{3.16 \times 10^{-3} \text{ s}} = 3.16 \text{ ms}$

CHECK and THINK
With the equation for the characteristic time of the *RC* and *RL* circuits, we are able to calculate the resistance and time constant.

14. (N) After being closed for a long time, the switch S in the circuit shown in Figure P33.14 is thrown open at $t = 0$. In the circuit, $\mathcal{E} = 24.0$ V, $R_A = 4.00$ kΩ, $R_B = 7.00$ kΩ, and $L = 589$ mH.
a. What is the emf across the inductor immediately after the switch is opened?
b. When does the current in the resistor R_B have a magnitude of 1.00 mA?

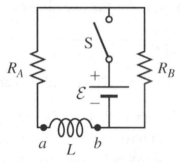

Figure P33.14

INTERPRET and ANTICIPATE
An inductor opposes changes in current. At steady-state, after a DC circuit is connected to an *RL* circuit for a long time, the current reaches a constant value and the inductor has no effect. Once the switch is opened, even with the voltage source disconnected, the inductor prevents the current from turning off immediate. (It decays exponentially instead.)

SOLVE **a.** With the switch closed, the left hand loop is an *RL* circuit with a voltage source connected, so the current increases according to Eq. 33.12. After the switch is closed for "a long time" (i.e. as time goes to infinity), the current reaches a constant value that can be found using Ohm's law. That is, after a DC voltage source is connected for a long time, the current is constant and the inductor has no effect on the circuit.	$I_0 = \dfrac{\mathcal{E}}{R}\left(1 - e^{-\frac{t}{\tau}}\right) = \dfrac{\mathcal{E}}{R}\left(1 - e^{-\infty}\right) = \dfrac{\mathcal{E}}{R}$
In this case, the resistance is that of the loop on the left since the voltage source imposes a constant voltage across the resistor and inductor on the left.	$I_0 = \dfrac{\mathcal{E}}{R} = \dfrac{24.0\text{ V}}{4.00 \times 10^3\ \Omega} = 6.00 \times 10^{-3}\text{ A} = 6.00\text{ mA}$
Immediately after the switch is opened, the 6.00 mA current will still flow around the outer loop of the circuit because the inductor prevents it from changing instantaneously. Apply Kirchhoff's loop rule going counter-clockwise. The voltage across the inductor is ΔV_L and across each resistor is $\Delta V = IR$ from Ohm's law.	$\Delta V_L - I_0 R_A - I_0 R_B = 0$ $\Delta V_L = \left(6.00 \times 10^{-3}\text{ A}\right)\left(4.00 \times 10^3\ \Omega + 7.00 \times 10^3\ \Omega\right)$ $\Delta V_L = \boxed{66.0\text{ V}}$
b. After the switch is opened, the outer loop is an *RL* circuit with initial current I_0 that decays according to Eq. 33.15 with time constant Eq. 33.13.	$I(t) = I_i e^{-t/\tau}$ and $\tau = \dfrac{L}{R}$ $I(t) = I_0 e^{-Rt/L}$

Solve for time. This will allow us to determine at what time the current has reached a value $I = 1.00$ mA given the initial current (from part *a*), inductance, and resistance.	$t = \left(\dfrac{L}{R}\right)\ln\left(\dfrac{I_0}{I}\right)$
We know that $I_0 = 6.00$ mA, $R = 11.0$ kΩ, and $L = 589$ mH.	$t = \left(\dfrac{0.589\text{ H}}{11.0\times 10^3\ \Omega}\right)\ln\left(\dfrac{6.00\text{ mA}}{1.00\text{ mA}}\right)$ $t = \boxed{9.59\times 10^{-5}\text{ s}} = 95.9\ \mu s$

CHECK and THINK

This problem focuses on the *RL* circuit. Whether connected to a battery or suddenly disconnected with a current running through it, the inductor prevents instantaneous changes in current.

17. In Figure 33.9A (page 1052), the switch is closed at *a* at $t = 0$.

a. (A) Find an expression for the total energy dissipated by the resistor in one time constant.

b. (A) After the switch is left at *a* for many time constants, it is switched to *b*. Find an expression for the total energy dissipated by the resistor in one time constant.

c. (C) Compare your results and comment.

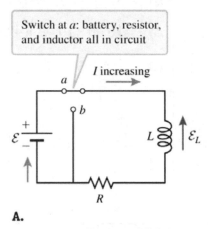

Figure 33.9A

INTERPRET and ANTICIPATE

When switched to *a*, the battery is connected to the *LR* circuit and the current exponentially approaches a steady-state value. When switched to *b*, the *LR* circuit initially has a current, but no voltage source, so the current exponentially decreases to zero. We can use these expressions for current to determine the power dissipated by the resistor.

SOLVE **a.** The current in an *LR* circuit connected to a voltage source is given by Eq. 33.12. It is initially zero but increases and exponentially approaches the steady-state value, $\mathcal{E}/R$.	$I = \dfrac{\mathcal{E}}{R}\left(1 - e^{-t/\tau}\right)$
The power dissipated by the resistor is I^2R. Since the current is increasing in time, so is the power dissipated.	$P = I^2 R$ $P = \left[\dfrac{\mathcal{E}}{R}\left(1 - e^{-t/\tau}\right)\right]^2 R$ $P = \dfrac{\mathcal{E}^2}{R}\left(1 - e^{-t/\tau}\right)^2 \qquad\qquad (1)$
We are asked about the energy dissipated after one time constant. Power is the *rate* at which energy is dissipated, so we can integrate the power over time to determine the energy dissipated.	$P = \dfrac{dE}{dt} \quad \rightarrow \quad \Delta E = \displaystyle\int P\,dt \qquad (2)$
Insert Eq. (1) into (2) and integrate from time 0 to τ, one time constant. It's easiest to multiply out the terms and then integrate.	$\Delta E = \displaystyle\int_0^\tau \dfrac{\mathcal{E}^2}{R}\left(1 - e^{-t/\tau}\right)^2 dt$ $\Delta E = \dfrac{\mathcal{E}^2}{R}\displaystyle\int_0^\tau \left(1 - 2e^{-t/\tau} + e^{-2t/\tau}\right)dt$
Perform the integral, using $\displaystyle\int dt = t$ and $\displaystyle\int e^{ct}\,dt = \dfrac{1}{c}e^{ct}$.	$\Delta E = \dfrac{\mathcal{E}^2}{R}\left[t + 2\tau e^{-t/\tau} - \dfrac{\tau}{2}e^{-2t/\tau}\right]_0^\tau$ $\Delta E = \dfrac{\mathcal{E}^2}{R}\left[\tau + 2\tau\left(\dfrac{1}{e} - 1\right) - \dfrac{\tau}{2}\left(\dfrac{1}{e^2} - 1\right)\right]$ $\boxed{\Delta E = \dfrac{\mathcal{E}^2\tau}{R}\left[\dfrac{2}{e} - \dfrac{1}{2e^2} - \dfrac{1}{2}\right]}$
b. The process is very similar, except the current for an *LR* circuit without a voltage source exponentially decreases to zero as in Eq. 33.15. Because the switch	$I(t) = I_{max}\,e^{-t/\tau} = \dfrac{\mathcal{E}}{R}e^{-t/\tau}$

has been closed for a long time leading up to this moment, the initial current is $I_{max} = \mathcal{E}/R$.	
Calculate the power as in part (a).	$$P = I^2 R = \left(\frac{\mathcal{E}}{R} e^{-t/\tau} \right)^2 R$$ $$P = \frac{\mathcal{E}^2}{R} e^{-2t/\tau} \qquad (3)$$
Now insert Eq. (3) into (2) and integrate from 0 to τ.	$$\Delta E = \frac{\mathcal{E}^2}{R} \int_0^\tau e^{-2t/\tau} \, dt$$ $$\Delta E = \frac{\mathcal{E}^2}{R} \left[-\frac{\tau}{2} e^{-2t/\tau} \right]_0^\tau$$ $$\Delta E = -\frac{\mathcal{E}^2 \tau}{2R} \left(e^{-2} - 1 \right)$$ $$\Delta E = \boxed{\frac{\mathcal{E}^2 \tau}{2R} \left(1 - \frac{1}{e^2} \right)}$$
c. One way to compare these is to find the ratio $\Delta E_b / \Delta E_a$. We find that the ratio is greater than one, so $\Delta E_b > \Delta E_a$, indicating that the resistor dissipates more energy in the first time constant when the current is decreasing from the steady state value than when it's increasing from zero. This makes sense because the current is higher initially when the current starts decreasing compared to when it starts increasing from zero. So, the average current is higher in part (b) compared to part (a) and the power dissipated depends on the current squared.	$$\frac{\Delta E_b}{\Delta E_a} = \frac{\dfrac{\mathcal{E}^2 \tau}{2R} \left(1 - \dfrac{1}{e^2} \right)}{\dfrac{\mathcal{E}^2 \tau}{R} \left[\dfrac{2}{e} - \dfrac{1}{2e^2} - \dfrac{1}{2} \right]} = \frac{e^2 - 1}{4e - 1 - e^2} \approx 2.57$$

Chapter 33 – Inductors and AC Circuits

CHECK and THINK

This required a bit of work in performing the integral, but conceptually, we just needed to use the current in the *RL* circuit to find the power dissipated by the resistor and integrate to find out how much energy was dissipated over a certain amount of time.

20. (N) If a high-voltage power line 25 m above the ground carries a current of 1.00×10^3 A, estimate the energy density of the magnetic field near the ground and compare it to the energy density of the Earth's magnetic field.

INTERPRET and ANTICIPATE

The energy density due to a magnetic field depends on the strength of the magnetic field, which we can calculate for this current-carrying wire.

SOLVE

The energy density due to a magnetic field B is given by Equation 33.17. We can calculate the magnetic field assuming that the high voltage line can be approximated as an infinitely long wire.

$$u_B = \frac{1}{2}\frac{B^2}{\mu_0} \quad \text{and} \quad B = \frac{\mu_0 I}{2\pi r}$$

$$u_B = \frac{\mu_0 I^2}{8\pi^2 r^2}$$

We can now plug in known values. The energy density of the earth's magnetic field was calculated in Example 33.3 and found to be around 10^{-3} J/m³. The field from the overhead power line is only a few percent of the value due to the Earth's field, so it is significantly smaller.

$$u_B = \frac{\left(4\pi \times 10^{-7}\,\frac{\text{T}\cdot\text{m}}{\text{A}}\right)(1000\,\text{A})^2}{8\pi^2 (25\,\text{m})^2}$$

$$u_B = \boxed{2.5 \times 10^{-5}\ \text{J/m}^3}$$

CHECK and THINK

The energy density of the magnetic field due to the high voltage power line was calculated assuming the field is that of a current-carrying wire. The energy density is smaller than that of the Earth's magnetic field calculated in Example 33.3.

26. (N) Figure P33.26 shows a circuit with $\mathcal{E} = 9.00$ V, $R = 6.00\ \Omega$, $L = 75.0$ mH, and $C = 2.55\ \mu$F. After a long time interval at the position *a* shown in the figure, the switch S is thrown to position *b* at time $t = 0$. What is the maximum **a.** charge on the capacitor and **b.** current in the inductor for $t > 0$? **c.** What is the frequency of oscillation of the resulting *LC* circuit for $t > 0$?

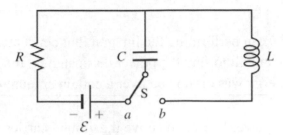

Figure P33.26

INTERPRET and ANTICIPATE

In position a, the left loop is closed and forms a charging RC circuit. After a long time, the capacitor is fully charged. When thrown to position b, the right loop is closed and is an LC circuit in which the capacitor is initially charged. The charge on the capacitor, the current in the circuit, and the energy stored in the capacitor and inductor oscillate in an LC circuit.

SOLVE	
a. After a is closed for a long time, the capacitor is fully charged. At this point, it's as if the battery is simply connected to the capacitor (since there is no current and therefore no voltage across the resistor) and $Q = C\Delta V$.	$Q = C\Delta V = C\mathcal{E}$ $Q = \left(2.55 \times 10^{-6} \text{ F}\right)\left(9.00 \text{ V}\right)$ $Q = \boxed{2.30 \times 10^{-7} \text{C}} = 23.0 \ \mu C$
b. The charge and current oscillate in time. While the *total* energy is constant, the energy is transferred back and forth between the capacitor and inductor, so the energy of each oscillates in time. (We can think of it as the energy sloshing back and forth between the capacitor and the inductor.) The energy is initially entirely stored in the capacitor $\left(\dfrac{1}{2}C\mathcal{E}^2\right)$ but is transferred back and forth such that at other points in time, all of the energy is stored in the inductor, $\left(\dfrac{1}{2}LI^2, \text{ Eq. 33.3}\right)$. The current is at its maximum value when all of the energy is stored in the inductor.	$\dfrac{1}{2}C\mathcal{E}^2 = \dfrac{1}{2}LI_{max}^2$

Solve for the maximum current and insert numbers.	$I_{max} = \mathcal{E}\sqrt{\dfrac{C}{L}}$ $I_{max} = (9.00 \text{ V})\sqrt{\dfrac{(2.55 \times 10^{-6} \text{ F})}{(75.0 \times 10^{-3} \text{ H})}}$ $I_{max} = 5.25 \times 10^{-2} \text{ A} = \boxed{52.5 \text{ mA}}$
c. The angular velocity is given by Eq. 33.21. We also use the fact that $\omega = 2\pi f$.	$\omega = \dfrac{1}{\sqrt{LC}} = 2\pi f$
Solve for the frequency and insert numbers.	$f = \dfrac{1}{2\pi\sqrt{LC}}$ $f = \dfrac{1}{2\pi\sqrt{(75.0 \times 10^{-3} \text{ H})(2.55 \times 10^{-6} \text{ F})}}$ $f = \boxed{364 \text{ Hz}}$

CHECK and THINK

In position a, we have an RC circuit that eventually fully charges the capacitor. When switched to b, the RL circuit is connected and the charge, current, and energy associated with the capacitor and inductor oscillate in time. Given the maximum energy in the capacitor, we can determine the maximum energy (and therefore maximum current) in the inductor. The frequency of oscillation depends on the capacitance and inductance, which are given.

31. (N) A fully charged capacitor and a 0.20-H inductor are connected to form a complete circuit. If the circuit oscillates with a frequency of 1.2×10^3 Hz, determine the capacitance of the capacitor.

INTERPRET and ANTICIPATE
This is an LC circuit, which exhibits a capacitor charge and current that oscillate in time. The oscillation frequency depends on both the inductance and the capacitance, which are given.

SOLVE The angular frequency for the LC circuit is given by Eq. 33.21. We also use the fact that $\omega = 2\pi f$.	$\omega = \sqrt{\dfrac{1}{LC}} = 2\pi f$

Solve for the capacitance.	$$C = \frac{1}{4\pi^2 L f^2}$$
Finally, insert numerical values.	$$C = \frac{1}{4\pi^2 (0.2 \text{ H})(1.2 \times 10^3 \text{ Hz})^2}$$ $$C = \boxed{8.8 \times 10^{-8} \text{ F}} = 88 \text{ nF}$$

CHECK and THINK

This is a straightforward application of the frequency of the LC circuit.

35. An AC generator delivers an alternating current $I(t) = (2.0 \text{ A})\sin\left[(120\pi \text{ rad/s})t\right]$ to a single resistor in series with the generator.

a. (N) What is the rms value of the current in the circuit?

b. (N) If the resistor has a resistance of 100.0 V, what is the rms value of the source emf?

c. (N) What is the maximum value of the source emf?

d. (A) Write a function that describes the source emf as a function of time.

INTERPRET and ANTICIPATE

An AC circuit with a single resistor can be modeled using the rms and maximum values of current and emf, related by Eq. 32.20, 32.21, and 33. 31. We write the potential difference across the resistor using Eq. 33.29, where we know the current through the resistor is in phase with the potential difference across the resistor. Note that the rms values of current and emf should be less than their maximum values.

SOLVE	
a. The maximum value of the current is identified as the coefficient in the equation for current as a function of time. Then, we can use Eq. 32.21 to find the rms current.	$$I_{rms} = I_{max}/\sqrt{2} = (2.0 \text{ A})/\sqrt{2} = \boxed{1.4 \text{ A}}$$
b. Find the maximum emf using Eq. 33.31.	$$\mathcal{E}_{max} = I_{max}R = (2.0 \text{ A})(100.0 \text{ }\Omega) = 2.0 \times 10^2 \text{ V}$$
Then, we use Eq. 32. 20 to find the rms emf.	$$\mathcal{E}_{rms} = \mathcal{E}_{max}/\sqrt{2} = (2.0 \times 10^2 \text{ V})/\sqrt{2} = \boxed{1.4 \times 10^2 \text{ V}}$$

c. We actually already found this when solving part (b).	$\mathcal{E}_{max} = I_{max}R = (2.0\ A)(100.0\ \Omega) = \boxed{2.0\times10^2\ V}$
d. Using Eq. 33.29, we are able to construct the expression for the potential difference across the resistor.	$\mathcal{E} = \mathcal{E}_{max}\sin\left[(120\pi\ \text{rad/s})t\right]$ $\boxed{\mathcal{E}(t) = (2.0\times10^2\ V)\sin\left[(120\pi\ \text{rad/s})t\right]}$

CHECK and THINK

If we were to write an equation for the current through the resistor as a function of time, we would have the same time dependence because the current is in phase with the potential difference across the resistor. The phasor diagram for this scenario would also show the current phasor in line with the phasor for V_R.

39. (N) An 8.00-mF capacitor is connected across an AC source with an rms voltage of 68.0 V oscillating with a frequency of 50.0 Hz. What are the **a.** capacitive reactance of, **b.** rms current in, and **c.** maximum current in this circuit?

INTERPRET and ANTICIPATE

The capacitive reactance depends on the capacitance and the angular frequency, with a smaller frequency leading to a larger reactance. Once we calculate this quantity, we can easily determine the current given a particular AC voltage.

SOLVE

a. The capacitive reactance is given by Eq. 33.39 with $\omega = 2\pi f$.	$X_C = \dfrac{1}{\omega C} = \dfrac{1}{2\pi f C}$
Insert values in metric base units. Reactance opposes AC currents and is measured in ohms.	$X_C = \dfrac{1}{2\pi(50.0\ \text{Hz})(8.00\times10^{-3}\ F)}$ $X_C = \boxed{0.398\ \Omega}$
b. Apply Eq. 33.41 to determine the current. According to Eq. 32.20 and 32.21, $\mathcal{E}_{rms} = \dfrac{\mathcal{E}_{max}}{\sqrt{2}}$ and $I_{rms} = \dfrac{I_{max}}{\sqrt{2}}$, so we can express Eq. 33.41 in terms of the rms values.	$I_{rms} = \dfrac{V_{C,rms}}{X_C}$ $I_{rms} = \dfrac{68.0\ V}{0.398\ \Omega} = \boxed{171\ A}$

c. Apply Eq. 32.21.	$I_{max} = \sqrt{2}\, I_{rms}$ $I_{max} = \sqrt{2}\,(171\text{ A}) = \boxed{242\text{ A}}$

CHECK and THINK

The capacitive reactance depends on the capacitance and angular frequency of the voltage source. This is the impedance of this circuit and with it we can determine the current given the voltage.

41. An AC generator delivers an rms current of 3.50 A to a 6.25-μF capacitor connected in series. The frequency of the source emf is 60.0 Hz.
a. (N) What is the capacitive reactance of the circuit?
b. (N) What is the rms emf of the generator?
c. (N) What are the maximum values of the current and the source emf?
d. (A) Write a function for the potential difference across the capacitor as a function of time.

INTERPRET and ANTICIPATE

An AC circuit with a single capacitor can be modeled using the rms and maximum values of current and emf, related by Eq. 32.20, 32.21, and 33.41, where the capacitive reactance is defined by Eq. 33.39. We know that in a capacitor, the current leads the potential difference across the capacitor by $\pi/2$ rad, though we write the potential difference across the capacitor using Eq. 33.34. Note that the rms values of current and emf should be less than their maximum values.

SOLVE	
a. The capacitive reactance can be found using Eq. 33.39 and the relationship between frequency and angular frequency, $\omega = 2\pi f$.	$X_C = 1/\omega C = 1/\left(2\pi(60.0\text{ Hz})(6.25\times10^{-6}\text{ F})\right)$ $X_C = \boxed{424\ \Omega}$
b. The rms values of current and emf can be related to their maximum values using Eq. 32.20 and Eq. 32.21.	$I_{max} = I_{rms}\sqrt{2}$ $\mathcal{E}_{max} = \mathcal{E}_{rms}\sqrt{2}$
Then, use these expressions with Eq. 33.41 and the values from the problem statement to find the rms emf.	$\mathcal{E}_{max} = I_{max}X_C$ $\mathcal{E}_{rms}\sqrt{2} = I_{rms}\sqrt{2}X_C$ $\mathcal{E}_{rms} = I_{rms}X_C$ $\mathcal{E}_{rms} = (3.50\text{ A})\left(1/\left(2\pi(60.0\text{ Hz})(6.25\times10^{-6}\text{ F})\right)\right)$ $\mathcal{E}_{rms} = \boxed{1.49\times10^{3}\text{ V}}$

c. Use Eq. 32.21 to find the maximum current.	$I_{max} = I_{rms}\sqrt{2} = (3.50\ A)\sqrt{2} = \boxed{4.95\ A}$
Then, use Eq. 33.41 to find the maximum emf.	$\mathcal{E}_{max} = I_{max} X_C$ $\mathcal{E}_{max} = (4.95\ A)\left(1/\left(2\pi(60.0\ Hz)(6.25\times10^{-6}\ F)\right)\right)$ $\mathcal{E}_{max} = \boxed{2.10\times10^3\ V}$
d. Using Eq. 33.34, we are able to construct the expression for the potential difference across the capacitor.	$V_C(t) = \mathcal{E}_{max}\sin(\omega t)$ $V_C(t) = (2.10\times10^3\ V)\sin(2\pi(60.0\ Hz)t)$ $V_C(t) = \boxed{(2.10\times10^3\ V)\sin((377\ rad/s)t)}$

CHECK and THINK

If we were to write an equation for the current through the capacitor as a function of time, we would have to apply a phase shift of $\pi/2$ rad inside the sine function expression, because it leads the potential difference across the capacitor by that amount. The phasor diagram for this scenario would also show the current phasor 90° ahead of the phasor for V_C, or $\pi/2$ rad ahead.

47. (N) A 68.0-mH inductor is connected across an AC source that has an rms voltage of 120.0 V oscillating at a frequency of 110.0 Hz. What are the **a.** inductive reactance of, **b.** rms current in, and **c.** maximum current in this circuit?

INTERPRET and ANTICIPATE

The inductive reactance depends on the inductance and angular frequency of the oscillating voltage source, with the inductor providing a higher reactance at higher frequencies. Once we know the reactance, we can determine the current using the analog of Ohm's law for AC circuits, $I = V/Z$.

SOLVE a. The inductive reactance is given by Eq. 33.61 with the angular frequency $\omega = 2\pi f$.	$X_L = \omega L = 2\pi f L$
Insert numerical values. The reactance tends to limit an AC current and has units of ohms.	$X_L = 2\pi(110.0\ Hz)(68.0\times10^{-3}\ H)$ $X_L = \boxed{47.0\ \Omega}$

b. The rms current can be found using Eq. 33.63. We also use Eqs. 32.20 and 32.21, $$\mathcal{E}_{rms} = \frac{\mathcal{E}_{max}}{\sqrt{2}} \text{ and } I_{rms} = \frac{I_{max}}{\sqrt{2}} \text{ to determine}$$ the rms current.	$$I_{max} = \frac{\mathcal{E}_{max}}{X_L} \quad \text{or} \quad I_{rms} = \frac{\mathcal{E}_{rms}}{X_L}$$
The emf is the rms voltage across the inductor or 120 V and the inductive reactance was calculated in part (a).	$$I_{rms} = \frac{120.0 \text{ V}}{47.0 \; \Omega} = \boxed{2.55 \text{ A}}$$
c. Use Eq. 32.21, mentioned above.	$$I_{max} = \sqrt{2} \, I_{rms} = \sqrt{2} \left(2.55 \text{ A} \right)$$ $$I_{max} = \boxed{3.61 \text{ A}}$$

CHECK and THINK

The inductive reactance depends on the inductance and angular frequency of the voltage source. This is the impedance of this circuit and with it we can determine the current given the voltage.

52. (N) When connected to a 90.0-Hz AC source, an inductor has an inductive reactance of 33.0 Ω. What is the maximum current in this inductor if it is connected to an AC source that has an rms voltage of 120.0 V and a frequency of 110.0 Hz?

INTERPRET and ANTICIPATE The inductive reactance is proportional to frequency, so given the value at 90 Hz, we can determine the value at 110.0 Hz. Then, given the voltage and the inductive reactance (the impedance for this circuit), we can calculate the current.	
SOLVE The inductive reactance is proportional to frequency (Eq. 33.61, with $\omega = 2\pi f$). We can use this to find the ratio for two different frequencies.	$$X_L = \omega L = 2\pi f L$$ $$\frac{X_{L,1}}{X_{L,2}} = \frac{2\pi f_1 L}{2\pi f_2 L} = \frac{f_1}{f_2}$$
Calculate the reactance for 110.0 Hz given the value for 90.0 Hz.	$$\frac{X_{L,\,110Hz}}{X_{L,\,90Hz}} = \frac{110.0 \text{ Hz}}{90.0 \text{ Hz}}$$ $$X_{L,\,110Hz} = \frac{110.0 \text{ Hz}}{90.0 \text{ Hz}} \left(33.0 \; \Omega \right) = 40.3 \; \Omega$$

The maximum current can now be calculated using Equation 33.63. We also use the definition of rms voltage, $V_{rms} = \dfrac{V_{max}}{\sqrt{2}}$.	$I_{max} = \dfrac{V_{max}}{X_L} = \dfrac{\sqrt{2}\,(V_{rms})}{X_L}$
	$I_{max} = \dfrac{\sqrt{2}\,(120.0\ \text{V})}{40.3\ \Omega} = \boxed{4.21\ \text{A}}$

CHECK and THINK

The inductive reactance depends on the inductance and frequency. Given the value at one frequency, we can determine the value at another. This then allows us to determine the current if a particular voltage source is connected.

57. (N) A series *RLC* circuit with a resistance of 120.0 Ω has a resonance angular frequency of 4.0×10^5 rad/s. At resonance, the voltages across the resistor and inductor are 60.0 V and 40.0 V, respectively. **a.** Determine the values of *L* and *C*. **b.** At what frequency does the current in the circuit lag the voltage by 45°?

INTERPRET and ANTICIPATE

At resonance, the inductive reactance and capacitive reactance are equal, such that the total impedance of the circuit is at its lowest possible value, equal to the resistance of the resistor. This will allow us to get started.

SOLVE **a.** The impedance is given by Eq. 33.81. At resonance, $\omega = \omega_0$ and $Z = R$. At this frequency, the current is simply given by Ohm's law.	$I = \dfrac{\mathcal{E}}{R} = \dfrac{60.0\ \text{V}}{120.0\ \Omega} = 0.500\ \text{A}$
The voltage across the inductor depends on the inductive reactance (Eq. 33.61) and current according to Eq. 33.63. We can solve this for the inductance since we are given the voltage across the inductor and we calculated the current above.	$I = \dfrac{V_L}{X_L} = \dfrac{V_L}{\omega_0 L}$ $L = \dfrac{V_L}{\omega_0 I}$
Insert numerical values.	$L = \dfrac{40\ \text{V}}{(4.0\times10^5\ \text{rad/s})(0.50\ \text{A})}$ $L = 2.0\times10^{-4}\ \text{H} = \boxed{0.20\ \text{mH}}$

The angular frequency of resonance depends on the inductance and capacitance according to Eq. 33.21. Solve to find the capacitance.	$\omega_0 = \dfrac{1}{\sqrt{LC}}$ $C = \dfrac{1}{L\omega_0^2}$
Insert values.	$C = \dfrac{1}{\left(2.0 \times 10^{-4}\ \text{H}\right)\left(4.0 \times 10^5\ \text{rad/s}\right)^2}$ $C = 3.1 \times 10^{-8}\ \text{F} = \boxed{31\ \text{nF}}$
b. The phase angle is given by Eq. 33.80 and depends on the competing effects of the capacitor (which causes the voltage to lag relative to the current) and the inductor (which causes it to lead).	$\varphi = \tan^{-1}\left(\dfrac{X_L - X_C}{R}\right)$ $\tan\varphi = \dfrac{X_L - X_C}{R}$
We want to determine when this phase angle is 45 degrees using the fact that tan(45°) = 1.	$\tan 45° = 1 = \dfrac{L\omega - \dfrac{1}{C\omega}}{R}$ $LC\omega^2 - RC\omega - 1 = 0$
This is a quadratic equation in terms of ω, so we can use the quadratic formula.	$\omega = \dfrac{RC \pm \sqrt{\left(RC\right)^2 + 4LC}}{2LC}$

Now insert numerical values.

$$\omega = \dfrac{\left(120\ \Omega\right)\left(3.1 \times 10^{-8}\ \text{F}\right) \pm \sqrt{\left(\left(120\ \Omega\right)\left(3.1 \times 10^{-8}\ \text{F}\right)\right)^2 + 4\left(2.0 \times 10^{-4}\ \text{H}\right)\left(3.1 \times 10^{-8}\ \text{F}\right)}}{2\left(2.0 \times 10^{-4}\ \text{H}\right)\left(3.1 \times 10^{-8}\ \text{F}\right)}$$

$$\omega = \frac{3.7 \times 10^{-6} \pm \sqrt{\left(3.7 \times 10^{-6}\right)^2 + 4\left(6.2 \times 10^{-12}\right)}}{2\left(6.2 \times 10^{-12}\right)} \text{ rad/s}$$

$$\omega = 8.0 \times 10^5 \text{ rad/s}$$

Then, the frequency is

$$f = \frac{\omega}{2\pi} = \frac{8.0 \times 10^5 \text{ rad/s}}{2\pi} = \boxed{1.3 \times 10^5 \text{ Hz}}$$

Note, we take the positive root as being physically meaningful. There is also a solution where $\omega = -2.0 \times 10^5$ rad/s, but a negative frequency does not make sense. So, at this angular frequency (800,000 Hz), the phase angle between the current and voltage is 45 degrees.

CHECK and THINK
When an RLC circuit is at resonance, the impedance is at its minimum value and equals the resistance. This allows us to calculate the current. Given the voltage across each circuit element and the current, we are able to calculate the inductance and capacitance.

60. (N) An AC source of angular frequency ω is connected to a resistor R and a capacitor C in series. The maximum current measured is I_{max}. While the same maximum emf is maintained, the angular frequency is changed to $\omega/3$. The measured current is now $I/2$. Determine the ratio of the capacitive reactance to the resistance at the initial frequency ω.

INTERPRET and ANTICIPATE
The problem concerns the RC circuit driven by an AC voltage. We can calculate the impedance to find the current.

SOLVE	
First, write the impedance for the RC circuit using Eq. 33.81.	$Z = \sqrt{\left(\dfrac{1}{C\omega}\right)^2 + R^2}$
The current is given by Eq. 33.82.	$I = \dfrac{\mathcal{E}}{\sqrt{\left(\dfrac{1}{C\omega}\right)^2 + R^2}}$ (1)

Now consider the fact that the current decreases to half the original value when the frequency is reduced by a factor of three.	$$\frac{1}{2}I = \frac{\mathcal{E}}{\sqrt{\left(\dfrac{1}{C\omega/3}\right)^2 + R^2}} = \frac{\mathcal{E}}{\sqrt{\left(\dfrac{3}{C\omega}\right)^2 + R^2}} \qquad (2)$$
Use Equation (1) to substitute the current in equation (2) and solve for the capacitive reactance, $X_C = \dfrac{1}{\omega C}$ (Eq. 33.39).	$$\frac{1}{2}\frac{\mathcal{E}}{\sqrt{\left(\dfrac{1}{C\omega}\right)^2 + R^2}} = \frac{\mathcal{E}}{\sqrt{\left(\dfrac{3}{C\omega}\right)^2 + R^2}}$$ $$\frac{9}{C^2\omega^2} + R^2 = 4\left(\frac{1}{C^2\omega^2} + R^2\right)$$ $$\frac{5}{C^2\omega^2} = 3R^2$$ $$\frac{1}{C^2\omega^2} = \frac{3}{5}R^2$$ $$X_C^2 = \frac{3}{5}R^2$$ $$\frac{X_C}{R} = \sqrt{\frac{3}{5}} = \boxed{0.77}$$

CHECK and THINK
The reactance is less than the resistance in this particular case.

64. **(N)** In an *RLC* circuit (Fig. 33.31, page 1072), the resistance is 325 Ω, the inductance is 126 mH, and the capacitance is 13.7 μF (13.7×10^{-6} F). The angular frequency is 377 rad/s and $\mathcal{E}_{max} = 18.8$ V. **a.** What is the impedance? **b.** What is the maximum current?
c. What is the phase constant φ of the power supply's emf with respect to the current?

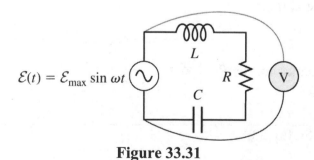

$\mathcal{E}(t) = \mathcal{E}_{max} \sin \omega t$

Figure 33.31

INTERPRET and ANTICIPATE

In an *RLC* circuit, the impedance is analogous to a resistance in that it determines the current in a circuit driven by an AC voltage. The properties of the current, including its maximum value and the phase relative to the driving voltage, depend on the values of the *RLC* components used, which are given. We don't have a way to predict the values, so we'll go right ahead and calculate!

SOLVE	
a. The impedance is given by Eq. 33.81.	$Z \equiv \sqrt{(X_L - X_C)^2 + R^2}$
The capacitive reactance and inductive reactance are given by Equations 33.39 and 33.61 respectively.	$X_L = \omega L \quad \text{and} \quad X_C = \dfrac{1}{\omega C}$
Insert these into the formula for the impedance.	$Z = \sqrt{\left(\omega L - \dfrac{1}{\omega C}\right)^2 + R^2}$

Now insert values. The unit for impedance is ohms.

$$Z = \sqrt{\left((377\,\text{Hz})(0.126\,\text{H}) - \frac{1}{(377\,\text{Hz})(13.7 \times 10^{-6}\,\text{F})}\right)^2 + (325\,\Omega)^2}$$

$$Z = \boxed{3.56 \times 10^2\,\Omega}$$

b. The maximum current can be found using the maximum voltage and the impedance, as in Eq. 33.82.	$I_{max} = \dfrac{\mathcal{E}_{max}}{Z}$
We are given the maximum voltage and know the impedance from part (a).	$I_{max} = \dfrac{18.8\,\text{V}}{356\,\Omega} = \boxed{5.28 \times 10^{-2}\,\text{A}}$
c. The phase angle depends on the competing effects of the capacitor (which causes the voltage to lag relative to the current) and the inductor (which causes it to lead). This is expressed by Eq. 33.80, where we again use the reactances from part (a).	$\varphi = \tan^{-1}\left(\dfrac{X_L - X_C}{R}\right) = \tan^{-1}\left(\dfrac{\omega L - \dfrac{1}{\omega C}}{R}\right)$

Insert numerical values.

$$\varphi = \tan^{-1}\left(\frac{(377\text{ Hz})(0.126\text{ H}) - \dfrac{1}{(377\text{ Hz})(13.7\times10^{-6}\text{ F})}}{325\ \Omega}\right) = -0.422\text{ rad} = \boxed{-24.2°}$$

CHECK and THINK

Given the values of the *R, L,* and *C*, we can determine the response of the circuit, both in terms of the maximum values of the current and its phase relative to the voltage. The negative value of the phase angle indicates that the voltage lags relative to the current.

70. (N) An ideal solenoid with 855 turns is 0.100 m long and has a cross-sectional area of 3.00×10^{-3} m². **a.** What is the inductance of this solenoid? **b.** If we connect this solenoid in series with a 12.0-V battery and a 185-Ω resistor, what is the maximum value of the current? **c.** Assuming the current is zero when $t = 0$ as the circuit elements are connected, at what time will the current reach 50% of its maximum value?

INTERPRET and ANTICIPATE

The circuit is really an *RL* series DC circuit, so we may use Eq. 33.12 to model the circuit's behavior. The time constant will be given by Eq. 33.13, but we must first find the inductance of the solenoid by using Eq. 33.5. Note that the maximum value of the current is the coefficient of Eq. 33.12, which is really Ohm's law for the circuit once the inductor is no longer affecting the current flow. We expect the time in the last part to be very little as most *RL* circuits do not take a long period of time to saturate and reach the maximum current, thought the time does depend on the values of *R* and *L*.

SOLVE

a. We can find the inductance of the solenoid by using Eq. 33.5.

$$L = \frac{\mu_0 N^2}{\ell} A$$

$$L = \frac{(4\pi\times10^{-7}\text{ T}\cdot\text{m/A})(855)^2}{0.100\text{ m}}(3.00\times10^{-3}\text{ m}^2)$$

$$L = \boxed{2.76\times10^{-2}\text{ H}}$$

b. In series with a battery, the maximum current would occur when the potential difference of the battery is equal to the potential difference across the resistor, as we can see from the coefficient of Eq. 33.12.

$$I_{max} = \mathcal{E}/R = (12.0\text{ V})/(185\ \Omega) = \boxed{6.49\times10^{-2}\text{ A}}$$

c. We can begin by expressing the time constant for the *RL* circuit using Eq. 33.13.	$\tau = L / R$
Using Eq. 33.12, we can find the time when the current is half of the maximum current. We choose the current $I = (1/2)(\mathcal{E}/R)$ Note that we must apply the natural log function to both sides of the equation to isolate t from the exponential function.	$I = \dfrac{\mathcal{E}}{R}\left(1 - e^{-t/\tau}\right)$ $\dfrac{1}{2}\dfrac{\mathcal{E}}{R} = \dfrac{\mathcal{E}}{R}\left(1 - e^{-t/(L/R)}\right)$ $0.5 = \left(1 - e^{-t/(L/R)}\right)$ $e^{-t/(L/R)} = 0.5$ $-t / (L/R) = \ln(0.5)$ $t = -\dfrac{L}{R}\ln(0.5)$ $t = -\dfrac{\left(2.76 \times 10^{-2}\ \text{H}\right)}{185\ \Omega}\ln(0.5)$ $t = \boxed{1.03 \times 10^{-4}\ \text{s}}$

CHECK and THINK

While it doesn't take a long amount of time, the inductor prevents the current from reaching its maximum value immediately when the circuit is closed. This allows for a gradual rise in current which can help keep physical parts of the circuit from burning out. So, if you want to control the rate of rise of current (or fall, when opening the circuit), use an inductor!

74. (N) A 22.5-μF capacitor is charged by a 6.00-V battery and then connected to a 75.0-mH inductor. At what frequency does the current oscillate, and what is the maximum current?

INTERPRET and ANTICIPATE	
The circuit described is an *LC* circuit with no resistance and an initially charged capacitor. The charge on the capacitor and current through the circuit oscillate in time with a frequency that depends on the inductance and capacitance.	
SOLVE The charge on the capacitor plates oscillates as given by Equation 33.20. The maximum charge when the capacitor is charged by the battery can also be expressed.	$Q(t) = Q_{max}\cos(\omega t + \varphi)$ $Q_{max} = CV$

The angular frequency of oscillation is given by Equations 33.21.	$\omega = \sqrt{\dfrac{1}{LC}}$
We can calculate the angular frequency using the inductance and capacitance. We could also find the frequency f for this oscillation, since $\omega = 2\pi f$.	$\omega = \sqrt{\dfrac{1}{\left(75.0\times10^{-3}\ \text{H}\right)\left(22.5\times10^{-6}\ \text{F}\right)}} = 7.70\times10^2\ \text{rad/s}$ $f = \dfrac{\omega}{2\pi} = \dfrac{7.70\times10^2\ \text{rad/s}}{2\pi} = \boxed{123\ \text{Hz}}$
The current is the rate of change of the charge on the capacitor plates and has a maximum value given by Equation 33.23. Since the sine function oscillates between plus and minus one, the maximum current equals the magnitude of the factor in front, ωQ_{max}. (You can also compare to Eq. 33.23 to see that $I_{max} = \omega Q_{max}$.)	$I(t) = \dfrac{dQ(t)}{dt} = -\omega Q_{max}\sin\left(\omega t + \varphi\right)$ $I_{max} = \omega Q_{max} = \omega C V$
We can now insert numerical values to find the maximum current.	$I_{max} = \left(7.70\times10^2\ \text{rad/s}\right)\left(22.5\times10^{-6}\ \text{F}\right)\left(6.00\ \text{V}\right)$ $I_{max} = \boxed{0.104\ \text{A}}$

CHECK and THINK
The frequency of the oscillating current depends on the inductance and capacitance. The maximum current can also be calculated given the initial charge on the capacitor.

77. (N) A 45.0-V battery with an internal resistance of 13.0 Ω is connected to a 7.40-H inductor. The current is zero at $t = 0$. At what rate is the current in the inductor increasing at **a.** $t = 0$ and **b.** $t = 2.00$ s?

INTERPRET and ANTICIPATE
The question describes and RL circuit connected to a voltage source. The current increases from zero and eventually reaches a steady-state value. We can determine the expression for the current and then find the *rate of change of the current* that we are asked to find.

SOLVE The current in this *RL* circuit increases from zero according to Eq. 33.12 and approaches a steady-state value that we can determine from Ohm's law. (After a long time, when the current eventually reaches a constant steady-state value, the inductor has no effect since it opposes *changes* in current.)	$I = I_{max}\left(1 - e^{-t/\tau}\right)$ with $I_{max} = \dfrac{\mathcal{E}}{R}$
The time constant is given by Eq. 33.13.	$\tau = \dfrac{L}{R}$
The rate of change of the current is how rapidly the current changes in time, so we can take the derivative of this expression for current with respect to time.	$\dfrac{dI}{dt} = -I_{max}\left(e^{-t/\tau}\right)\left(-\dfrac{1}{\tau}\right)$ $\dfrac{dI}{dt} = \dfrac{\mathcal{E}}{L}e^{-t/\tau}$
a. First use this expression at *t* = 0. Since all quantities are in metric base units, the answer for "current per time" must be as well: Amps per second.	$\dfrac{dI}{dt} = \dfrac{\mathcal{E}}{L}e^{0} = \dfrac{\mathcal{E}}{L}$ $\dfrac{dI}{dt} = \dfrac{45.0\text{ V}}{7.40\text{ H}} = \boxed{6.08\text{ A/s}}$
b. Now, apply the expression at *t* = 2.00 s. The time constant from above is: $\tau = \dfrac{L}{R} = \dfrac{7.40\text{ H}}{13.0\ \Omega} = 0.569\text{ s}$	$\dfrac{dI}{dt} = \dfrac{\mathcal{E}}{L}e^{-t/\tau}$ $\dfrac{dI}{dt} = \dfrac{45.0\text{ V}}{7.40\text{ H}}e^{-2.00/0.569} = \boxed{0.181\text{ A/s}}$

CHECK and THINK

The current increases from zero and exponentially approaches a steady-state value. Initially, the current is changing most rapidly. After a long time, the current is constant and so the *rate of change* is eventually zero. Indeed, we see that the rate of change is smaller after two seconds than its initial value.

83. (N) An electronic device is composed of an *RLC* circuit with frequency *f* = 60.0 Hz. The capacitor in the circuit goes bad and must be replaced, but the engineer does not know the capacitance. If she wants the impedance to be 1000.0 Ω but knows only that *R* = 500.0 Ω and *L* = 2.50 H, find the two values of capacitance that will solve her dilemma. (There are two answers to this problem!)

INTERPRET and ANTICIPATE

Because we know the impedance and have enough information to find the capacitive and inductive reactance, we need to set up Eq. 33.81 using Eq. 33.39 and Eq. 33.61, and solve for the unknown capacitance. As to why we should expect two values for C, note that after we square both sides of Eq. 33.81, we will eventually have to take the square root of both sides of the equation later in order to solve for the capacitive reactance. When we take the square root of an equation while solving for an unknown quantity, we should always consider that the other side of the equation could have been positive or negative. This may lead to two possible answers for the variable. Sometimes, only one of those mathematical answers will be physically reasonable, but in this case, as the problem suggests, there should be two physically reasonable answers.

SOLVE	
First, write Eq. 33.81, and substitute Eq. 33.39 and Eq. 33.61 for the capacitive and inductive reactances.	$$Z = \sqrt{\left(X_L - X_C\right)^2 + R^2} = \sqrt{\left(\omega L - \left(1/\omega C\right)\right)^2 + R^2}$$
Now, we begin the process of solving for the capacitance. Note that when we take the square root of both sides of the equation while trying to solve for C, we must include the possibility that the result could have been negative or positive. Then, in the panels, below, we will solve each of these possibilities separately.	$$Z^2 = \left(\omega L - \left(1/\omega C\right)\right)^2 + R^2$$ $$Z^2 - R^2 = \left(\omega L - \left(1/\omega C\right)\right)^2$$ $$\sqrt{Z^2 - R^2} = \pm\left(\omega L - \left(1/\omega C\right)\right)$$
Solving now, using the $+\left(\omega L - \left(1/\omega C\right)\right)$, we find one value of capacitance.	$$\sqrt{Z^2 - R^2} = \omega L - \left(1/\omega C\right)$$ $$\left(1/\omega C\right) = \omega L - \sqrt{Z^2 - R^2}$$ $$\omega C = \frac{1}{\omega L - \sqrt{Z^2 - R^2}}$$ $$C = \left(\frac{1}{\omega}\right)\frac{1}{\omega L - \sqrt{Z^2 - R^2}}$$

Insert the values from the problem statement with $\omega = 2\pi f$ to find one value for C.	$C = \left(\dfrac{1}{2\pi(60.0 \text{ Hz})}\right)\dfrac{1}{2\pi(60.0 \text{ Hz})(2.50 \text{ H}) - \sqrt{(1000.0 \ \Omega)^2 - (500.0 \ \Omega)^2}}$ $C = \boxed{3.47 \times 10^{-5} \text{ F}}$
Solving now, using the $-(\omega L - (1/\omega C))$, we find the other value of capacitance.	$\sqrt{Z^2 - R^2} = -\left(\omega L - (1/\omega C)\right)$ $(1/\omega C) = \sqrt{Z^2 - R^2} + \omega L$ $\omega C = \dfrac{1}{\sqrt{Z^2 - R^2} + \omega L}$ $C = \left(\dfrac{1}{\omega}\right)\dfrac{1}{\sqrt{Z^2 - R^2} + \omega L}$
Insert the values from the problem statement with $\omega = 2\pi f$ to find the other value for C.	$C - \left(\dfrac{1}{2\pi(60.0 \text{ Hz})}\right)\dfrac{1}{\sqrt{(1000.0 \ \Omega)^2 - (500.0 \ \Omega)^2} + 2\pi(60.0 \text{ Hz})(2.50 \text{ H})}$ $C = \boxed{1.47 \times 10^{-6} \text{ F}}$

CHECK and THINK

In this case, both solutions turned out to be physically reasonable (they are both positive, as C must be, and real). In another scenario with a different resistor, inductor and required impedance, it is possible that only one of the two possible solutions would be possible.

34

Maxwell's Equations and Electromagnetic Waves

3. (N) A circular coil of radius 0.50 m is placed in a time-varying magnetic field $B(t) = (5.80 \times 10^{-4}) \sin\left[(12.6 \times 10^2 \, \text{rad/s})t\right]$ where B is in teslas. The magnetic field is perpendicular to the plane of the coil. Find the magnitude of the induced electric field in the coil at $t = 0.001$ s and $t = 0.01$ s.

INTERPRET and ANTICIPATE	
We can use Faraday's law, which relates the change in magnetic flux to the induced electric field.	

SOLVE Faraday's law is given by Equation 34.5 or, after inserting the definition of flux, Eq. 34.6.	$\oint \vec{E} \cdot d\vec{\ell} = -\dfrac{d\Phi_B}{dt}$ $\oint \vec{E} \cdot d\vec{\ell} = -\dfrac{d}{dt} \int \vec{B} \cdot d\vec{A}$
The induced electric field along the coil is constant, so the integral on the left is just the magnitude of the field times the length around the coil. On the right, the area of the circular coil is πr^2. The area is constant in time but the magnetic field is changing. This allows us to solve for E. The sign simply indicates the direction around the ring, but since we are only asked to find the magnitude, we drop the negative sign.	$E(2\pi r) = -\dfrac{d}{dt}(B\pi r^2)$ $E(2\pi r) = -\pi r^2 \dfrac{dB}{dt}$ $|E| = \dfrac{r}{2}\dfrac{dB}{dt}$

Chapter 34 – Maxwell's Equations and Electromagnetic Waves

We can now calculate the derivative of the magnetic field.	$E = \dfrac{r}{2}\dfrac{d}{dt}\left[B_0 \sin(\omega t)\right]$
	$E = \dfrac{1}{2} r\omega B_0 \cos(\omega t)$

Calculate the factor in front, noting that the frequency is $\omega = 12.6 \times 10^2$ rad/s.

$$\frac{1}{2}r\omega B_0 = \frac{1}{2}(0.50 \text{ m})(12.6\times10^2 \text{ rad/s})(5.80\times10^{-4} \text{ T}) = 0.18 \text{ V/m}$$

So, with t measured in seconds,

$$E = 0.18 \cos\left[\left(12.6\times10^2\right)t\right] \text{ V/m}$$

Find the induced electric field at $t = 0.001$ s. Remember that the number after the cosine is in radians and not degrees.	$E(t = 0.001 \text{ s}) = 0.18 \cos\left[\left(12.6\times10^2\right)(0.001 \text{ s})\right]$ V/m
	$E(t = 0.001 \text{ s}) = \boxed{0.056 \text{ V/m}}$
Next at $t = 0.01$ s.	$E(t = 0.01 \text{ s}) = 0.18 \cos\left[\left(12.6\times10^2\right)(0.01 \text{ s})\right]$ V/m
	$E(t = 0.01 \text{ s}) = \boxed{0.18 \text{ V/m}}$

CHECK and THINK
Notice that the flux is changing due to the changing magnetic field. The flux and the induced electric field are oscillating, which means the induced current is an alternating current.

9. (N) A capacitor with square plates, each with an area of 36.0 cm^2 and plate separation $d = 2.54$ mm, is being charged by a 265-mA current. **a.** What is the change in the electric flux between the plates as a function of time? **b.** What is the magnitude of the displacement current between the capacitor's plates?

INTERPRET and ANTICIPATE
As the charge increases, the electric field between the plates increases and therefore the flux increases. The change of flux determines the displacement current.

SOLVE **a.** We can first express the electric field between the capacitor plates.	$E = \dfrac{\sigma}{\varepsilon_0} = \dfrac{Q}{A\varepsilon_0}$

285

The electric flux is equal to the electric field times the area.	$$\Phi_E = EA = \frac{Q}{\varepsilon_0}$$	
We can now determine the derivative with respect to time. Note that the rate of change of the charge in time, dQ/dt, is the current.	$$\frac{d\Phi_E}{dt} = \frac{dQ/dt}{\varepsilon_0} = \frac{I}{\varepsilon_0}$$	(1)
Insert numbers.	$$\frac{d\Phi_E}{dt} = \frac{(0.265 \text{ A})}{8.85 \times 10^{-12} \text{ C}^2/\text{N}\cdot\text{m}^2}$$ $$\frac{d\Phi_E}{dt} = \boxed{2.99 \times 10^{10} \text{ V}\cdot\text{m/s}}$$	
b. We write the formula for displacement current and then use Eq. (1). The displacement current, associated with the changing electric flux between the capacitor plates, equals the current carrying charge into the capacitor.	$$I_d = \varepsilon_0 \frac{d\Phi_E}{dt}$$ $$I_d = \varepsilon_0 \left(\frac{I}{\varepsilon_0}\right) = I = \boxed{0.265 \text{ A}}$$	

CHECK and THINK
We are able to calculate the electric flux and then the displacement current associated with the change in flux. We see that it equals the current into the capacitor.

14. (N) A circular conductor encloses a uniform magnetic field that is perpendicular to the page (not shown), pointing inward and increasing. The magnetic field outside the circular loop is zero. Suppose the changing magnetic flux induces a 1.0-V emf in the conductor and the conductor's resistance is 1.0 Ω. Imagine connecting two voltmeters to the conductor (Fig. P34.13). The clockwise distance from A to B is one-fourth the circumference. Because the resistance of a wire is proportional to its length, this portion of the loop has resistance $R_1 = 0.25$ Ω. Likewise, the clockwise distance from B to A is three-fourths the circumference, so this portion's resistance is $R_2 = 0.75$ Ω. It may seem incredible, but what a voltmeter measures depends on its placement; in Problem 13, we showed that $V_2 - V_1 = \mathcal{E}$. Find the voltage measured by both meters. *Hint*: Kirchhoff's loop rule (and other rules from Chapter 29) may be applied to a loop that does not enclose the changing magnetic flux.

Chapter 34 – Maxwell's Equations and Electromagnetic Waves

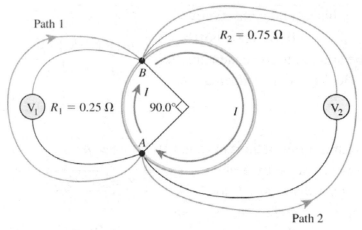

Figure P34.13

INTERPRET and ANTICIPATE The key to doing this problem is knowing when you can use Kirchhoff's loop rule and when you cannot. If we consider a loop that does not enclose the magnetic field then we can use Kirchhoff's loop rule. It is helpful to note that the current in the conducting loop is same everywhere in the loop since there are no branches.	
SOLVE Using the hint, apply Kirchhoff's loop rule to the loop that includes voltmeter 1 and the short portion of the conductor from A to B with Ohm's law. As the loop does not include a changing magnetic field, there is no induced electric field that produces a nonconservative field. Since the current is from the black to the red lead of the voltmeter, the voltmeter reads a negative potential difference.	$V_1 = -IR_1 \qquad (1)$
Likewise, we can use Kirchhoff's loop rule to the loop that runs through voltmeter 2 and larger portion of the conductor from A to B. Since the current is from the red to the black lead in this case, voltmeter 2 reads a positive potential difference.	$V_2 = IR_2 \qquad (2)$
In Problem 13, we prove that $V_2 - V_1 = \mathcal{E}$. Substitute Equations (1) and (2) into this expression to find the current in the conductor.	

Notice that because this is a nonconservative electric field, the $V_2 - V_1 \neq 0$. So Kirchhoff's loop rule cannot be applied to the conductor, as this is an induced electric field producing the voltage.	$V_2 - V_1 = \mathcal{E}$ $IR_2 - (-IR_1) = \mathcal{E}$ $I(R_1 + R_2) = \mathcal{E}$ $I = \dfrac{\mathcal{E}}{R_1 + R_2} = \dfrac{1.0 \text{ V}}{(0.25 + 0.75) \ \Omega} = 1.0 \text{ A}$
Substitute the current in to Equations (1) and (2) to find the measurement made by each voltmeter. The results reflect the 1 V total around the conductor and the fact that V_1 spans 1/4 of the ring and V_2 3/4 of the ring.	$V_1 = -IR_1 = -(1.0 \text{ A})(0.25 \ \Omega)$ $V_1 = \boxed{-0.25 \text{ V}}$ $V_2 = IR_2 = (1.0 \text{ A})(0.75 \ \Omega)$ $V_2 = \boxed{0.75 \text{ V}}$

CHECK and THINK

This problem gives us a chance to build up our intuition about nonconservative forces. The induced electric field is nonconservative and that means that where you put your voltmeter matters. Both voltmeters are connected at points A and B. You would expect to get the same reading. That is because your expectation is based on your laboratory work with conservative electric fields. A conservative electric field results from the excess of charge in some region such as the excess charge that you find on plates of capacitor. The nonconservative electric field results from a changing magnetic field.

15. (N) What is the acceleration of a proton moving with velocity $\vec{v} = 15.0\,\hat{j}$ m/s through a region that has magnetic field $\vec{B} = 1.20\hat{i}$ T and electric field $\vec{E} = \left(3.00\,\hat{j} + 2.00\hat{k}\right)$ V/m ?

INTERPRET and ANTICIPATE

The force on a charge in an electric and magnetic field is the Lorentz force. Given a charge, the electric force depends on the electric field and the magnetic force depends on the velocity and magnetic field. We can then calculate the acceleration with Newton's second law.

SOLVE The force on a charge in an electric and magnetic field is called the Lorentz force and is given by Eq. 30.21.	$\vec{F} = q\vec{E} + q\vec{v} \times \vec{B}$

Using Newton's second law, we can determine the acceleration. We also insert the charge of the proton, $q = +e$.	$\vec{a} = \dfrac{\vec{F}}{m} = \dfrac{e}{m}\left[\vec{E} + \vec{v}\times\vec{B}\right]$
Calculate the cross product.	$\vec{v}\times\vec{B} = \begin{vmatrix} \hat{i} & \hat{j} & \hat{k} \\ 0 & 15.0 & 0 \\ 1.20 & 0 & 0 \end{vmatrix} = -18.0\,\hat{k}$
Insert values into the formula for acceleration above.	$\vec{a} = \dfrac{\left(1.60\times10^{-19}\right)}{1.67\times10^{-27}}\left[\left(3.00\hat{j} + 2.00\hat{k}\right) + \left(-18.0\hat{k}\right)\right]$ $\vec{a} = \left(9.58\times10^{7}\right)\left[3.00\hat{j} - 16.0\hat{k}\right]$ $\vec{a} = \boxed{\left(2.87\times10^{8}\,\hat{j} - 15.3\times10^{8}\,\hat{k}\right)\ \text{m/s}^2}$

CHECK and THINK

Given the charge, velocity, and fields we can determine the Lorentz force and then the acceleration.

20. (N) An electromagnetic wave is given by $E(x,t) = 3.75\sin\left(0.60x - \omega t\right)\,\text{V/m}$ in SI units.

a. What is the angular frequency?

INTERPRET and ANTICIPATE	
The equation given for the electric field is a wave equation that describes a wave traveling in space. The wavenumber $k = 0.60$ rad/m and the angular frequency ω are related to the speed of the wave, i.e. the speed of light. We would expect the angular frequency to be a large number in SI units for an electromagnetic wave.	

SOLVE Eq. 34.20 relates the wave number and angular frequency to the speed of light for an electromagnetic wave.	$c = \dfrac{\omega}{k} \quad \rightarrow \quad \omega = ck$
Insert numerical values.	$\omega = \left(3.00\times10^{8}\ \text{m/s}\right)\left(0.60\ \text{rad/m}\right)$ $\omega = \boxed{1.8\times10^{8}\ \text{rad/s}}$

Chapter 34 – Maxwell's Equations and Electromagnetic Waves

CHECK and THINK	
The angular frequency is a very large number, as expected. In the electromagnetic spectrum (Fig. 34.11), we see that that familiar electromagnetic waves have angular frequencies in the range of 10^8 to 10^{21} Hz.	

b. What is the magnetic field at $x = 2.0$ m and $t = 3.0$ s?

INTERPRET and ANTICIPATE	
The magnetic field and electric fields for an electromagnetic wave are related, with the magnetic field magnitude a factor of the speed of light smaller.	
SOLVE Apply. Eq. 34.24.	$E(x,t) = cB(x,t) \quad \rightarrow \quad B(x,t) = \dfrac{E(x,t)}{c}$
Insert numerical values with $x = 2.0$ m and $t = 3.0$ s. We use SI units in each case and therefore the answer will be in the SI unit for magnetic field (tesla).	$B(x,t) = \dfrac{3.75}{3.00 \times 10^8} \sin\big((0.60)(2.0) - (1.8 \times 10^8)(3.0)\big)$ $B(x,t) = \boxed{-5.9 \times 10^{-9} \text{ T}}$

CHECK and THINK	
The magnetic field is proportional to the electric field, which we can calculate using the expression provided.	

24. (A) Write equations for both the electric and magnetic fields for an electromagnetic wave in the red part of the visible spectrum that has a wavelength of 710 nm and a peak electric field magnitude of 2.5 V/m.

INTERPRET and ANTICIPATE	
The electric and magnetic fields are traveling waves. Equations can be written in terms of the wavelength and frequency as well as the peak electric and magnetic fields. The speed of light can be used to determine the frequency based on the wavelength and the magnetic field magnitude in terms of the electric field magnitude.	
SOLVE The electric and magnetic fields can be described using Equations 34.18 and 34.19.	$E(x,t) = E_{max} \sin(kx - \omega t)$ $B(x,t) = B_{max} \sin(kx - \omega t)$

We can determine the angular frequency and wave number (refer back to Chapter 17 if you would like to review these equations). The frequency is related to the wavelength by the speed of light (Eq. 34.20).	$k = \dfrac{2\pi}{\lambda} = \dfrac{2\pi}{710 \text{ nm}} = 8.8 \times 10^6 \text{ m}^{-1}$ $f = \dfrac{c}{\lambda} = \dfrac{3 \times 10^8 \text{ m/s}}{710 \text{ nm}} = 4.2 \times 10^{14} \text{ Hz}$ $\omega = 2\pi f = 2\pi \left(\dfrac{3 \times 10^8 \text{ m/s}}{710 \text{ nm}} \right) = 2.7 \times 10^{15} \text{ Hz}$
We can also relate the peak electric field to the peak magnetic field with Eq. 34.23.	$E_{max} = cB_{max} = 2.5 \ \dfrac{\text{V}}{\text{m}}$ $B_{max} = \dfrac{E_{max}}{c} = \dfrac{2.5 \ \dfrac{\text{V}}{\text{m}}}{3.00 \times 10^8 \text{ m/s}} = 8.3 \times 10^{-9} \text{ T}$

We now have all the pieces to write equations for the electric and magnetic fields.

$$E(x,t) = \boxed{2.5 \sin\left(\left(8.8 \times 10^6 \text{ m}^{-1}\right)x - \left(2.7 \times 10^{15}\text{ Hz}\right)t \right) \text{ V/m}}$$

$$B(x,t) = \boxed{8.3 \times 10^{-9} \sin\left(\left(8.8 \times 10^6 \text{ m}^{-1}\right)x - \left(2.7 \times 10^{15}\text{ Hz}\right)t \right) \text{ T}}$$

CHECK and THINK

Given the wavelength and electric field magnitude, we were able to write equations for the electric and magnetic waves that correspond to these values.

28. (N) Suppose the magnetic field of an electromagnetic wave is given by

$$B = \left(1.5 \times 10^{-10}\right) \sin\left(kx - \omega t\right) \text{ T}.$$

a. What is the maximum energy density of the magnetic field of this wave?
b. What is the maximum energy density of the electric field?

INTERPRET and ANTICIPATE

The maximum energy density in the magnetic field can be found using Eq. 33.17. Note that the coefficient in the equation for the magnetic field as a function of time is the maximum value of the magnetic field. By Eq. 34.26 the maximum energy density of the electric field is the same as the maximum energy density of the magnetic field, which will allow us to answer part (b) quickly after answering part (a).

SOLVE

a. Use Eq. 33.17 to find the maximum energy density in the magnetic field.	$u_B = \dfrac{B^2}{2\mu_0} = \dfrac{\left(1.5 \times 10^{-10} \text{ T}\right)^2}{2\left(4\pi \times 10^{-7} \text{ T} \cdot \text{m/A}\right)} = \boxed{9.0 \times 10^{-15} \text{ J/m}^3}$

b. Then, based on Eq. 34.26 and the properties of an electromagnetic wave, the maximum energy density of the electric field must be the same as that of the magnetic field.	$u_E = u_B = \boxed{9.0 \times 10^{-15} \text{ J/m}^3}$

CHECK and THINK

The energy is stored, and shared, equally in both fields that comprise an electromagnetic wave. As the fields oscillate, the amount of energy changes, though we can say that the average total energy density in the wave is equal to one-half the maximum total energy density, or either the maximum energy density of the magnetic field, or the maximum energy density of the electric field.

32. (N) An electromagnetic wave is traveling in the positive x direction with the electric field oscillating along the z direction. If the wavelength is 555 nm, $E_{max} = 0.050$ V/m, and the wave is in a vacuum, write equations describing the electric and magnetic fields that make up the electromagnetic wave as functions of x and t.

INTERPRET and ANTICIPATE

We will use Eq. 34.18 and Eq. 34.19 to write the electric and magnetic fields of the wave as functions of time. Both fields of the wave must have the same time dependence, and thus the same wave number and angular frequency. We will begin by using the given wavelength with Eq. 34.20 and $\omega = 2\pi f$ to find the angular frequency and wave number of each field. The speed of the wave in vacuum will be c. Also, note that because the fields must be perpendicular, the magnetic field must oscillate along the y direction, since the electric field oscillates along the z direction and the wave travels in the x direction.

SOLVE	
Use Eq. 34.20 to find and expression for the frequency of the wave.	$c = \lambda f$ $f = c/\lambda = (3.00 \times 10^8 \text{ m/s})/(555 \times 10^{-9} \text{ m})$
Then, use $\omega = 2\pi f$ to find the angular frequency.	$\omega = 2\pi f$ $\omega = 2\pi((3.00 \times 10^8 \text{ m/s})/(555 \times 10^{-9} \text{ m}))$ $\omega = 3.40 \times 10^{15}$ rad/s
Now, go back to Eq. 34.20 to find the wave number, k.	$k = \omega/c$ $k = (2\pi((3.00 \times 10^8 \text{ m/s})/(555 \times 10^{-9} \text{ m})))/(3.00 \times 10^8 \text{ m/s})$ $k = 1.13 \times 10^7 \text{ m}^{-1}$

Given that we know the maximum electric field strength, we can now use Eq. 34.18 to write the electric field as a function of time. We note that the wave is traveling along the positive x direction, so that is the spatial variable along which the wave travels. The sign of the kx and ωt terms must be opposite since the wave travels in the positive direction. Also, the electric field component is oscillating along the z direction.	$\vec{E}(x,t) = \left[(0.050 \text{ V/m}) \sin\left((1.13 \times 10^7 \text{ m}^{-1})x - (3.40 \times 10^{15} \text{ rad/s})t \right) \right] \hat{k}$
The time dependence of the magnetic field must be the same as that for the electric field. However, we need to find the magnitude of the maximum magnetic field using Eq. 34.23.	$B_{max} = E_{max}/c = (0.050 \text{ V/m})/(3.00 \times 10^8 \text{ m/s})$ $B_{max} = 1.7 \times 10^{-10} \text{ T}$
Now, we can write the function for the magnetic field using Eq. 34.19.	$\vec{B}(x,t) = \left[(1.7 \times 10^{-10} \text{ T}) \sin\left((1.13 \times 10^7 \text{ m}^{-1})x - (3.40 \times 10^{15} \text{ rad/s})t \right) \right] \hat{j}$

CHECK and THINK
The magnetic and electric fields that comprise an electromagnetic wave must be in phase with each other (have their maximums always occur at the same moment in time), be perpendicular to each other, travel with the same velocity, and have the same frequency, wavelength, angular frequency, and wave number. Knowing that the maximum values of the fields are related to each other via the speed of light in vacuum allows us to more easily construct the equations that describe the fields as functions of time and space.

41. (N) Find the intensity of the electromagnetic wave described in each case: An electromagnetic wave with **a.** a wavelength of 710 nm and a peak electric field magnitude of 2.5 V/m and **b.** an angular frequency of 7.5×10^{18} rad/s and a peak magnetic field magnitude of 10^{-10} T.

INTERPRET and ANTICIPATE

The intensity of the electromagnetic wave can be determined given the peak value of the electric or magnetic field—the larger the field, the higher the intensity.

SOLVE The intensity of an electromagnetic wave depends on the peak electric and magnetic field as expressed in Equation 34.30.	$I = \dfrac{E_{max}^2}{2\mu_0 c} = \dfrac{c}{2\mu_0} B_{max}^2$
a. The peak electric field for this wave is 2.5 V/m.	$I = \dfrac{E_{max}^2}{2\mu_0 c} = \dfrac{\left(2.5\ \dfrac{V}{m}\right)^2}{2\left(4\pi \times 10^{-7}\ \dfrac{N}{A^2}\right)\left(3\times 10^8\ \dfrac{m}{s}\right)}$ $I = \boxed{0.0083\ \text{W/m}^2}$
b. The peak magnetic field for this wave is 10^{-10} T.	$I = \dfrac{c}{2\mu_0} B_{max}^2 = \dfrac{\left(3\times 10^8\ \dfrac{m}{s}\right)\left(10^{-10}\ T\right)^2}{2\left(4\pi \times 10^{-7}\ \dfrac{N}{A^2}\right)}$ $I = \boxed{1.2\times 10^{-6}\ \text{W/m}^2}$

CHECK and THINK

The amplitude of the magnetic or electric field can be used to determine the intensity of the light. We see that the wavelength and angular frequency actually don't come into the calculation at all. The intensity of other types of waves (for instance sound) behaves similarly actually in that it depends on the amplitude of the wave.

43. (N) You wish to send a probe to the Moon. The probe has a radio transmitter that you test in the laboratory. When the probe is 10 m from your receiver, the intensity is I_0. When the probe is on the Moon, it sends radio waves to you. If you want to receive radio waves of the same intensity, how much stronger must the probe's electric field be when it is on the Moon?

INTERPRET and ANTICIPATE We will assume that the radio transmitter radiates in all directions. At points further from the transmitter, the energy is spread over a larger area. Therefore, in order to have the same intensity (power per unit area), the transmitter that is on the moon (and much further away) must have a much larger power.	
SOLVE Similar to Example 34.2, we take the transmitter to be radiating in all directions such that the intensity falls off as $1/r^2$, as in Eq. 17.23. We also know that the intensity is related to the electric field according to Eq. 34.30.	$$I = \frac{P_{av}}{4\pi r^2} = \frac{E_{max}^2}{2\mu_0 c}$$
We're told that the intensity is I_0 at a distance of 10 m. We want to increase the power from P_0 to a larger value P_{moon} so that we detect the same intensity when the transmitter is on the moon.	$$I_0 = \frac{P_0}{4\pi (10 \text{ m})^2} = \frac{P_{moon}}{4\pi \left(r_{Earth\text{-}moon}\right)^2}$$
Find the ratio of the high power transmitter placed on the moon to the lower power transmitter measured in the lab. The moon is 3.84×10^8 m from Earth. Not surprisingly, we need a *much* higher power when it's on the moon to measure the same intensity on Earth.	$$\frac{P_{moon}}{P_0} = \left(\frac{r_{Earth\text{-}moon}}{10 \text{ m}}\right)^2$$ $$\frac{P_{moon}}{P_0} = \left(\frac{3.84 \times 10^8 \text{ m}}{10 \text{ m}}\right)^2 = 1.5 \times 10^{15}$$
For the lower power transmitter, we can express the electric field using Eq. 34.30.	$$\frac{P_0}{4\pi (10 \text{ m})^2} = \frac{E_{max,0}^2}{2\mu_0 c} \qquad (1)$$
Now, we imagine that we are on the moon, 10 m from the higher power transmitter and express the same relationship.	$$\frac{P_{moon}}{4\pi (10 \text{ m})^2} = \frac{E_{max,moon}^2}{2\mu_0 c} \qquad (2)$$
Finally, take the ratio of Equations (1) and (2) and solve for the ratio of the electric field near the higher power transmitter on the moon relative to the lower power one on the Earth.	$$\frac{P_{moon}}{P_0} = \frac{E_{max,moon}^2}{E_{max,0}^2}$$ $$\frac{E_{max,moon}}{E_{max,0}} = \sqrt{\frac{P_{moon}}{P_0}} = \sqrt{1.5 \times 10^{15}} \approx \boxed{4 \times 10^7}$$

295

CHECK and THINK
The electric field is indeed much stronger—over 10 billion times stronger!

49. (C) Determine the direction of energy flow for the following four cases, where $\vec{E}$ and $\vec{B}$ are the electric and magnetic fields, respectively:

a. $\vec{E} = E\hat{j}, \vec{B} = B\hat{k}$;

b. $\vec{E} = -E\hat{i}, \vec{B} = B\hat{k}$;

c. $\vec{E} = -E\hat{i}, \vec{B} = -B\hat{j}$;

d. $\vec{E} = -E\hat{i}, \vec{B} = -B\hat{k}$

INTERPRET and ANTICIPATE
Energy flows in the same direction as the wave propagation and can be found using the Poynting vector.

SOLVE	
Energy flows in the direction of the Poynting vector given by Eq. 34.27. In each case, we can use the right hand rule to determine the direction of $\vec{S}$.	$\vec{S} \equiv \dfrac{1}{\mu_0}\left(\vec{E} \times \vec{B}\right)$
a. Imagine a coordinate system with x pointing to the right, y upward, and z out of the page. To apply the right hand rule, point your fingers upward and curl them out of the page, in which case your thumb will point to the right along the $+x$ direction. This is the direction of energy flow.	$\vec{S} = \dfrac{1}{\mu_0}\left(E\hat{j} \times B\hat{k}\right) = \dfrac{EB}{\mu_0}\left(\hat{j} \times \hat{k}\right)$ $\vec{S} = \dfrac{EB}{\mu_0}\boxed{\hat{i}}$
b. Repeat this procedure for the other three. In this case, the energy flows in the $+y$ direction.	$\vec{S} = \dfrac{1}{\mu_0}\left(-E\hat{i} \times B\hat{k}\right) = \dfrac{-EB}{\mu_0}\left(\hat{i} \times \hat{k}\right)$ $\vec{S} = -\dfrac{1}{\mu_0}EB\left(-\hat{j}\right) = \dfrac{1}{\mu_0}EB\boxed{\hat{j}}$
c. The energy flows in the $+z$ direction.	$\vec{S} = \dfrac{1}{\mu_0}\left(\left(-E\hat{i}\right) \times \left(-B\hat{j}\right)\right) = \dfrac{EB}{\mu_0}\left(\hat{i} \times \hat{j}\right)$ $\vec{S} = \dfrac{EB}{\mu_0}\boxed{\hat{k}}$

d. The energy flows in the –x direction.	$\vec{S} = \dfrac{1}{\mu_0}\left(\left(-E\,\hat{i}\right)\times\left(-B\,\hat{k}\right)\right) = \dfrac{EB}{\mu_0}\left(\hat{i}\times\hat{k}\right)$ $\vec{S} = \dfrac{EB}{\mu_0}\left(-\hat{j}\right) = \dfrac{EB}{\mu_0}\left(\boxed{-\hat{j}}\right)$

CHECK and THINK

Notice that in each case, the wave travels perpendicular to *E* and *B*. That is, the electric and magnetic fields oscillate perpendicular to the direction that the light propagates.

54. (N) The intensity of sunlight on the Earth is about 1.00×10^3 W/m². The average orbital radius of Mercury is about 40% of the Earth's orbital radius. Using this information, determine the approximate intensity of the sunlight and the radiation pressure on Mercury.

INTERPRET and ANTICIPATE

Assuming that the sun is a point source, the intensity of light falls off as one over the square of the distance from the sun. Using this fact, we can determine the intensity and the radiation pressure on Mercury due to the sun.

SOLVE The intensity of light due to a point source is given by Equation 17.23.	$I = \dfrac{P_{av}}{4\pi r^2}$
We can use this to find the ratio of the intensity on Mercury compared to earth.	$\dfrac{I_M}{I_E} = \dfrac{r_E^2}{r_M^2} \quad\rightarrow\quad I_M = \left(\dfrac{r_E}{r_M}\right)^2 I_E$
Plug in numbers. The distance from the sun to the Earth is 1 AU (astronomical unit), so the distance from the sun to Mercury is 0.4 AU.	$I_M = \left(\dfrac{1\,\text{AU}}{0.4\,\text{AU}}\right)^2 \left(1.00\times10^3\ \text{W/m}^2\right) = \boxed{6.25\times10^3\ \text{W/m}^2}$
The radiation pressure, assuming the light is absorbed, can be determined using the intensity in Equation 34.38.	$P = \dfrac{I}{c}$ $P = \dfrac{6.25\times10^3\ \text{W/m}^2}{3.00\times10^8\ \text{m/s}} = \boxed{2.08\times10^{-5}\ \text{N/m}^2}$

Chapter 34 – Maxwell's Equations and Electromagnetic Waves

CHECK and THINK
Since the Earth is about 2.5 times further from the sun than Mercury, the intensity and radiation pressure are $(2.5)^2$ or about 6 times smaller on Earth.

57. (N) The accepted value of the intensity of sunlight at the Earth's surface is 1.36 kW/m^2. **a.** Determine the pressure exerted by sunlight if it strikes a perfect absorber. **b.** Determine the pressure exerted by sunlight if it strikes a perfect reflector.

INTERPRET and ANTICIPATE
In both cases, the pressure depends on the intensity of the light. A perfect reflector experiences a higher pressure than an absorber since the light experiences a greater change in momentum on reflection.

SOLVE	
a. For a perfect absorber, we use Eq. 34.38. The final unit for pressure is pascals (or, equivalently, N/m^2).	$P = \dfrac{I}{c}$ $$P = \dfrac{1.36 \times 10^3 \text{ W/m}^2}{3.00 \times 10^8 \text{ m/s}}$$ $P = \boxed{4.53 \times 10^{-6} \text{ Pa}}$
b. For a perfect reflector, we use Eq. 34.39.	$P = 2\dfrac{I}{c}$ $$P = 2\left(\dfrac{1.36 \times 10^3 \text{ W/m}^2}{3.00 \times 10^8 \text{ m/s}}\right)$$ $P = \boxed{9.07 \times 10^{-6} \text{ Pa}}$

CHECK and THINK
This is quite a small pressure of course—we wouldn't expect a large pressure from sunlight.

60. (N) A budding magician holds a 5.00-mW laser pointer, wondering whether he could use it to keep an object floating in the air with the radiation pressure. This might be an idea for a new trick! Assuming the laser pointer has a circular beam 3.00 mm in diameter and the magician rigs up a totally reflecting sail on which to shine the laser, what is the maximum weight the magician could suspend with this technique?

INTERPRET and ANTICIPATE

The laser intensity can be used to calculate the radiation pressure and the force exerted by the laser pointer. The maximum weight that could be suspended using this technique is equal to this total force.

SOLVE The radiation pressure on a body that is completely reflecting the radiation depends on the intensity of the light according to Equation 34.39.	$$P = 2\frac{I}{c}$$
The intensity is the power per unit area. (We write out power in this formula just to distinguish it from pressure in the equation above.)	$$I = \frac{\text{power}}{A} = \frac{5.00 \times 10^{-3} \text{ W}}{\pi\left(1.50 \times 10^{-3} \text{ m}\right)^2}$$
Now calculate the pressure.	$$P = 2\left(\frac{\dfrac{5.00 \times 10^{-3} \text{ W}}{\pi\left(1.50 \times 10^{-3} \text{ m}\right)^2}}{3.00 \times 10^8 \ \dfrac{\text{m}}{\text{s}}}\right) = \boxed{4.72 \times 10^{-6} \text{ Pa}}$$
The radiation pressure (which is a force per unit area) is exerted on an area equal to the area of the beam. Therefore, we can determine the total force as pressure times the cross-sectional area of the beam. This is the maximum weight that can be balanced by the radiation pressure.	$$F = PA = \left(2\frac{\dfrac{5.00 \times 10^{-3} \text{ W}}{\pi\left(1.50 \times 10^{-3} \text{ m}\right)^2}}{3.00 \times 10^8 \ \dfrac{\text{m}}{\text{s}}}\right)\left(\pi\left(2.00 \times 10^{-3} \text{ m}\right)^2\right)$$ $$F = \boxed{5.93 \times 10^{-11} \text{ N}}$$

CHECK and THINK

Even the intensity of a reasonably bright light source, like the laser pointer, leads to a pressure that is *really* small. Given the area, the weight that could be suspended is

5.93×10^{-11} N—corresponding to a mass of around 1 nanogram! This must include the weight of the sail, so this is *far* too small to suspend anything that would produce an impressive magic trick!

63. (N) The light in Problem 62 travels toward a workbench and encounters two polarizing filters.
a. If the light passes first through a filter with its transmission axis at an angle of 45.0° from the z direction, and then through a filter with its transmission axis at an angle of 90.0° from the z direction, what is the final intensity of the transmitted light?
b. If, instead, the light passes first through a filter with its transmission axis at an angle of 90.0° from the z direction, and then through a filter with its transmission axis at an angle of 45.0° from the z direction, what is the final intensity of the transmitted light?

INTERPRET and ANTICIPATE
In the first case (part (a)), we must apply Eq. 34.40 twice; once for the first filter where the angle is 45.0°, and then again for the light that exits the first filter and hits the second. In the second case, the angle will again be 45.0° because the light is polarized along the transmission axis after passing through the first filter. The angle between the polarization of the light after the first filter and the next transmission axis is $90.0° - 45.0° = 45.0°$. Thus, we expect a nonzero final intensity.

We do not expect a nonzero final intensity in the second case (part (b)) because the incident light's polarization is 90.0° from the transmission axis of the first filter.

SOLVE	
a. Apply Eq. 34.40 to the light as it passes through the first filter.	$I_{f1} = I_i \cos^2 \varphi = \left(1409 \text{ W/m}^2\right)\cos^2\left(45.0°\right)$
Then, light with that resulting intensity is incident on the second filter, such that it's polarization is 45.0° from the transmission axis of the second filter. Apply Eq. 34.40 to that filter and get the final intensity of the light.	$I_{f2} = I_{f1} \cos^2 \varphi$ $I_{f2} = \left[\left(1409 \text{ W/m}^2\right)\cos^2\left(45.0°\right)\right]\cos^2\left(45.0°\right)$ $I_{f2} = \boxed{352 \text{ W/m}^2}$
b. Apply Eq. 34.40 to the incident light on the first polarizer at an angle of 90.0°. We find there is no need to continue and determine the effect of the second polarizer because the intensity of the light has become 0.	$I_f = I_i \cos^2 \varphi = \left(1409 \text{ W/m}^2\right)\cos^2\left(90.0°\right) = \boxed{0}$

CHECK and THINK
The polarization of the light is always parallel to the transmission axis after it passes through a polarizing filter. In order to apply Eq. 34.40, we must reevaluate the angle between the polarization of the wave and the next transmission axis after the light passes through each filter.

69. (N) What is the amplitude of the magnetic field a distance of 3.50 km away from a radio station that is broadcasting isotropically (in all directions) with a power of 43.0 kW?

INTERPRET and ANTICIPATE
The intensity is power per unit area. Since the power of the source is fixed, the intensity decreases as one over distance squared, as the energy is spread over a larger area (the area of a spherical surface $4\pi r^2$). The intensity can also be expressed in terms of the maximum electric or magnetic field of the wave. We can set these equal to each other to determine the magnetic field that corresponds to the intensity for this power and distance.

SOLVE	
The intensity can be expressed in terms of the maximum magnetic field (which is what we want to find) using Eq. 34.30.	$I = \dfrac{c}{2\mu_0} B_{max}^2$
It can also be expressed in terms of power and distance from the emitter with Eq. 17.23.	$I = \dfrac{P}{4\pi r^2}$
Set these expressions equal to each other and solve for B_{max},	$\dfrac{c}{2\mu_0} B_{max}^2 = \dfrac{P}{4\pi r^2}$ $B_{max} = \sqrt{\left(\dfrac{P}{4\pi r^2}\right)\left(\dfrac{2\mu_0}{c}\right)}$
Now insert numerical values.	$B_{max} = \sqrt{\dfrac{(43.0\times10^3\ \text{W})}{4\pi(3.50\times10^3\ \text{m})^2}\dfrac{2(4\pi\times10^{-7}\ \text{T}\cdot\text{m/A})}{(3.00\times10^8\ \text{m/s})}}$ $B_{max} = \boxed{1.53\times10^{-9}\ \text{T}} = 1.53\ \text{nT}$

CHECK and THINK

The intensity can be expressed in terms of the power per unit area or in terms of the magnetic field strength. Therefore, we are able to determine the maximum magnetic field associated with the desired intensity.

---

70. (N) The magnetic field of an electromagnetic wave is given by

$$B = 1.5 \times 10^{-10} \sin(kx - \omega t)\,\text{T}.$$

a. If the wavelength is 752 nm, what are the frequency, angular frequency, and wave number of this wave? **b.** What is the maximum total energy density of this wave?

INTERPRET and ANTICIPATE

Using the wavelength and Eq. 34.20 with $\omega = 2\pi f$, we can find the frequency of the wave, the angular frequency, and the wave number. The coefficient in the equation for the magnetic field represents the maximum magnetic field. We can use this maximum field strength to find the total energy density of the electromagnetic wave. We will use the speed of light in vacuum, c, for the speed of the wave.

SOLVE	
a. Using the wavelength and Eq. 34.20, we can find the frequency of the wave.	$f = c/\lambda = (3.00 \times 10^8 \text{ m/s})/(752 \times 10^{-9} \text{ m}) = \boxed{3.99 \times 10^{14} \text{ Hz}}$
Then, we can use the relationship between angular frequency and frequency.	$\omega = 2\pi f = 2\pi(3.99 \times 10^{14} \text{ Hz}) = \boxed{2.51 \times 10^{15} \text{ rad/s}}$
Now, use Eq. 34.20 again to find the wave number.	$k = \omega/c = 2\pi(3.99 \times 10^{14} \text{ Hz})/(3.00 \times 10^8 \text{ m/s})$ $k = \boxed{8.36 \times 10^6 \text{ m}^{-1}}$
b. Use Eq. 34.31 and the maximum magnetic field given to find the energy density of the wave.	$u = \dfrac{1}{2\mu_0} B_{max}^2 = \dfrac{1}{2(4\pi \times 10^{-7} \text{ T·m/A})}(1.5 \times 10^{-10} \text{ T})^2$ $u = \boxed{9.0 \times 10^{-15} \text{ J/m}^3}$

CHECK and THINK

The energy density of the wave can be found using either the maximum magnetic field, or the maximum electric field. The energy is stored in the both fields. Also, Eq. 34.20 provides a set of useful relationships between the properties of the electromagnetic wave, which could be used to help us complete the wave description of the electric, or magnetic fields.

71. (N) Household solar panels typically have an efficiency of 15.0%. If the intensity of sunlight is assumed to be a constant 9.80×10^2 W/m^2, what should be the total area of rooftop solar panels on a house in order to supply an average of 33.0 kW each day? Efficiency denotes the percentage of incident solar energy that is converted to electricity by the solar panel.

INTERPRET and ANTICIPATE

We are given the average solar intensity and what percentage the panels can extract as electrical power, so it should be straightforward to determine the area needed to achieve the desired power level.

SOLVE

If we could extract *all* of the energy from the sunlight, we could get a power out that equals the intensity times the area. (Intensity is measured in kW/m^2, so multiplying by an area results in the power output.) The power that is *actually* converted into electrical energy is 15% of the total available, so we multiply by a factor of 0.15.	$P_{electrical} = 0.15\,IA$
Rearrange and solve for the area.	$A = \dfrac{P_{electrical}}{0.15I} = \dfrac{33000 \text{ W}}{0.15\left(980\ \dfrac{\text{W}}{\text{m}^2}\right)} = \boxed{224 \text{ m}^2}$

CHECK and THINK

This area is quite large compared to the rooftops of most houses!

---

74. (E) Case Study The region around the Earth is filling up with space junk such as old satellites. One idea for cleaning up space involves using sails that create drag (Fig. P34.74). Perhaps one day satellites will be equipped with sails that are deployed at the end of their missions. The NASA mission NanoSail-D was launched in 2010 to test this idea. This problem compares the drag on a solar sail due to the Earth's upper atmosphere with the force exerted on the sail by the Sun's radiation and with the gravitational force exerted on the satellite by the Earth. Nano-Sail-D orbited at an altitude of 6.5×10^5 m, and its solar sail had an area of 10 m^2. Assume the satellite's total mass was about 4 kg and it was in a circular orbit. Also assume the Earth's atmosphere at that altitude has a density of about 5×10^{-14} kg/m^3 and a drag coefficient $C \approx 1$, and the Earth's gravitational force provided the centripetal acceleration. Finally, assume the sail perfectly reflected the sunlight. Estimate the magnitude of the three forces exerted on the satellite.

Figure P34.74

INTERPRET and ANTICIPATE This problem involves material from Chapter 6 on drag and from Chapter 7 on gravity. You may want to refer back to those chapters while completing this problem. It will be interesting to see how these three forces compare!	

SOLVE Let's start with the gravitational force. We need the distance from the center of the Earth up to the satellite. We can then apply Newton's universal law of gravitation. The resulting force is around 30 N.	$r = R_\oplus + h$ $r = 6.38 \times 10^6 \text{ m} + 6.5 \times 10^5 \text{ m} = 7.0 \times 10^6 \text{ m}$ $F_G = G \dfrac{M_\oplus m}{r^2}$ $F_G = \left(6.67 \times 10^{-11} \ \dfrac{\text{N} \cdot \text{m}^2}{\text{kg}^2}\right) \dfrac{\left(5.97 \times 10^{24} \text{ kg}\right)\left(4 \text{ kg}\right)}{\left(7.0 \times 10^6 \text{ m}\right)^2}$ $F_G = 33 \text{ N} \approx \boxed{30 \text{ N}}$
Assume that the sail works something like a parachute, such that the drag force is proportional to the velocity squared.	$F_D = \dfrac{1}{2} C\rho A v^2$
We need to know the satellite's speed. If the gravitational force is responsible for the centripetal acceleration, we can find the speed that allows the satellite to stay in circular orbit.	$a_c = \dfrac{v^2}{r} = \dfrac{F_G}{m}$ $v = \sqrt{\dfrac{F_G r}{m}} = \sqrt{\dfrac{\left(33 \text{ N}\right)\left(7.0 \times 10^6 \text{ m}\right)}{4 \text{ kg}}}$ $v = 7.6 \times 10^3 \text{ m/s}$

Now calculate the drag force. We find that it is *much* smaller than the gravitational force.	$F_D = \frac{1}{2}(1)\left(5\times10^{-14}\text{ kg/m}^3\right)\left(10\text{ m}^2\right)\left(7.6\times10^6\text{ m/s}\right)^2$ $F_D \approx \boxed{1.4\times10^{-5}\text{ N}}$
Finally, to find the force exerted by the Sun's radiation, use Example 34.6 as a guide. First, apply Eq. 34.37 for the pressure P when radiation is reflected from a surface and Eq. 17.23, which relates the power of the source (the sun) to the intensity at a given distance. In this case, r is the distance between the Sun and the satellite or about 1 AU $= 1.5 \times 10^{11}$ m (the distance from the sun to the Earth). The power P_{av} is the luminosity of the sun, $L_\odot = 4\times10^{26}\text{ W}$.	$P = 2\dfrac{I}{c}$ $\qquad$ (Eq. 34.37) $I = \dfrac{P_{av}}{4\pi r^2} = \dfrac{L_\odot}{4\pi r^2}$ $\qquad$ (Eq. 17.23)
The force equals the pressure times the area A of the sail.	$F_{rad} = PA = 2\dfrac{I}{c}A$ $F_{rad} = \dfrac{2}{c}\left(\dfrac{L_\odot}{4\pi r^2}\right)A$
Now, insert values. This value is also really small compared to the force of gravity. It is actually comparable to the drag force.	$F_{rad} = \dfrac{2}{3\times10^8\text{ m/s}}\left(\dfrac{4\times10^{26}\text{ W}}{4\pi\left(1.5\times10^{11}\text{m}\right)^2}\right)10\text{ m}^2$ $F_{rad} \approx \boxed{9\times10^{-5}\text{ N}}$

CHECK and THINK

So the gravitational force exerted on this lightweight satellite is *much* stronger than the force exerted by drag or radiation on the sail. On Earth, the satellite's weight is about 40 N. It may actually be surprising that the weight in orbit is as high as it is, but the orbit is at an altitude that is about 10% of the radius of the Earth. The drag force and the force exerted by radiation on the sail are comparable in this case.

Chapter 34 – Maxwell's Equations and Electromagnetic Waves

80. (A) Consider two circular regions of the same cross-sectional area A with uniform magnetic fields, one with the field pointing into the page and the other pointing out of the page (Fig. P34.80). Initially, both fields have magnitude B_0, and the magnitude of each is increasing at the same rate dB/dt. Calculate $\oint \vec{E} \cdot d\vec{\ell}$ using

a. circular path 1 and
b. circular path 2.

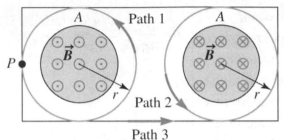

Figure P34.80

INTERPRET and ANTICIPATE
This problem is really about using Faraday's law. We need to find the derivative of the magnetic flux to find the path integral in each case.

SOLVE	
a. Use Faraday's law, Eq. 34.6. For path 1, $d\vec{A}$ is out of the page and $\vec{B}$ is out of the page, so the dot product is positive. Lenz's law tells us the path integral in negative, as indicated by the negative sign in formula.	$\oint \vec{E} \cdot d\vec{\ell} = -\dfrac{d}{dt}\int \vec{B} \cdot d\vec{A}$ $\oint \vec{E} \cdot d\vec{\ell} = -\dfrac{d}{dt}\int B\,dA = -\dfrac{dB}{dt}\int dA$ $\oint \vec{E} \cdot d\vec{\ell} = \boxed{-A\dfrac{dB}{dt}}$
b. For path 2, $d\vec{A}$ is out of the page and $\vec{B}$ is into the page, so the dot product is negative. Lenz's law tells us the path integral in positive.	$\oint \vec{E} \cdot d\vec{\ell} = -\dfrac{d}{dt}\int \vec{B} \cdot d\vec{A}$ $\oint \vec{E} \cdot d\vec{\ell} = -\dfrac{d}{dt}\int -B\,dA = \dfrac{dB}{dt}\int dA$ $\oint \vec{E} \cdot d\vec{\ell} = \boxed{A\dfrac{dB}{dt}}$

CHECK and THINK
The relative sign of our answers makes sense because the only difference between part (a) and part (b) is the direction of the magnetic field.

35

Diffraction and Interference

8. (N) Monochromatic light is incident on a pair of slits that are separated by 0.200 mm. The screen is 2.50 m away from the slits.
a. If the distance between the central bright fringe and either of the adjacent bright fringes is 1.67 cm, find the wavelength of the incident light.
b. At what angle does the next set of bright fringes appear?

INTERPRET and ANTICIPATE We can use Eq. 35.1 and the geometry described in Figure 35.15 to find the wavelength. According to the geometry, if the distance between the slits and the screen is x and the distance between the nth fringe and the central bright fringe is y_n, then they can be related by the angle at which the nth fringe occurs. We expect the wavelength to assuredly be less than the distance between the slits. Once we know the wavelength, we can use Eq. 35.1 to find the angel at which the next bright fringe occurs ($n = 2$).	
SOLVE **a.** Using the geometry of Figure 35.15, we express the relationship between the distance from the slits to the screen, the angular position of the fringe, and the distance between the nth fringe and the central bright fringe.	$\tan\theta_n = y_n/x$
Since the fringe is adjacent to the central bright fringe, $n = 1$. Also, in the small-angle approximation, which should be valid given the distances in the problem statement, it can be shown that the sine of the angular	$\sin\theta_n \approx \tan\theta_n = y_n/x$ $\sin\theta_1 \approx \tan\theta_1 = y_1/x$

position is approximately equal to the tangent of the angular position.	
Now, use Eq. 35.1 to find the wavelength of the incident light, using the approximation for the sine of the angular position.	$d\sin\theta_n = n\lambda \rightarrow d\sin\theta_1 = (1)\lambda$ $d(y_1/x) = (1)\lambda$ $\lambda = \dfrac{dy_1}{x} = \dfrac{(0.200\times10^{-3}\ \text{m})(1.67\times10^{-2}\ \text{m})}{(2.50\ \text{m})} = \boxed{1.34\times10^{-6}\ \text{m}}$
b. The next set of bright fringes ($n = 2$) occurs at a different angle, which can be found using Eq. 35.1. Write this equation, using the answer from part (a).	$d\sin\theta_n = n\lambda$ $d\sin\theta_2 = (2)\lambda$ $\sin\theta_2 = (2)(1.34\times10^{-6}\ \text{m})/(0.200\times10^{-3}\ \text{m})$ $\theta_2 = \sin^{-1}\left[(2)(1.34\times10^{-6}\ \text{m})/(0.200\times10^{-3}\ \text{m})\right]$ $\theta_2 = \boxed{0.768°}$

CHECK and THINK

The wavelength is less than the slit separation which allows for the formation of an interference pattern, as expected. The small-angle approximation will be valid when the distance from the slits to the screen is much larger (say, two or more orders of magnitude) than the distance from the central fringe to the location of another fringe on the screen.

10. (N) In a Young's double-slit experiment with microwaves of wavelength 3.00 cm, the distance between the slits is 5.00 cm and the distance between the slits and the screen is 100.0 cm. Determine the number of bright fringes on the screen and their distances from the central bright fringe on the screen. Assume the screen is long enough to at least show the first order bright fringes.

INTERPRET and ANTICIPATE

There will be at least one maximum in the center. We can determine the location of the other maxima using the equation for the double slit experiment.

SOLVE We first sketch the situation (which looks like Figure 35.13).	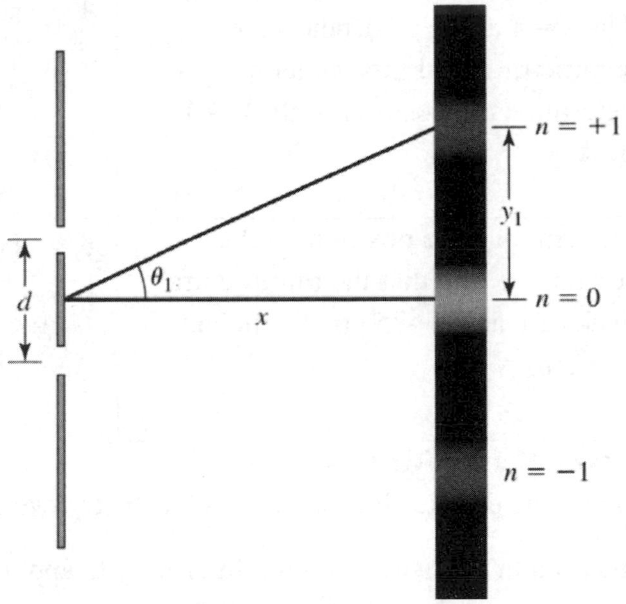 **Figure P35.10ANS**
The angles of the maxima for a double slit are given by Eq. 35.1, where n is the index for the maximum, λ is the wavelength, and d the separation between the slits. We can rearrange this expression to solve for the angle.	$d \sin \theta = n\lambda$ $\theta = \sin^{-1}\left(\dfrac{n\lambda}{d}\right)$
When we look at the values given in the problem statement, this expression becomes that seen here on the right.	$\theta = \sin^{-1}(0.600n)$
Note, though, that it must be the case that $0.600n \leq 1$ because the sine of the angle cannot be bigger than 1. Thus, we can only have the first order fringes possibly appear on the screen, because they are the only ones created in this scenario! If $n = 2$ or greater, there is no solution.	$0.600n \leq 1$
We can also use geometry to relate the distance on the screen from the central maximum y based on the angle of the first maximum. The distance to the screen is x.	$\tan \theta = \dfrac{y}{x} \quad \rightarrow \quad y = x \tan \theta$

The central maximum is at $\theta = 0°$. The $n = 1$ and $n = -1$ peaks are symmetric about this, so let's determine the position of the $n = 1$ peak.	$\theta_{n=1} = \sin^{-1}\left(\dfrac{\lambda}{d}\right) = \sin^{-1}\left(\dfrac{3.00 \text{ cm}}{5.00 \text{ cm}}\right) = 36.9°$
Determining the position on the screen, we find that the three central peaks are at $y = -75$ cm, 0 cm, and $+75$ cm.	$y_{n=1} = (100.0 \text{ cm})\tan(36.9°) = 75.0 \text{ cm}$ $y_{n=-1,0,+1} = \boxed{-75.0 \text{ cm}, \ 0 \text{ cm}, \ +75.0 \text{ cm}}$

CHECK and THINK

Using the double slit formula and geometry, we can determine the locations of the maxima in intensity. You might also try to apply Eq. 35.4, $y = \dfrac{nx\lambda}{d}$, to calculate the position directly. The one problem though, is that in deriving this equation, we used the small angle approximation, where we assume $\sin\theta \approx \theta$ and $\tan\theta \approx \theta$ for small θ. However, the angle we found (around 37 degrees), is not that small. If we use this equation, we find

$$y_{n=\pm1} = \pm\frac{(100.0 \text{ cm})(3.00 \text{ cm})}{5.00 \text{ cm}} = \pm60.0 \text{ cm}$$

So, instead of $= -75.0$ cm, 0 cm, and $+75.0$ cm, we get -60.0 cm, 0, and $+60.0$ cm—not horrible, but off by 20%, so we are justified in not using the small angle approximation here.

13. The wavelength of light emitted by a particular laser is 633 nm. This laser light illuminates two slits that are 50.0 μm apart, and a screen is 1.50 m from the slits.
a. (N) What is the (linear) distance on the screen between the central maximum and the $n = 2$ maximum?
b. (N) What is the (linear) distance on the screen between the central maximum and the $n = 20$ maximum?
c. (C) Are the maxima evenly spaced? Explain.

INTERPRET and ANTICIPATE

We are asked about maxima, so we must apply the condition for constructive interference as in Example 35.1.

SOLVE **a.** Find θ_2 using Eq. 35.1. This is the angular position of the $n =2$ maximum. The angle depends on the separation of the slits d and the wavelength λ. We also use the fact that 1 μm = 1000 nm, though of course you can also use the fact that 1 nm = 10^{-9} m and 1 μm = 10^{-6} m.	$d \sin \theta_n = n\lambda$ $\theta_n = \sin^{-1}\left(\dfrac{n\lambda}{d}\right)$ $\theta_2 = \sin^{-1}\left(\dfrac{2(633 \text{ nm})}{50.0 \times 10^3 \text{ nm}}\right) = 1.45°$
Find the distance from the central maximum to the $n = 2$ maximum (y_2) using geometry (see Example 35.1). The distance to the screen is x.	$y_2 = x \tan \theta_2$ $y_2 = (1.50 \text{ m}) \tan 1.45°$ $y_2 = 0.0380 \text{ m} = \boxed{3.80 \text{ cm}}$
b. Now find θ_{20} as in part (a). This is the angular position of the $n =20$ maximum.	$\theta_{20} = \sin^{-1}\left(\dfrac{20(633 \text{ nm})}{50.0 \times 10^3 \text{ nm}}\right) = 14.7°$ $y_{20} = x \tan \theta_{20}$ $y_{20} = (1.50 \text{ m}) \tan 14.7°$ $y_{20} = 0.393 \text{ m} = \boxed{39.3 \text{ cm}}$
c. If the fringes were evenly spaced, we would expect the ratio y_{20}/y_2 to be 10.0. But we find this ratio to be greater than 10.0, so the fringes are not evenly spaced.	$\dfrac{y_{20}}{y_2} = \dfrac{39.3 \text{ cm}}{3.80 \text{ cm}} = 10.3$ Fringes are $\boxed{\textbf{\textit{not} evenly spaced}}$.

CHECK and THINK

The fringes near the central maximum look evenly spaced, but they are not quite since the sine and tangent are nonlinear expressions. Notice that we had to consider the 20th fringe to see this slight discrepancy. Fringes that occur at small angles appear equally spaced, but the angular separation between adjacent maxima increases as n increases.

15. (N) Light from a sodium vapor lamp ($\lambda = 589$ nm) forms an interference pattern on a screen 0.80 m from a pair of slits in a double-slit experiment. The bright fringes near the center of the pattern are 0.35 cm apart. Determine the separation between the slits. Assume the small-angle approximation is valid here.

INTERPRET and ANTICIPATE

The small angle approximation was used to derive Equation 35.4, the formula for the location of the intensity maxima. The double slit formula in terms of angle, $d\sin\theta = n\lambda$, is exact, but in deriving the formula for the position of the maxima on the screen, y_n, we used the facts that $\sin\theta \approx \theta$ and $\tan\theta \approx \theta$ when θ is small.

SOLVE Apply Eq. 35.4, where n is the index for the maximum, λ is the wavelength, x is the distance to the screen, and d is the separation between the slits.	$y_n \simeq \dfrac{nx\lambda}{d} \quad n = 0, \pm 1, \pm 2, \pm 3 \ldots$
We now find the distance between the adjacent fringes.	$\Delta y = y_{n+1} - y_n$ $\Delta y = \dfrac{x\lambda}{d}(n+1) - \dfrac{x\lambda}{d}n$ $\Delta y = \dfrac{x\lambda}{d}$
Now we solve for d to obtain the separation between the slits and insert numbers.	$d = \dfrac{x\lambda}{\Delta y}$ $d = \dfrac{(0.80 \text{ m})(589 \times 10^{-9}\text{m})}{0.35 \times 10^{-2}\text{m}}$ $d = \boxed{1.3 \times 10^{-4}\text{ m}} = 0.13 \text{ mm}$

CHECK and THINK

The separation between the slits is quite small in this case, which is consistent with similar examples we've seen.

22. (N) Red light with a wavelength of 715.5 nm and violet light with a wavelength of 412.5 nm are used simultaneously in a double-slit experiment. The first maximum of the violet light is 1.975 mm from the central maximum. What is the distance between the central maximum and the first maximum of the red light?

INTERPRET and ANTICIPATE

We are not given the slit separation, but we can set up a ratio using the data for the violet light to determine the distance for the red light.

SOLVE Let's apply the small angle approximation to the first maximum ($n = 1$). The position on the screen is then given by Eq. 35.4.	$$y_n \approx \frac{nx\lambda}{d} \quad \rightarrow \quad y_1 \approx \frac{x\lambda}{d}$$
The distances x and d are the same for both the violet and the red light. Write a ratio for the position of their first maxima in terms of their wavelengths. Solve for y_{red}.	$$\frac{y_{red}}{\lambda_{red}} = \frac{y_{violet}}{\lambda_{violet}} = \frac{x}{d}$$ $$y_{red} = \frac{\lambda_{red}}{\lambda_{violet}} y_{violet}$$
Insert values. Since red light has a larger wavelength, the diffraction pattern created is larger.	$$y_{red} = \left(\frac{715.5 \text{ nm}}{412.5 \text{ nm}}\right)1.975 \text{ mm} = \boxed{3.426 \text{ mm}}$$

CHECK and THINK

The first maximum of the red light is farther from the central maximum than that of the violet light. This also gives us an idea of how slits can be used to separate white light into a color spectrum.

27. (N) A thread must have a uniform thickness of 0.525 mm. To check the thickness of the thread, you can illuminate it with a laser of wavelength 625.8 nm. A diffraction pattern like the one produced by a single slit forms on a screen.
a. If the screen is 3.00 m from the thread, how far apart are the fifth-order minima from one another?
b. If the thread's thickness increases by 20%, how far apart will the fifth-order minima be?

INTERPRET and ANTICIPATE

If the thread acts as a single slit, we can use the single slit diffraction equation to determine the angle of the 5th order peak and then the separation of the 5th order peaks on the screen. Since diffraction effects are sensitive to the size of the slit, we would expect that the two different thread thicknesses will give us a measureable difference on the screen.

SOLVE **a.** Solve Equation 35.5 for $\sin \theta_5$.	$$w\sin\theta_m = m\lambda$$ $$\sin\theta_5 = 5\frac{\lambda}{d}$$

Insert values. The result on the right is very small, corresponding to a very small angle, so we use the small angle approximation, $\sin\theta \approx \theta$, keeping in mind that the angle calculated is in radians.	$\sin\theta_5 = 5\left(\dfrac{625.8\times10^{-9}\text{ m}}{0.525\times10^{-3}\text{ m}}\right) = 5.96\times10^{-3}$ $\theta_5 \approx 5.96\times10^{-3}$ rad
We can calculate the distance from the central maximum to the 5th order peak using geometry (see Example 35.1, for instance). We again use the small angle approximation, $\tan\theta \approx \theta$.	$y_n = x\tan\theta_n \approx x\theta_n$ $y_5 = x\theta_5 = (3.00\text{ m})(5.96\times10^{-3}\text{ rad})$ $y_5 = 0.0179\text{ m} = 1.79$ cm
This is the distance from the center to the 5th order peak, so the distance between the two 5th order peaks (on either side of the central peak) is twice this value.	$\Delta y = 2y_5 = \boxed{3.58\text{ cm}}$
b. If the size of the thread is 20% larger, it is a factor of 1.2 times the thickness above.	$1.2(0.525\text{ mm}) = 0.630$ mm
Now, we repeat the calculation from part (a).	$\sin\theta_5 \approx \theta_5 = 5\left(\dfrac{625.8\times10^{-9}\text{ m}}{0.630\times10^{-3}\text{ m}}\right)$ $\theta_5 = 4.97\times10^{-3}$ rad $y_5 = x\theta_5 = (3.00\text{ m})(4.97\times10^{-3}\text{ rad})$ $y_5 = 0.0149\text{ m} = 1.49$ cm $\Delta y = 2y_5 = \boxed{2.98\text{ cm}}$

CHECK and THINK

Indeed, we could definitely measure this difference on the screen. In fact, the pattern is about 20% smaller since the thread is 20% bigger. Diffraction effects allow us to measure this very small difference in thickness of only 100 μm by easily measuring a distance on the screen of around 0.6 cm!

33. **(N)** A single slit is illuminated by light consisting of two wavelengths. One wavelength is 540.0 nm. If the first minimum of one color is located at the second minimum of the other color, what are the possible wavelengths of the other color? Is the light visible?

INTERPRET and ANTICIPATE We can use the single slit formula to find the location of the first dark spot for each wavelength and require that the first order dark fringe for one is at the same angle as the second order for the other.	

SOLVE The first minimum of one color is at the second minimum of the other color, so $\theta_1 = \theta_2$, by which we mean θ_1 is the angle of the first dark fringe ($m = 1$) for one of the wavelengths λ_1 and θ_2 is the second dark fringe ($m = 2$) for the other wavelength λ_2.	$\theta_1 = \theta_2$
We can now write Eq. 35.5 in both cases.	$w \sin \theta_m = m\lambda$ $\sin \theta_1 = \dfrac{\lambda_1}{w}$ and $\sin \theta_2 = \dfrac{2\lambda_2}{w}$
Since the angles are the same, so are the sine of the angles, so we can set these two equations equal to each other.	$\sin \theta = \dfrac{\lambda_1}{w} = \dfrac{2\lambda_2}{w} \quad \rightarrow \quad \lambda_1 = 2\lambda_2$
We have not specified which wavelength is which, so the 540 nm light could be *either* λ_1 or λ_2. Calculate the other wavelength with each assumption.	If $\lambda_2 = 540.0$ nm, $\lambda_1 = 2\lambda_2 = \boxed{1.080 \times 10^{-6} \text{ m}}$. If $\lambda_1 = 540$ nm, $\lambda_2 = \lambda_1/2 = \boxed{2.700 \times 10^{-7} \text{ m}}$.
The resulting wavelengths are outside those that are visible (around 390 – 700 nm).	$\boxed{\text{The light is not visible in either case.}}$

CHECK and THINK In this case, we simply need to use the formula for the single slit dark fringes. We find that there are two possible wavelengths since the given one could be the wavelength at the first or second minimum.

Chapter 35 – Diffraction and Interference

36. (N) A diffraction pattern is produced by passing He-Ne laser light of wavelength 632.8 nm through a single slit. The pattern shown is viewed on a screen 2.0 m behind the slit, where the second-order minima are separated by 15.2 cm. Find the width of the slit.

INTERPRET and ANTICIPATE Knowing the separation of the second minima will allow us to determine the location of this minimum relative to the center and θ_2. From this, we can determine the width of the slit.	

SOLVE The fact that the two second-order minima are separated by 15.2 cm means that the location of the second-order minimum relative to the center of the pattern is half this distance, or 7.60 cm.	2nd minimum · · · 2nd minimum ⟵ 15.2 cm ⟶ $y = \dfrac{15.2}{2} = 7.6$ cm **Figure P35.36ANS**
Apply Eq. 35.5 for the second minimum, $m = 2$, where w is the width of the slit.	$w\sin\theta_2 = 2\lambda$
The angle can be determined from geometry (see Example 35.1). The position on the wall relative to the central maximum, y_2, can be expressed in terms of the distance to the screen x and the angle θ_2.	$y_2 = x\tan\theta_2 \quad \rightarrow \quad \theta_2 = \tan^{-1}\left(\dfrac{y_2}{x}\right)$ $\theta_2 = \tan^{-1}\left(\dfrac{0.076 \text{ m}}{2.0 \text{ m}}\right) = \tan^{-1}(0.038) = 2.2°$
Now, solve for the width of the slit.	$w = \dfrac{2\lambda}{\sin\theta_2}$ $w = \dfrac{2(632.8\times10^{-9} \text{ m})}{\sin(2.2°)}$ $w = \boxed{33\times10^{-6} \text{ m}} = 33 \ \mu\text{m}$

CHECK and THINK The width of the slit is quite small, as expected, and can be determined from the diffraction pattern.

316

© 2016 Cengage Learning. All Rights Reserved. May not be scanned, copied or duplicated, or posted to a publicly accessible website, in whole or in part.

39. (N) A double-slit experiment is conducted, and at some point on the screen the intensity is found to be 75.0% of its maximum. What is the minimum phase difference (in radians and degrees) that produces this result?

INTERPRET and ANTICIPATE	
The relative intensity depends directly on the phase difference, so knowing the intensity, we can determine the phase difference.	

SOLVE	
The intensity depends on phase according to Eq. 35.8.	$I = I_{max} \cos^2 \dfrac{\varphi}{2}$
Solve this for the phase φ,	$\dfrac{I}{I_{max}} = \cos^2 \dfrac{\varphi}{2}$ $\cos \dfrac{\varphi}{2} = \sqrt{\dfrac{I}{I_{max}}}$ $\varphi = 2 \cos^{-1} \sqrt{\dfrac{I}{I_{max}}}$
We can now insert numerical values.	$\varphi = 2 \cos^{-1} \sqrt{0.75}$ $\varphi = \boxed{1.05 \text{ rad}}$ or $\boxed{60.0°}$

CHECK and THINK	
Remember, this is the minimum phase difference, since a phase difference that is different by any multiple of 2π radians will result in the same value of I. So, the calculation provides a phase difference and does not give you a specific location on the screen.	

44. (N) Light of wavelength 455 nm is incident on two slits separated by $d = 3.75$ mm, forming an interference pattern on a screen placed 1.85 m away. At what distance y from the central maximum is the intensity exactly half that of the maximum?

INTERPRET and ANTICIPATE	
We have a formula for the intensity of the double slit diffraction pattern. We can use this and seek places with intensities that are half the maximum value.	

SOLVE The intensity of the fringes in a double slit is given by Equation 35.8, with φ defined by Eq. 35.9.	$I = I_{max} \cos^2\left(\dfrac{\varphi}{2}\right)$ and $\varphi = \dfrac{2\pi d \sin\theta}{\lambda}$
We will assume that the small angle approximation is valid. With this assumption, $\sin\theta \approx \tan\theta \approx \theta$. We can also relate the angle to y and the distance to the screen x, as in Example 35.1.	$\sin\theta \approx \tan\theta = \dfrac{y}{x}$
Combine these expressions.	$I = I_{max} \cos^2\left(\dfrac{\pi d \sin\theta}{\lambda}\right) = I_{max} \cos^2\left(\dfrac{\pi d y}{\lambda x}\right)$
Solve for y.	$y = \dfrac{\lambda x}{\pi d} \cos^{-1}\sqrt{\dfrac{I}{I_{max}}}$
Use the fact that $I = 0.500\, I_{max}$ and solve. Remember that the angle calculated using the arccosine is in radians.	$y = \dfrac{(455\times10^{-9}\ \text{m})(1.20\ \text{m})}{\pi(3.75\times10^{-3}\ \text{m})} \cos^{-1}\sqrt{0.500}$ $y = \boxed{3.64\times10^{-5}\ \text{m}} = 36.4\ \mu\text{m}$

CHECK and THINK

The distance turns out to be incredibly small compared to the distance to the screen. In fact, the angle is around 0.001°. We were certainly well justified in using the small angle approximation!

47. (N) In a Young's double-slit experiment, 586-nm-wavelength light is sent through the slits. The intensity at an angle of 2.50° from the central bright fringe is 80.0% of the maximum intensity on the screen. What is the spacing between the slits?

INTERPRET and ANTICIPATE

In order to do this, we must first use Eqs. 35.8 and 35.9 to find an expression for the intensity in terms of the given variables. The slit separation should depend on the maximum intensity, the intensity, the wavelength, and the angular position. Given that we have these quantities, we can then substitute and find the slit separation. We expect that this should be greater than the wavelength, assuming that an interference pattern has formed.

SOLVE First, write Eq. 35.9.	$$\varphi = \frac{2\pi}{\lambda} d \sin\theta_0$$
Now, substitute this into Eq. 35.8.	$$I = I_{max} \cos^2\left[\frac{2\pi}{2\lambda} d \sin\theta_0\right] = I_{max} \cos^2\left(\frac{\pi}{\lambda} d \sin\theta_0\right)$$
Now, solve for d.	$$I = I_{max} \cos^2\left(\frac{\pi}{\lambda} d \sin\theta_0\right)$$ $$I/I_{max} = \cos^2\left(\frac{\pi}{\lambda} d \sin\theta_0\right)$$ $$\sqrt{I/I_{max}} = \cos\left(\frac{\pi}{\lambda} d \sin\theta_0\right)$$ $$\frac{\pi}{\lambda} d \sin\theta_0 = \cos^{-1}\left(\sqrt{I/I_{max}}\right)$$ $$d = \frac{\lambda}{\pi \sin\theta_0} \cos^{-1}\left(\sqrt{I/I_{max}}\right)$$
Use this result with the given info to find the slit separation.	$$d = \frac{\lambda}{\pi \sin\theta_0} \cos^{-1}\left(\sqrt{I/I_{max}}\right)$$ $$d = \frac{586 \times 10^{-9} \text{ m}}{\pi \sin(2.50°)} \cos^{-1}\left(\sqrt{0.800}\right)$$ $$d = \boxed{1.98 \times 10^{-6} \text{ m}}$$

CHECK and THINK

The slit separation is greater than the wavelength. The intensity will fall off as we move away from the center of the central bright region, and then oscillate up and down as we track across the interference pattern. Note that the angular position given is not a location of a maxima or minima; it is just a given angular position.

52. (N) Light with a wavelength of 426 nm is incident on a single slit with a width of 4.5 μm. If the intensity of the central bright fringe is 655 W/m^2, find the intensity at these angular positions from the fringe: **a.** 20.0°, **b.** 40.0°, and **c.** 60.0°.

INTERPRET and ANTICIPATE

We must use Eq. 35.10 in conjunction with Eq. 35.11 in each case to find the intensity. It is hard to have an expectation for the intensity at each angular position, but it must be less than 655 W/m^2. The angles could be near a minimum or a maximum. We would have to use Eq. 35.5, if we wanted to try to find nearby minima.

SOLVE First, substitute Eq. 35.11 into Eq. 35.10.	$I = I_{max} \left(\dfrac{\sin\left(\dfrac{\pi w}{\lambda}\sin\theta\right)}{\dfrac{\pi w}{\lambda}\sin\theta} \right)^2$
a. Now, substitute the given values and solve for the intensity.	$I = I_{max} \left(\dfrac{\sin\left(\dfrac{\pi w}{\lambda}\sin\theta\right)}{\dfrac{\pi w}{\lambda}\sin\theta} \right)^2$ $I = \left(655 \text{ W/m}^2\right) \left[\dfrac{\sin\left(\dfrac{\pi\left(4.5\times10^{-6}\text{ m}\right)}{\left(426\times10^{-9}\text{ m}\right)}\sin\left(20.0°\right)\right)}{\dfrac{\pi\left(4.5\times10^{-6}\text{ m}\right)}{\left(426\times10^{-9}\text{ m}\right)}\sin\left(20.0°\right)} \right]^2$ $I = \boxed{4.5 \text{ W/m}^2}$
b. Again, substitute the given values and solve.	$I = I_{max} \left(\dfrac{\sin\left(\dfrac{\pi w}{\lambda}\sin\theta\right)}{\dfrac{\pi w}{\lambda}\sin\theta} \right)^2$ $I = \left(655 \text{ W/m}^2\right) \left[\dfrac{\sin\left(\dfrac{\pi\left(4.5\times10^{-6}\text{ m}\right)}{\left(426\times10^{-9}\text{ m}\right)}\sin\left(40.0°\right)\right)}{\dfrac{\pi\left(4.5\times10^{-6}\text{ m}\right)}{\left(426\times10^{-9}\text{ m}\right)}\sin\left(40.0°\right)} \right]^2$ $I = \boxed{0.54 \text{ W/m}^2}$

c. Again, substitute the given values and solve.	$$I = I_{max}\left(\dfrac{\sin\left(\dfrac{\pi w}{\lambda}\sin\theta\right)}{\dfrac{\pi w}{\lambda}\sin\theta}\right)^2$$ $$I = \left(655\ \text{W/m}^2\right)\left[\dfrac{\sin\left(\dfrac{\pi\left(4.5\times10^{-6}\ \text{m}\right)}{\left(426\times10^{-9}\ \text{m}\right)}\sin\left(60.0°\right)\right)}{\dfrac{\pi\left(4.5\times10^{-6}\ \text{m}\right)}{\left(426\times10^{-9}\ \text{m}\right)}\sin\left(60.0°\right)}\right]^2$$ $$I = \boxed{0.16\ \text{W/m}^2}$$

CHECK and THINK

Recall that in single-slit diffraction, it is common that the bright fringes are highly localized at higher angles (see Figure 35.22, for example). This provides some explanation for the likelihood of finding three intensities that are extremely low.

---

54. (N) Monochromatic light of wavelength 414 nm is incident on a single slit of width 32.0 μm. The distance from the slit to the screen is 2.60 m. Consider a point at $y = 16.5$ mm from the center of the central maximum. What is the ratio of the intensity at that point to the maximum intensity?

INTERPRET and ANTICIPATE

The intensity depends on the phase shift, which we can calculate. Since we're asked for a ratio of this intensity to the maximum, our answer will definitely be less than one.

SOLVE The intensity for a single slit diffraction pattern is given by Eq. 35.10, with α defined by Eq. 35.11. The width of the slit is w and the wavelength of light is λ.	$$I = I_{max}\left(\dfrac{\sin\alpha}{\alpha}\right)^2 \quad \text{with} \quad \alpha = \dfrac{\pi w}{\lambda}\sin\theta$$
We need to calculate the angle to this spot, θ. Since the distance to this point on the spectrum is $y = 16.5$ mm, which is much smaller than the distance to the screen of $x = 2.60$ m, the angle from the center is clearly very small and we can use the small angle approximation. Relate the angle to x and y using geometry as in	$$\tan\theta \approx \theta \approx \dfrac{y}{x}$$ $$\theta \approx \dfrac{16.5\times10^{-3}\ \text{m}}{2.60\ \text{m}} = 6.35\times10^{-3}\ \text{rad}$$

Example 35.1. Remember that the angle calculated is in radians.	
Insert values into Eq. 35.11.	$\alpha = \dfrac{\pi w}{\lambda}\sin\theta$ $\alpha = \dfrac{\pi\left(32.0\times10^{-6}\text{ m}\right)}{\left(414\times10^{-9}\text{ m}\right)}\sin\left(6.35\times10^{-3}\text{ rad}\right)$ $\alpha = 1.54\text{ rad}$
Solve Eq. 35.10 for the ratio of the intensity to the maximum value.	$\dfrac{I}{I_{max}} = \left(\dfrac{\sin\alpha}{\alpha}\right)^2$ $\dfrac{I}{I_{max}} = \left(\dfrac{\sin(1.54\text{ rad})}{1.54\text{ rad}}\right)^2 = \boxed{0.421}$

CHECK and THINK
The intensity at this point is 42% of the maximum value at the center.

59. (A) Two slits are separated by distance d and each has width w. If $d = 2w$, how many bright fringes are within the central maximum of the diffraction pattern?

INTERPRET and ANTICIPATE	
A double slit pattern consists of an envelope due to the diffraction pattern of each individual slit with smaller bright and dark spots due to the interference pattern from the double slit (see Figures 35.26 and 35.27 for examples). We can determine the location of the first minimum in intensity due to the single slit diffraction pattern. Then, we can determine how many double slit peaks fit into this envelope.	

SOLVE	
The first peak of the central maximum stretches from the center of the pattern to the first minimum in the single slit diffraction pattern, θ_1. Use Eq. 35.5 with $m = 1$, where w is the width of the single slit and λ the wavelength of light.	$\sin\theta_1 = \dfrac{\lambda}{w}$

The bright fringes in the double-slit interference pattern are given by Eq. 35.1.	$d \sin \theta_n = n\lambda \quad n = 0, \pm 1, \pm 2...$ $\sin \theta_n = \dfrac{n\lambda}{d}$
Since there are multiple peaks from the double slit in the central single slit peak, the question is what order n of the double slit pattern lines up with the first minimum of the single slit? So, what $\theta_n = \theta_1$? That will tell us how many double slit peaks are on each side of the center and still fit inside the single slit maximum.	$\theta_n = \theta_1 \quad$ so $\quad \sin \theta_n = \sin \theta_1$ $\dfrac{n\lambda}{d} = \dfrac{\lambda}{w}$ $n = \dfrac{d}{w}$
If $d = 2w$, $n = 2$. This means that the second peak is at the edge of the central bright spot. Therefore, the peaks that are actually *inside* the bright spot are $n = -1$, 0, and +1. So, *three fringes are within the central peak.*	$n = \dfrac{2w}{w} = 2$ Three fringes are *within* the central peak, $n = -1$, 0, and +1

CHECK and THINK

The double slit intensity corresponds to a small pattern from the double slit inside a larger central peal due to the single slit diffraction pattern. Therefore, we can relate the spacing for each to find out how many spots are in the central maximum.

63. (N) Light of wavelength 570.0 nm incident on a single slit forms a diffraction pattern on a screen placed 1.10 m from the slit. What is the width of the slit if the first and fifth dark fringes are separated by 5.30 mm?

INTERPRET and ANTICIPATE The pattern of bright and dark fringes depends on the wavelength of light, distance to the screen, and the width of the slit. Given data about the fringes, we can determine the slit width. Since this is diffraction of visible light, we expect the slit width to be quite small, perhaps a fraction of a millimeter.	
SOLVE The angle of the dark fringes for a single slit pattern are given by Eq. 35.5, where m is the fringe order, λ the wavelength of light, and w the width of the slit.	$\sin \theta_m = \dfrac{m\lambda}{w} \qquad m = \pm 1, \pm 2, \pm 3,...$

We can relate the angle to the position of the fringes on the screen y_m and the distance from the slit to the screen x using the geometry in Example 35.1.	$\tan\theta_m = \dfrac{y_m}{x}$
Since the separation on the screen is only 5.3 mm compared to the distance to the screen of around 1 m, we can safely use the small angle approximation, $\sin\theta \approx \tan\theta \approx \theta$. That allows us to set the above equations equal and solve for the width. Since we are looking to relate a distance on the screen to a range of fringes indicated by m, we express this in terms of $\Delta y = y_5 - y_1$ and $\Delta m = 5 - 1$.	$\dfrac{\Delta m\lambda}{w} = \dfrac{\Delta y}{x}$ $w = \dfrac{\Delta m\lambda x}{\Delta y}$
The distance between the fringes is $\Delta y = 5.30\times10^{-3}$ m and corresponds to $\Delta m = 5-1 = 4$.	$w = \dfrac{4\left(570\times10^{-9}\ \text{m}\right)\left(1.10\ \text{m}\right)}{\left(5.30\times10^{-3}\ \text{m}\right)}$ $w = \boxed{4.73\times10^{-4}\ \text{m}}$

CHECK and THINK
The width is quite small as expected.

66. **(N)** The two slits of a double-slit apparatus each have a width of 0.120 mm and their centers are separated by 0.720 mm. What orders are missing in the diffraction pattern?

INTERPRET and ANTICIPATE Missing orders occur when the condition for a maximum of interference and for a minimum of diffraction are both fulfilled for the same value of the angle θ.	
SOLVE The condition for a maximum intensity due to interference between the two slits is given by Equation 35.1 where n is an integer and d is the separation between the two slits.	$d\sin\theta_n = n\lambda \qquad n = 0,\ \pm1,\ \pm2,\ \pm3\ ...$
The condition for a minimum of intensity due to diffraction of each slit is given by Equation 35.5 where m is an integer and w is the width of each slit.	$w\sin\theta_m = m\lambda \qquad m = \pm1,\ \pm2,\ \pm3\ ...$

Missing orders occur when the condition for a maximum of interference and for a minimum of diffraction are both fulfilled for the same value of the angle $\theta_n = \theta_m = \theta$.	$\dfrac{d}{n} = \dfrac{\lambda}{\sin\theta} = \dfrac{w}{m} \quad \rightarrow \quad \dfrac{d}{w} = \dfrac{n}{m}$ $\dfrac{n}{m} = \dfrac{0.720\ \text{mm}}{0.120\ \text{mm}} = 6 \quad \text{or} \quad \boxed{n = 6m}$ Since m must be an integer, the missing orders are those for which n is a multiple of 6.

CHECK and THINK

The ratio of n/m determines the orders that are missing. Using this data, we could also use Eq. 35.1 or 35.5 to determine the corresponding angles.

---

74. (N) Sound with a wavelength of 2.29 m is incident on an opening in a wall that acts as a single slit with a width of 4.59 m. The maximum intensity of the sound after passing through the opening is 1.00×10^{-6} W/m². The locations of absolute silence would be found in a way similar to finding the angular positions of dark fringes for light passing through a single slit. What is the angular position for the first-order ($m = 1$) location of silence in this example?

INTERPRET and ANTICIPATE

We can use Eq. 35.5 to find the location of the first-order dark fringe. Even though this is a sound wave, it can still diffract like any other wave. Note that, in reality, there is a limit to human hearing that would cause the "dark" fringe, or location of silence to be broader than a single point. Here, we are really finding the center of that region.

SOLVE First, write Eq. 35.5, noting that we are looking at the $m = 1$ fringe.	$w\sin\theta_1 = (1)\lambda$
Use the given information in the problem statement and solve for the angular position.	$\sin\theta_1 = \lambda/w = (2.29\ \text{m})/(4.59\ \text{m})$ $\theta_1 = \boxed{29.9°}$

CHECK and THINK

Given the large angle at which the center of the first region of silence occurs for this wavelength, the sound has spread forward through the opening in a fairly large cone forward (about 30° on either side). A lower wavelength, or higher frequency would have resulted in the first region of silence occurring at a lesser angle, meaning that higher frequencies are more forward-scattered, and do not diffract, or bend, out of the opening as well as lower frequencies. This is why you might be able to hear someone talking in a room, even when you are not in front of the doorway, but their voice sounds "bassier", or you hear the lower frequency tones more prominently.

Chapter 35 – Diffraction and Interference

77. (N) Consider two monochromatic sources A and B with wavelength λ such that A is initially ahead of B in phase by 66°. The waves interfere at a certain point after having traveled different paths, where B travels $\lambda/4$ farther than A. Determine the phase difference between the waves when they interfere at this point.

INTERPRET and ANTICIPATE We need to determine the total phase difference, which includes both the initial 66 degree difference and the additional phase shift due to the fact that ray B travels farther than A.	
SOLVE A phase shift of $\lambda/4$ means one quarter of a wave. Since an entire wave is 360°, one quarter of the wave corresponds to a phase shift of 90°.	$\varphi_{\lambda/4} = \dfrac{360°}{\lambda}\Delta x = \dfrac{360°}{\lambda}\dfrac{\lambda}{4} = 90°$
We are told that A is initially ahead of B, or in other words that B is behind A. Since B then travels further than A (and it oscillates for a longer time), then B lags even further behind A. That is, the phase shifts add together and B is ultimately 156° behind A.	$\varphi = 66° + 90° = \boxed{156° = 2.72 \text{ rad}}$

CHECK and THINK
The total phase difference is that due to both the initial phase difference and the shift due to the difference in path length. We just need to be careful about whether they add together or offset each other.

82. (N) A pair of slits separated by 0.340 mm and placed 2.10 m away from a viewing screen is illuminated by a source that produces both 580-nm and 620-nm light, resulting in overlapping interference patterns. What is the shortest distance from the central maximum for which bright fringes for the two wavelengths coincide?

INTERPRET and ANTICIPATE We can use the formula for the location of the bright fringes for a double slit. The locations of the peaks depend on the wavelength, so we can express this relationship for each wavelength and determine how they are related.

SOLVE Let's assume that we can use the small angle approximation, in which case we can use Eq. 35.4 to express the locations of the intensity peaks in terms of the integer index n, distance to the screen x, wavelength λ, and the separation of the slits d.	$y \approx \dfrac{nx\lambda}{d}$
Write this equation for the yellow light (λ_1 = 580 nm). The distance to the screen and the separation between the slits is the same in both cases.	$y_1 = \dfrac{n_1 x \lambda_1}{d}$
And now for the red light (λ_2 = 620 nm).	$y_2 = \dfrac{n_2 x \lambda_2}{d}$
We want to find which peaks (indexed by n_1 and n_2) will coincide at the same position ($y_1 = y_2 = y$).	$\dfrac{n_1 x \lambda_1}{d} = \dfrac{n_2 x \lambda_2}{d}$
Rearrange this expression.	$\dfrac{n_2}{n_1} = \dfrac{\lambda_1}{\lambda_2} = \dfrac{580 \text{ nm}}{620 \text{ nm}} = \dfrac{29}{31}$
The smallest integers satisfying the equation are $n_1 = 31$ and $n_2 = 29$. Use either to find the y location on the screen. Here we use the yellow light, for which $n_1 = 31$ and $\lambda_1 = 580$ nm.	$y = \dfrac{31\left(2.10 \text{ m}\right)\left(580 \times 10^{-9} \text{ m}\right)}{3.40 \times 10^{-4} \text{ m}} = \boxed{0.111 \text{ m}}$

CHECK and THINK

We can recalculate for the location of the red light and confirm that we get the same value.

$$y = \frac{29\left(2.10 \text{ m}\right)\left(620 \times 10^{-9} \text{ m}\right)}{3.40 \times 10^{-4} \text{ m}} = 0.111 \text{ m}$$

83. (N) Monochromatic light waves of wavelengths 400.0 nm and 560.0 nm are incident simultaneously and normally on a Young's double-slit apparatus. The separation between the slits is 0.100 mm, and the distance between the slits and the screen is 1.00 m. Determine the distance between the first two adjacent dark fringes on the screen.

INTERPRET and ANTICIPATE

For a point on a screen to be completely dark, we need a minimum in intensity for *both* waves. We can determine the location of the minima for each and see where they line up.

SOLVE We use Equation 35.2 to determine the location of the minima in a double slit pattern.	$d \sin \theta_n = \left(n + \dfrac{1}{2} \right) \lambda \quad n = 0, \pm 1, \ \pm 2, \ \pm 3 \ldots$
We now find when the minima are at the same locations. This occurs when θ is the same. Since d is the same in both cases, the left hand side of Eq. 35.2 is the same and we can set the right hand side equal for the two different wavelengths. We need to keep in mind that the index is associated with light of a particular wavelength, so we call the index in Eq. 35.2 m for the 400.0 nm light and n for the 560.0 nm light.	$\left(m + \dfrac{1}{2} \right)(400 \text{ nm}) = \left(n + \dfrac{1}{2} \right)(560 \text{ nm})$
Solve this to find a relationship between n and m. We now try to find values that satisfy this equation. After some trial and error (perhaps by considering $m = 1, 2, 3, \ldots$ and trying to see whether n is an integer), $m = 3, n = 2$ and $m = 10, n = 7$ work.	$\dfrac{2m+1}{2n+1} = \dfrac{560}{400} = \dfrac{7}{5}$ $10m + 5 = 14n + 7$ $10m = 14n + 2$
We can now determine the spacing between these two minima. For instance, the minima for $\lambda = 560.0 \text{ nm}$ occur at $n = 2$ and $n = 7$. Use Eq. 35.2 to find the angle in each case.	$d \sin \theta_n = \left(n + \dfrac{1}{2} \right) \lambda \ \rightarrow \ \theta_n = \sin^{-1} \left(\dfrac{\left(n + \dfrac{1}{2} \right) \lambda}{d} \right)$

$$\theta_2 = \sin^{-1}\left(\frac{2.5\left(560.0\times10^{-9} \text{ m}\right)}{0.100\times10^{-3} \text{ m}}\right) = 0.802°$$

$$\theta_7 = \sin^{-1}\left(\frac{7.5\left(560.0\times10^{-9} \text{ m}\right)}{0.100\times10^{-3} \text{ m}}\right) = 2.41°$$

We now find the position on the screen in both cases using the geometry in Fig. 35.15. In this case, x is the distance to the screen and y is the distance from the central maximum on the screen.	$y_n = x\tan\theta_n$ $y_2 = \left(1.00 \text{ m}\right)\tan\left(0.802°\right) = 0.0140 \text{ m}$ $y_7 = \left(1.00 \text{ m}\right)\tan\left(2.41°\right) = 0.0420 \text{ m}$ $\Delta y = y_7 - y_2 = \boxed{2.80\times10^{-2} \text{ m}}$

CHECK and THINK

The key is that for a dark spot, both wavelengths must produce a minimum. By finding where the minimum occurs for each wavelength, we can determine when they are equal. The separation is quite small—in this case, we could have used the small angle approximation, where we take $\sin\theta \approx \theta$ and $\tan\theta \approx \theta$.

36

Applications of the Wave Model

3. **(N)** The "hydrogen line" at 1420.4 MHz corresponds to the natural frequency of neutral hydrogen atoms and plays an important role in radio astronomy. What size dish is required so that a radio telescope receives this frequency with an angular resolution of 0.0500°?

INTERPRET and ANTICIPATE	
This is an example of Rayleigh's criterion, which allows us to determine the smallest angular separation that we can resolve from a circular aperture, including this telescope. We can relate the desired angular separation with the size of the aperture, which in this case is the diameter of the telescope.	

SOLVE	
We will use Rayleigh's criterion (Eq. 36.4), which indicates the minimum angular separation (in radians) that can be resolved when an aperture of diameter d is used to collect light of wavelength λ.	$\theta_{min} \approx 1.22 \dfrac{\lambda}{d}$
Calculate the wavelength of the hydrogen line (which is also known as the 21-centimeter line, which we see refers to the wavelength).	$\lambda = \dfrac{c}{f} = \dfrac{3.00 \times 10^8 \text{ m}}{1420.4 \times 10^6 \text{ Hz}}$ $\lambda = 0.211 \text{ m} = 21.1 \text{ cm}$
It's important to remember that the angle in Rayleigh's criterion is in radians, so we need to convert from degrees to radians.	$\theta_{min} = 0.0500° \left(\dfrac{\pi}{180°} \right) = 8.73 \times 10^{-4} \text{ rad}$

Solve Eq. 36.4 for the diameter d and insert values.	$d = 1.22 \dfrac{\lambda}{\theta_{min}}$ $d = 1.22 \left(\dfrac{0.211 \text{ m}}{8.73 \times 10^{-4} \text{ rad}} \right)$ $d = \boxed{295 \text{ m}}$

CHECK and THINK

This is a *really* large dish. Only the Arecibo radio telescope in Puerto Rico is large enough to observe the 21-cm line at this resolution. However, radio telescopes are often combined electronically to increase the "baseline" or effective observing size of the telescope.

4. (N) Post-Impressionist Georges Seurat is famous for a painting technique known as pointillism. His paintings are made up of different-colored dots, and each dot is roughly 2 mm in diameter. When viewed from far enough away, the dots blend together and only large-scale images are seen. Assuming that the human pupil is about 2.5 mm in diameter, determine an approximate minimum distance you need to stand from one of Seurat's paintings to see the dots blended together.

INTERPRET and ANTICIPATE

If we stand far enough away, our eyes cannot resolve the dots. This is a question of what is the minimum angular separation of two objects that we can resolve, which is ultimately asking about the diffraction limited resolution in Rayleigh's criterion.

SOLVE Estimate the resolution of the human eye using the equation for diffraction limited resolution (Rayleigh's criterion, Eq. 36.4). We can choose the center of the visible spectrum, green light at around 550 nm, for our estimate. This provides the smallest angular separation that we can resolve.	$\theta_{min} = 1.22 \dfrac{\lambda}{d}$ $\theta_{min} = 1.22 \left(\dfrac{550 \times 10^{-9} \text{ m}}{2.5 \times 10^{-3} \text{ m}} \right)$ $\theta_{min} = 2.68 \times 10^{-4} \text{ rad}$
Let's consider two dots that touch, such that their center-to-center distance y is equal to one diameter. Two dots will just barely be resolved if their separation is equal to θ_{min}. Using the small angle	$\dfrac{y}{x} = \tan \theta_{min} \approx \theta_{min}$ $x \approx \dfrac{y}{\theta_{min}} = \dfrac{2 \times 10^{-3} \text{ m}}{2.68 \times 10^{-4} \text{ rad}} \approx \boxed{7.5 \text{ m}}$

approximation, we can relate the angle to the distance between dots on the painting y and the distance from your eye to the painting x. We find that the distance (to one significant figure) is around 7.5 meters.

CHECK and THINK
So if we stand further than around 7.5 m (about 25 ft) from the painting we would not be able to see the individual dots. Of course, if you see one of these paintings in a museum, you are likely to want to approach even closer to see how much work went into creating the dots that blend together from a distance!

10. (N) Later in this textbook, you will explore the intriguing idea that matter can exhibit wavelike behavior when moving at velocities near the speed of light. One example where this is used is in the imaging of materials in an electron microscope. A common wavelength for the electrons in a transmission electron microscope is 2.5×10^{-12} m. Suppose the diameter of the aperture is 5.0 μm. What is the diffraction-limited resolution of this microscope?

INTERPRET and ANTICIPATE
The diffraction-limited resolution is given by Eq. 36.3 or Eq. 36.4. Given the relative values of the wavelength and the diameter (about 6 orders of magnitude apart), we can safely use Eq. 36.4, which is the small-angle approximation of Eq. 36.3. Given that the TEM is used to image crystals and see effects due to the arrangement of planes of atoms in the crystal, this angle should be very small. That way, the aperture can be placed near the specimen to achieve imaging things that are nanometers apart.

SOLVE
Use Eq. 36.4 to find the diffraction-limited resolution.

$$\theta_{min} \approx 1.22\frac{\lambda}{d} = 1.22\frac{\left(2.5\times10^{-12}\text{ m}\right)}{\left(5.0\times10^{-6}\text{ m}\right)} = \boxed{6.1\times10^{-7}\text{ rad}}$$

CHECK and THINK
The resolution defines the angular separation required between two objects, as measured from the aperture, so that they might be resolved, given the shape and size of the aperture.

14. (N) When you spread oil (n_{oil} = 1.50) on water (n_{water} = 1.33) and have white light incident from above, you might observe different reflected colors that depend on the thickness of the oil on the water at each point. Consider the wavelength of red light (750.0 nm) and the wavelength of violet light (380.0 nm), covering the visible spectrum. What is the minimum possible thickness of oil that would give rise to seeing **a.** the red light reflected and **b.** the violet light reflected?

INTERPRET and ANTICIPATE	
We must first determine whether we should use Eq. 36.12, or Eq. 36.14, in order to model the desired interference in the thin film. In this problem the light will undergo a phase shift when reflecting off of the first surface because the index of refraction of the oil is greater than that of air, but not when it reflects off of the second surface because the index of refraction of the water is less than that of the oil. Given that it is desired to see the color reflected in each case, we are looking at a case of constructive interference. Thus, for constructive interference in this scenario, we must use Eq. 36.12, where the index of refraction of the thin film is 1.50. We expect the minimum thickness must be greater for red than violet because the wavelength of red light is greater than that of the violet light.	
SOLVE **a.** Use Eq. 36.12 to find the minimum thickness ($m = 0$).	$$2w = (m+1/2)\frac{\lambda_0}{n_f} = 2w = (0+1/2)\frac{\lambda_0}{n_f}$$ $$2w = (1/2)\frac{\lambda_0}{n_f}$$
Then, solve for the minimum thickness of the oil film, w.	$$w = (1/4)\frac{\lambda_0}{n_f}$$ $$w_{red} = (1/4)\frac{750.0\times10^{-9}\ \text{m}}{1.50} = \boxed{1.25\times10^{-7}\ \text{m}}$$
b. Use Eq. 36.12 to find the minimum thickness ($m = 0$). This time, we use the wavelength of the violet light in air.	$$w = (1/4)\frac{\lambda_0}{n_f}$$ $$w_{violet} = (1/4)\frac{380.0\times10^{-9}\ \text{m}}{1.50} = \boxed{6.33\times10^{-8}\ \text{m}}$$
CHECK and THINK	
The minimum thickness required is larger for the red light as expected. Note that there are other thicknesses for which we would see constructive interference for either color. To find them, we would just need to change m and solve again.	

17. (N) An oil slick on water displays a variety of iridescent colors due to interference effects. Assuming the film has the minimum thickness to produce the colors observed, what thickness of the oil film will appear green ($\lambda = 550.0$ nm) or red ($\lambda = 700.0$ nm)? The index of refraction of the oil is 1.47.

INTERPRET and ANTICIPATE We can use the condition for constructive interference for the wavelengths given. Since the index of refraction is higher than the index of refraction for both air and water, we can use the formulas for a thin film that were derived in the chapter. Light reflecting off the top surface of the oil (where the index of refraction n increases from 1 to 1.47) experiences a 180° phase shift while light reflecting off the bottom surface (where n decreases from 1.47 to something like 1.33 if it's on water) does not. Therefore, for constructive interference, the minimum thickness is ¼ of the wavelength in the oil λ_f, since the total path difference (the extra segment for the ray that goes down and back up through the oil) is $\lambda/2$ and with the additional half wavelength phase shift produces constructive interference. Therefore, we expect that the minimum width is ¼ of the wavelength in the oil ($\lambda_f = \lambda_0/n$).	

SOLVE The condition for constructive interference is given by Equation 36.12, where w is the width of the film, λ_0 is the wavelength of light in air, n_f is the index of refraction of the fluid, and $m = 0$ corresponds to the smallest thickness film leading to constructive interference.	$2w = \left(m + \dfrac{1}{2} \right) \dfrac{\lambda_0}{n_f} \quad m = 0,\ 1, 2, 3...$
Solve for the oil layer thickness. Note, this formula agrees with the discussion in the Interpret and Anticipate section above.	$w = \dfrac{1}{4} \dfrac{\lambda_0}{n_f}$
Plug in numbers for each color.	$w_{\text{green}} = \dfrac{1}{4} \cdot \dfrac{550 \text{ nm}}{1.47} = 93.5 \text{ nm} = \boxed{9.35 \times 10^{-8} \text{ m}}$ $w_{\text{red}} = \dfrac{1}{4} \cdot \dfrac{700 \text{ nm}}{1.47} = 119 \text{ nm} = \boxed{1.19 \times 10^{-7} \text{ m}}$

CHECK and THINK
We found the minimum thickness of the oil layer needed to appear green or red. In both cases, a width of ¼ of the wavelength in the fluid leads to a total extra path length of a

half wavelength. Along with the half wavelength shift on reflection from the oil (since $n_{oil} > n_{air}$), this produces an interference maximum. The smaller the wavelength, the thinner the film leading to constructive interference.

23. (N) An oil slick of thickness 325 nm and index of refraction $n_{oil} = 1.50$ forms on the surface of a still pond ($n_{water} = 1.33$). When it is viewed straight from above, what is the wavelength of visible light that is **a.** most strongly reflected and **b.** most strongly transmitted?

INTERPRET and ANTICIPATE

Note that this is different than the situation described in Example 36.3A, since the oil in this case has an index of refraction higher than the water, but it is similar to Problem 17 above. For instance, the light rays that reflect off the front surface (where n increases from 1 to 1.5) experience a 180° phase shift while the light that reflects off the back surface (where n decreases from 1.5 to 1.33) does not. This determines which of the interference equations we use for constructive or destructive interference.

SOLVE

a. Since only the ray reflected from the front surface experience a phase reversal, there will be constructive interference for reflection if the thickness is a quarter wavelength. This would lead to a path length difference of half a wavelength as the ray traveling through the oil traverses it on the way in and out and therefore, including the extra 180° phase shift, the total phase difference would be 1 wavelength). This suggests that Eq. 36.14 is appropriate here, since for $m = 0$, $w = \dfrac{1}{4} \cdot \dfrac{\lambda_0}{n_f}$,

where λ_0 is the wavelength in air and λ_0/n_f is the wavelength in the fluid.

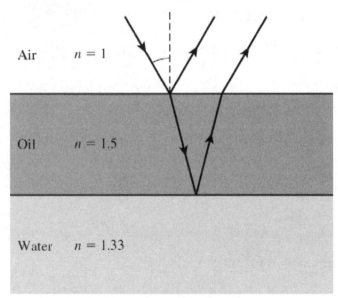

Figure P36.23aANS

$$2w = \left(m + \frac{1}{2} \right) \frac{\lambda_0}{n_f} \qquad \text{(Eq. 36.14)}$$

Solve for the wavelength.	$$\lambda_0 = \frac{2wn_f}{m+1/2}$$
Insert numbers. We see that the $m = 1$ constructive interference wavelength is in the visible spectrum, in the red range.	$$\lambda_0 = \frac{2(3.25\times10^{-7}\text{ m})(1.50)}{0.5} = 1.95\times10^{-6}\text{ m} \quad (m=0)$$ $$\lambda_0 = \frac{2(3.25\times10^{-7}\text{ m})(1.50)}{1.5} = \boxed{6.50\times10^{-7}\text{ m}} \quad (m=1)$$ $$\lambda_0 = \frac{2(3.25\times10^{-7}\text{ m})(1.50)}{2.5} = 3.90\times10^{-7}\text{ m} \quad (m=2)$$
b. For transmitted light (see Figure 36.15), neither ray experiences a phase shift and they arrive in phase if the thickness is a multiple of half a wavelength, such that the total phase shift is a multiple of a whole wavelength. That is, $w = m \cdot \dfrac{1}{2} \cdot \dfrac{\lambda_0}{n_f}$, which is Eq. 36.12.	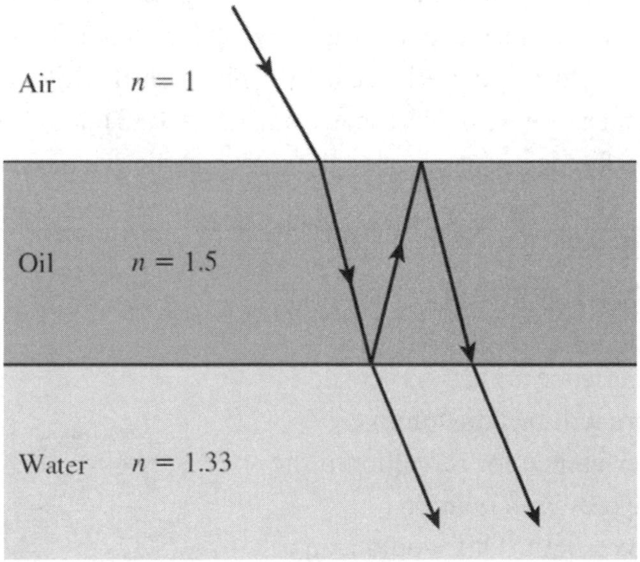 Air $\quad n = 1$ Oil $\quad n = 1.5$ Water $\quad n = 1.33$ **Figure P36.23bANS** $$2w = m\frac{\lambda_0}{n_f}$$
Again, solve for the wavelength.	$$\lambda_0 = \frac{2wn_f}{m}$$
Insert numbers. We find that the $m = 2$ constructive interference wavelength is in the visible spectrum, in the blue range.	$$\lambda_0 = \frac{2(3.25\times10^{-7}\text{ m})(1.50)}{1} = 9.75\times10^{-7}\text{ m} \quad (m=1)$$ $$\lambda_0 = \frac{2(3.25\times10^{-7}\text{ m})(1.50)}{2} = \boxed{4.88\times10^{-7}\text{ m}} \quad (m=2)$$ $$\lambda_0 = \frac{2(3.25\times10^{-7}\text{ m})(1.50)}{3} = 3.25\times10^{-7}\text{ m} \quad (m=3)$$

CHECK and THINK

For thin film interference, there are basically two equations we can use, Eqs. 36.12 and 36.14. We need to think through the individual situations, which vary based on the relative index of refraction of the layers (and therefore where phase reversals occur on reflection), to determine which one is appropriate.

25. (N) Light of wavelength 566 nm is incident on a grating. Its third-order maximum is at 15.7°. What is the density of rulings on this grating?

INTERPRET and ANTICIPATE

The angle of the diffraction peaks depends on the wavelength of light and the distance between the rulings in the grating. With the information given, we can calculate this distance and then the density of rulings. We would expect a very small distance and a very large density.

SOLVE	
The location of the lines in a diffraction pattern can be calculated using Eq. 36.15.	$d \sin \theta = m\lambda$
Solve for the distance between the rulings on the diffraction grating.	$d = \dfrac{m\lambda}{\sin \theta}$
Insert values.	$d = \dfrac{3\left(566 \times 10^{-9} \text{ m}\right)}{\sin 15.7°} = 6.27 \times 10^{-6} \text{ m}$
d is the number of meters between lines. The density is the number of lines per meter, so it's one over d.	$n = \dfrac{1}{d} = \dfrac{1}{6.27 \times 10^{-6} \text{ m}}$ $n = \boxed{1.59 \times 10^{5} \text{ rulings/m}}$

CHECK and THINK

As expected, the spacing is quite small, at around 6 microns, so the density is quite large.

26. (N) A diffraction grating with a ruling separation of 3.00 μm is illuminated with light of wavelength 525 nm. What is the highest-order maximum visible?

INTERPRET and ANTICIPATE	
We must use Eq. 36.15 to model the limit of m. Based on the question, there must be some maximum allowed value of m. This will come about because it must be true that $\sin\theta \leq 1$. Our answer must be a whole number and greater than or equal to 0.	
SOLVE Write Eq. 36.15 and solve for $\sin\theta$, using the values from the problem statement.	$d\sin\theta = m\lambda$ $\sin\theta = \dfrac{m\lambda}{d} = m\dfrac{525\times10^{-9}\text{ m}}{3.00\times10^{-6}\text{ m}}$ $\sin\theta = m(0.175)$
Examining the above result, because it must be true that $\sin\theta \leq 1$, the biggest m can be is 5. If we plug in $m = 6$, we get the result shown to the right, but this has no solution. The last value for which an angle exists is $\boxed{5}$.	$\sin\theta = (6)(0.175) = 1.05$ $\theta = \text{does not exist}$
CHECK and THINK	
Note that this has nothing to do with the location of a screen or other kind of geometry. This limit is imposed by the ratio of the wavelength and the ruling separation.	

30. (N) A diffraction grating is 5.00 cm wide and has 515 rulings per millimeter. If light of **a.** 450.0 nm and **b.** 650.0 nm is incident on the grating, what is the half-width of the central maximum in each case?

INTERPRET and ANTICIPATE	
The half-width is an angle that describes the separation between the edge of the maximum relative to the center of the spot. From the chapter, we have a formula for the half-width of the lines, so the calculation should be straightforward.	
SOLVE **a.** For both parts, it is helpful to remember that the number of lines N times the spacing per line d equals the total length of the grating ℓ.	$Nd = \ell$
The half-width of each line is given by Eq. 36.16, where λ is the wavelength of light and θ the half-width angle.	$\Delta\theta_{hw} = \dfrac{\lambda}{Nd\cos\theta} = \dfrac{\lambda}{\ell\cos\theta}$

Insert values for the 450.0 nm light. The central maximum ($m = 0$) occurs at $\theta = 0$.	$\Delta\theta_{hw} = \dfrac{450 \times 10^{-9} \text{ m}}{\left(5.00 \times 10^{-2} \text{ m}\right)\cos 0}$
	$\Delta\theta_{hw} = \boxed{9.00 \times 10^{-6} \text{ rad}}$
b. Now, insert values for the 650.0 nm light.	$\Delta\theta_{hw} = \dfrac{650 \times 10^{-9} \text{ m}}{\left(5.00 \times 10^{-2} \text{ m}\right)\cos 0}$
	$\Delta\theta_{hw} = \boxed{1.30 \times 10^{-5} \text{ rad}}$

CHECK and THINK

The larger wavelength leads to a wider central maximum.

37. (N) A grating has 3.200×10^4 rulings and is 5.00 cm wide. Light of wavelength 555 nm is incident on the grating. What is its dispersion for the first- and second-order lines?

INTERPRET and ANTICIPATE

The dispersion indicates how strongly the grating separates wavelengths (or colors in the case of visible light, as in this problem).

SOLVE The dispersion of a grating is given by Eq. 36.19, where m is the order of the line, d is the spacing between rulings, and θ is the angle of the line.	$D \equiv \dfrac{\Delta\theta}{\Delta\lambda} = \dfrac{m}{d\cos\theta}$
We are given the total width of the grating ℓ and the number of rulings N, from which we can determine the distance between rulings.	$d = \dfrac{\ell}{N} = \left(0.0500 \text{ m}\right) \big/ \left(3.200 \times 10^4\right)$
We can now determine the angles for each line using Eq. 36.15.	$d\sin\theta = m\lambda$ $\theta_m = \sin^{-1}\left(\dfrac{m\lambda}{d}\right)$

Calculate the angle for the first and second orders.	$\theta_1 = \sin^{-1}\left(\dfrac{\lambda}{d}\right) = \sin^{-1}\left(\dfrac{555 \times 10^{-9}\ \text{m}}{(0.0500\ \text{m})/(3.200 \times 10^4)}\right)$ $\theta_2 = \sin^{-1}\left(\dfrac{2\lambda}{d}\right) = \sin^{-1}\left(\dfrac{2(555 \times 10^{-9}\ \text{m})}{(0.0500\ \text{m})/(3.200 \times 10^4)}\right)$

Finally, we can calculate the dispersion for each order

$$D_1 = \frac{1}{\left((0.0500\ \text{m})/(3.200 \times 10^4)\right)\cos\left(\sin^{-1}\left(\dfrac{555 \times 10^{-9}\ \text{m}}{(0.0500\ \text{m})/(3.200 \times 10^4)}\right)\right)} = \boxed{6.85 \times 10^5\ \text{rad/m}}$$

$$D_2 = \frac{2}{\left((0.0500\ \text{m})/(3.200 \times 10^4)\right)\cos\left(\sin^{-1}\left(\dfrac{2(555 \times 10^{-9}\ \text{m})}{(0.0500\ \text{m})/(3.200 \times 10^4)}\right)\right)} = \boxed{1.82 \times 10^6\ \text{rad/m}}$$

CHECK and THINK

The second order line has a larger dispersion, which means that it separates colors more strongly than the first order peak.

---

39. (N) A diffraction grating is exposed to light from a source that consists of two different wavelengths: 575 nm and 585 nm. What is the minimum number of rulings necessary to be able to resolve the **a.** $m = 1$ lines, **b.** $m = 2$ lines, and **c.** $m = 3$ lines?

INTERPRET and ANTICIPATE
Use Eq. 36.20 to first find the resolving power for each scenario. This does not vary as we look for the number of rulings necessary in order to resolve each order of lines. Given the resolving power required, we can then use Eq. 36.21 to find the number of rulings necessary to resolve a particular order. We expect that higher-order lines require fewer rulings than lower-order lines. This is because the dispersion or higher order lines is greater, making them more distinct and easier to resolve than lower-order lines.

SOLVE	
First, find the resolving power, using the wavelengths given and Eq. 36.20.	$R = \dfrac{\lambda_{av}}{\Delta\lambda} = \dfrac{(\lambda_1 + \lambda_2)/2}{\lambda_2 - \lambda_1} = \dfrac{(575\ \text{nm} + 585\ \text{nm})/2}{585\ \text{nm} - 575\ \text{nm}} = 58$
a. Now, use the resolving power with Eq. 36.21 to find the number of rulings necessary for the $m = 1$ lines.	$R = Nm$ $N = R/m$ $N = 58/1 = \boxed{58}$

b. Use the resolving power with Eq. 36.21 to find the number of rulings necessary for the $m = 2$ lines.	$R = Nm$ $N = R/m$ $N = 58/2 = \boxed{29}$
c. Use the resolving power with Eq. 36.21 to find the number of rulings necessary for the $m = 3$ lines. Here, we need 20 lines because 19 is not quite enough, as we see when we get 19.3, when solving.	$R = Nm$ $N = R/m$ $N = 58/3 = 19.3 \rightarrow \boxed{20}$

CHECK and THINK

As expected, the number of rulings required with a necessary resolving power of 58 goes down as the order number increases. The dispersion of the lines in each order matters, though we did not actually find the dispersion in this scenario.

---

42. (A) A diffraction grating with n rulings per unit length and total length ℓ is illuminated by light with several similar wavelengths. Find an expression for the resolving power of this diffraction grating for the first-order maxima of two similar wavelengths of light in terms of n and ℓ.

INTERPRET and ANTICIPATE

Note that the number of rulings per unit length, n, could be expressed as $n = N/\ell$, where N is the number of rulings, and ℓ is the length of the grating. Then, we can use Eq. 36.21 to express the resolving power. The resolving power should also depend on the order number, m, since it is easier to resolve higher-order lines, due to dispersion. In other words, the resolving power of a grating goes up for higher-order maxima.

SOLVE First, express Eq. 36.21.	$R = Nm$
Then, using the fact that $n = N/\ell$, we can solve for N and substitute into the equation for the resolving power to express it in terms of n, ℓ, and m.	$R = Nm = \boxed{n\ell m}$

CHECK and THINK

The number of rulings on the grating and its length are not changing. This means that the resolving power of the grating is higher for higher-order maxima, m, as expected. This, again, as in other examples, is due to dispersion.

45. (N) CASE STUDY Michelson's interferometer played an important role in improving our understanding of light, and it has many practical uses today. For example, it may be used to measure distances precisely. Suppose the mirror labeled 1 in Figure 36.30 (page 1176) is movable. If the laser light has a wavelength of 632.5 nm, how many fringes will pass across the detector if mirror 1 is moved just 1.000 mm? If you can easily detect the passage of just one fringe, how accurately can you measure the displacement of the mirror?

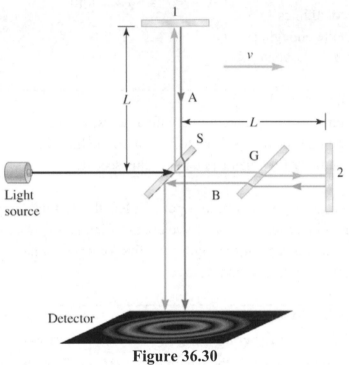

Figure 36.30

INTERPRET and ANTICIPATE

Moving the mirror on the order of the wavelength of light increases the path length of one arm of the interferometer leading to alternating constructive and destructive interference. Since it is easy to count these fringes, it means we have a method to reliably measure *really* small displacements that are on the order of the wavelength of light.

SOLVE

For a displacement of the mirror Δd, the path length increases by twice that since the light travels out and back along the arm this extra distance. The number of wavelengths that the path length changes by (which equals the number of fringes) is this extra distance divided by the wavelength of light, since changing the path length by one wavelength corresponds to a

$$n = \frac{2\Delta d}{\lambda}$$

Chapter 36 – Applications of the Wave Model

full cycle of constructive to destructive to constructive interference.	
Insert numbers.	$n = \dfrac{2\left(1.000\times10^{-3}\text{ m}\right)}{632.5\times10^{-9}\text{ m}} = \boxed{3162\text{ fringes}}$
Because we can easily measure one fringe ($n = 1$), we can measure a displacement of the mirror as small as half a wavelength, the distance the mirror moves for $n = 1$.	$\Delta d = \dfrac{\lambda}{2} = \dfrac{632.5\text{ nm}}{2}$ $\Delta d = \boxed{316.3\text{ nm}}$

CHECK and THINK

Simply by counting fringes, we can measure an incredibly small displacement of the mirror—around 0.3 μm!

48. (N) CASE STUDY The Michelson–Morley interferometer can also be used to determine the wavelength of light. Given light with an unknown wavelength, the position of one of the mirrors is shifted by 1.000 mm and the interference pattern shifts by exactly 4500 fringes. What is the wavelength of the light?

INTERPRET and ANTICIPATE	
The displacement of the mirror changes the optical path length. We can relate the change in phase with the displacement of the mirror.	

SOLVE A displacement of the mirror changes the optical path length by twice the displacement distance Δx, since the light has to travel out and back this extra distance. Moving by one fringe indicates a phase difference corresponding to a path length difference of one wavelength.	$4500\text{ fringes} = 4500\lambda = 2\Delta x$
Solve for the wavelength and insert numerical values.	$\lambda = \dfrac{2\Delta x}{4500}$ $\lambda = \dfrac{2\left(1.000\times10^{-3}\text{ m}\right)}{4500}$ $\lambda = \boxed{4.4\times10^{-7}\text{ m}} = 440\text{ nm}$

343

Chapter 36 – Applications of the Wave Model

50. (E) Optical flats are flat pieces of glass used to determine the flatness of other optical components. They are placed at an angle above the component as shown in Figure P36.49A, and monochromatic light is incident and observed from above, leading to interference fringes. Figure P36.49C shows the results of one of these tests. What is the approximate difference in the gap thickness between the left and right sides of the optical flat and the component? Is it possible to determine from this figure alone which side has the greater gap thickness (left or right)?

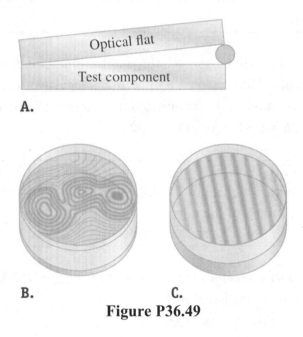

Figure P36.49

Moving from left to right, each bright line indicates constructive interference. Neighboring bright lines have gaps that differ in total path length by one wavelength. This path difference is a difference between light rays that bounce off the top surface of the air gap versus rays that make an extra down-and-up reflection in the air gap, bouncing off the top of the lower glass piece. (It is also possible to use optical flats with transmitted light, but the argument is similar.) Neighboring bright lines have gaps that differ by half a wavelength to give a total path difference of one wavelength. Since there are around 8 bright lines across the sample, the gap thickness changes by around 4λ from left to right. For visible light, roughly around 500 nm, this would correspond to about 2 microns. See Problem 53 below for a quantitative example.

We are not able to determine which side has the larger gap thickness. There is nothing to indicate that a bright fringe corresponds to a phase difference of 2π versus 20π or whether the phase increases or decreases by 2π as you move from one bright fringe to another. We can only determine the rate of change of the gap thickness as we move along the gap.

53. (N) Figure P36.53 shows two thin glass plates separated by a wire with a square cross section of side length w, forming an air wedge between the plates. What is the edge length w of the wire if 42 dark fringes are observed from above when 589-nm light strikes the wedge at normal incidence?

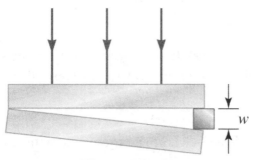

Figure P36.53

INTERPRET and ANTICIPATE

As we look along the plates from left to right, the air gap increases. Since the air gap will lead to interference and we know that the thickness of the gap determines whether there will be constructive or destructive interferences, we would expect to see alternating bright and dark fringes from left to right as the gap thickness steadily increases. The number of fringes will then be related to the change in size of the gap.

SOLVE	
Consider two rays: one that reflects off the top of the air gap and one that reflects off the top of the bottom piece of glass. The second one experiences a 180° phase shift (since n increases from air to glass). For destructive interference, we require that the gap thickness is a multiple of half a wavelength so that the total path length difference (due to the extra length down and up through the air gap) is a multiple of a whole wavelength. Then, with the extra 180° phase shift off the lower surface, the two rays will destructively interfere. This is described by Eq. 36.14 where each dark spot corresponds to a different index m.	$2w = m\dfrac{\lambda_0}{n_f}$

In this case, the fluid is just air, so $n_f = 1$. If we assume that one edge of the plate corresponds the to $m = 0$ (zero gap), then the other edge is at $m = 41$ (so that there are 42 dark fringes across the glass plate including both $m = 0$ and $m = 41$), we can determine the width w of that side of the gap.	$w = \dfrac{m\lambda}{2}$
Plug in values. The result is the width of the right side of the gap and therefore the thickness of the square wire.	$w = \dfrac{41\left(589 \times 10^{-9} \text{ m}\right)}{2}$ $w = \boxed{1.21 \times 10^{-5} \text{ m}} = 12.1 \ \mu\text{m}$

CHECK and THINK

This is actually a very sensitive way to measure a small change in the gap thickness. Since it uses interference of light, we are able to resolve differences in height of a fraction of a wavelength.

57. (N) What is the radius of the beam of an argon laser with wavelength 454.6 nm when viewed 50.0 km away from the laser if the laser's aperture has a radius of 3.00 mm?

INTERPRET and ANTICIPATE

When a beam is emitted from a circular aperture, it has a central maximum that spreads outward as a cone, producing a cross-sectional pattern called the Airy disk. We can use the expression for the angular spread of this beam to determine how wide the beam is at 50 km.

SOLVE The beam undergoes diffraction from a circular opening of diameter $d = 2r$ = 6.00 mm and spreads as a cone with an angular radius (i.e. half-angle of the opening) given by Eq. 36.2. This is the Airy disk.	$\theta_R = 1.22 \dfrac{\lambda}{d}$
Insert values to find the angle at which the beam spreads.	$\theta_R = 1.22 \left(\dfrac{454.6 \times 10^{-9} \text{ m}}{0.006\,00 \text{ m}} \right)$ $\theta_R = 9.24 \times 10^{-5} \text{ rad}$

The angle in radians can be expressed as the radius of the beam r_{beam} at a distance L. We might also describe the radius of the beam is an arc length that subtends the angle.	**Figure P36.57ANS**
	$$\theta_{min} = \frac{r_{beam}}{L}$$
Solve for the radius of the beam at a distance L.	$$r_{beam} = \theta_{min} L$$ $$r_{beam} = (9.24 \times 10^{-5}\ \text{rad})(5.00 \times 10^4\ \text{m})$$ $$r_{beam} = \boxed{4.62\ \text{m}}$$

CHECK and THINK

We think of a laser as being a straight beam of light, but it spreads out slowly. At a very large distance of 50 km, the beam has spread to over 9 meters in diameter!

---

60. (N) How many rulings must a diffraction grating have if it is just to resolve the sodium doublet (589.592 nm and 588.995 nm) in the second-order spectrum?

INTERPRET and ANTICIPATE
The resolution of a grating depends on the number of rulings on the grating and the order of the diffraction that we are using. It can also be expressed in terms of the average wavelength divided by the difference in wavelength that we can resolve.

SOLVE The resolving power of grating is directly related to the number of rulings N through Equation 36.21, where m is the order of the spectrum and λ_{avg} and $\Delta\lambda$ are the average wavelength and difference between the two.	$$R = \frac{\lambda_{avg}}{\Delta\lambda} = Nm$$
For the sodium doublet, we first calculate the average wavelength λ_{avg} and the difference $\Delta\lambda$.	$$\lambda_{avg} = \frac{589.592 + 588.995}{2}\ \text{nm} = 589.294\ \text{nm}$$ $$\Delta\lambda = 588.995\ \text{nm} - 589.592\ \text{nm} = 0.597\ \text{nm}$$

Now use Eq. 36.21 to solve for N for the second order ($m = 2$) spectrum.	$N = \dfrac{\lambda_{\text{avg}}}{m\Delta\lambda} = \dfrac{589.294 \text{ nm}}{2(0.597 \text{ nm})} = \boxed{494}$

CHECK and THINK

The grating has 494 rulings. The resolving power is quite high $R = Nm = 988$.

64. (N) Later in this book, you will explore the intriguing idea that matter can exhibit wavelike behavior when moving at velocities near the speed of light. One example where this is used is in the imaging of materials in an electron microscope. A common wavelength for the electrons in a transmission electron microscope is 2.5×10^{-12} m. Assume the diameter of the aperture is 500.0 μm. What is the required distance between the aperture and the specimen if we want to resolve two atoms separated by approximately 1.0×10^{-9} m, assuming the resolution is not dependent on other factors?

INTERPRET and ANTICIPATE

We can use Eq. 36.4 to express the diffraction-limited resolution, given that the diameter of the aperture is so much greater than the distance between the objects we are trying to resolve (the two atoms). In order to think about the distance between the aperture and the specimen, however, a sketch is helpful (Figure P36.64ANS). In the sketch, we would like to find L, the distance between the aperture and specimen. This can be done, using geometry once we know the minimum angular resolution.

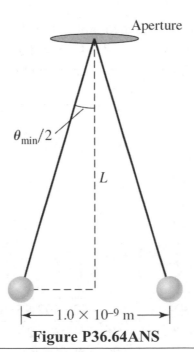

Figure P36.64ANS

SOLVE	
Use Eq. 36.4 and the values given in the problem statement, in order to find the angular resolution.	$\theta_{min} \approx 1.22\dfrac{\lambda}{d} = 1.22\dfrac{2.5 \times 10^{-12} \text{ m}}{5.000 \times 10^{-4} \text{ m}} = 6.10 \times 10^{-9} \text{ rad}$
From the sketch, we can see that this angle would be bisected by L, the distance from the aperture to the specimen. Also, the length of the opposite side from this half-angle would be $\dfrac{1}{2}\left(1.0 \times 10^{-9} \text{ m}\right)$. Thus, using the tangent function, we get the relationship shown on the right.	$\tan\left(\theta_{min}/2\right) = \dfrac{\dfrac{1}{2}\left(1.0 \times 10^{-9} \text{ m}\right)}{L}$
Use the angular resolution and the given distance between the atoms to find the distance between the aperture and the specimen.	$\tan\left(6.10 \times 10^{-9} \text{ rad}\right) = \dfrac{\dfrac{1}{2}\left(1.0 \times 10^{-9} \text{ m}\right)}{L}$ $L = \dfrac{\dfrac{1}{2}\left(1.0 \times 10^{-9} \text{ m}\right)}{\tan\left(6.10 \times 10^{-9} \text{ rad}\right)} = \boxed{8.2 \times 10^{-2} \text{ m}}$

CHECK and THINK

In a real TEM, there are other parameters that affect resolution, but if the issue were only dependent on diffraction limited resolution, then a distance of about 8.2 cm would be appropriate.

67. (N) The fringe width β is defined as the distance between two consecutive maxima (or consecutive minima). In a Young's double-slit experiment, the fringe width obtained from a source of wavelength 500.0 nm is 3.90 mm. If the apparatus is immersed in a liquid that has index of refraction $n = 1.30$, what is the new fringe width? Assume the apparatus was originally in air and that the small-angle approximation applies to this situation.

INTERPRET and ANTICIPATE	
The double slit was covered in Chapter 35. The distance between maxima depends on the spacing between the slits and the wavelength of light, which changes when the apparatus is submerged in water.	

SOLVE	
The position of the n-th maxima is given by equation 35.4 for small angles.	$y_n \approx \dfrac{nx\lambda}{d} \qquad n = 0, \pm 1, \pm 2, \pm 3...$

We now find the distance between the adjacent fringes, which is the fringe width.	$\beta = \Delta y = y_{n+1} - y_n$ $\beta = (n+1)\dfrac{x\lambda}{d} - n\dfrac{x\lambda}{d}$ $\beta = \dfrac{x\lambda}{d}$
The fringe width depends on the wavelength. As the apparatus is immersed in a liquid of refractive index n, the wavelength will decrease from the original value λ to $\lambda_n = \lambda/n$.	$\lambda_n = \dfrac{\lambda}{n}$
We find that the submerged fringe width is equal to that in air divided by the index of refraction.	$\beta_n = \dfrac{x\lambda_n}{d} = \dfrac{x\lambda}{nd} = \dfrac{\beta}{n}$
Now, insert values.	$\beta_n = \dfrac{3.90 \text{ mm}}{1.30} = 3.00 \text{ mm} = \boxed{3.00 \times 10^{-3} \text{ m}}$

CHECK and THINK

The new fringe width is smaller since the wavelength decreased when submerged in water.

37

Reflection and Images Formed by Reflection

3. An image is formed from an object using a camera obscura. The image is one-fifth the height of the object and is oriented upside-down compared to the orientation of the object.

a. (N) What is the magnification of the image? Answer using two significant figures.

b. (A) Write an expression for the image distance in terms of the object distance.

INTERPRET and ANTICIPATE	
Knowing that the image is inverted compared to the object's orientation, we can say that the magnification is negative. We also know that the image is smaller than the object. This means the magnification should be less than 1. We can relate the magnification to the image and object distances using Eq. 37.2. This will allow us to develop an expression relating the two distances, once we have answered part (a).	
SOLVE	
a. Use Eq. 37.1 to find the magnification, knowing that the image is 1/5 the size of the object, and inverted. Again, the image height negative since it is inverted.	$$M = \frac{h_i}{h_o} = \frac{(-1/5)h_o}{h_o} = -1/5 = \boxed{-0.20}$$
b. Now, we can use Eq. 37.2 to express the image distance in terms of the object distance.	$$M = -\frac{d_i}{d_o}$$ $$-0.20 = -\frac{d_i}{d_o}$$ $$d_i = \boxed{(0.20)d_o}$$
CHECK and THINK	
You must be very careful with the sign conventions for the object and image heights and distances. The formulae that you will encounter in this chapter and the next are set up to be self-consistent and aide you in thinking about the numerical results, but only if the information you feed the equations is expressed properly. Images or objects that are inverted have negative values for heights.	

351

5. (N) We have all enjoyed sitting in the shade of a leafy tree. If the leaves of the shade tree are not too densely packed, light from the Sun is able to filter through the tiny openings created by overlapping leaves. These openings essentially form tiny pinholes that project potentially hundreds of circular images of the Sun onto the forest floor. If one of the projected images of the Sun you observe on the forest floor is 4.1 cm in diameter, how high above the ground in the tree canopy is the pinhole that projected this image? *Hint*: You may need to look up the diameter of the Sun and the distance from the Sun to the Earth.

INTERPRET and ANTICIPATE This is an example of a pinhole camera or camera obscura, as in Figure 37.4. The pinhole (the opening in the tree canopy in this case) creates an image of the object (the Sun) on the ground (image). We are not given the distance to the Sun or the diameter of the Sun, so we will have to look those values up.	

SOLVE We can relate the ratio of the size of the Sun to its image to the ratio of the distance to the sun and the distance to the image using Eq. 37.2.	$$M = \frac{h_i}{h_o} = -\frac{d_i}{d_o}$$
In this equation, h_i is the size of the image (the height of the image if your looking straight at it) and d_i is the distance of the image from the pinhole (or the leaves in this case)—so this is the quantity we're seeking. h_o is the size of the object (the diameter of the Sun) and d_o is the distance from the pinhole to the object (that is, the distance from the Earth to the Sun). We have to look up the last couple values.	$h_i = 4.1 \text{ cm}$ $h_o = 1.391 \times 10^9 \text{ m}$ (Sun diameter) $d_o = 1.496 \times 10^{11} \text{ m}$ (Earth-Sun distance) $d_i = ?$
Solve Eq. 37.2 for the distance from the tree canopy to the image on the ground and insert values.	$$d_i = -d_o \frac{h_i}{h_o}$$ $$d_i = -\left(1.496 \times 10^{11} \text{ m}\right)\left(\frac{-4.1 \text{ cm}}{1.391 \times 10^9 \text{ m}}\right)$$ $$d_i = 4.4 \times 10^2 \text{ cm} = \boxed{4.4 \text{ m}}$$

8. (A) Two mirrors are perpendicular as shown in Figure P37.8. A narrow beam of light is incident on one mirror. The angle of incidence is θ_i. Find an expression for the angle of reflection θ_r from the second mirror.

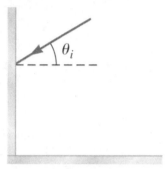

Figure P37.8

INTERPRET and ANTICIPATE

Since this problem concerns only reflections off of plane mirrors, we will use the law of reflection and geometry to determine the answer.

SOLVE	
Consider the first reflection. According to the law of reflection (Eq. 37.3), $\theta_i = \theta_r$.	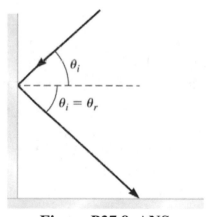 **Figure P37.8aANS**
Now consider the shaded region. The enclosed angles of the triangle must add up to 180°. Therefore, the incident angle for the second reflection must have a value $180° - \theta_i - 90°$ or $90° - \theta_i$ (where θ_i still refers to the initial angle). With	

this, the sum of the three angles is 180° as needed.

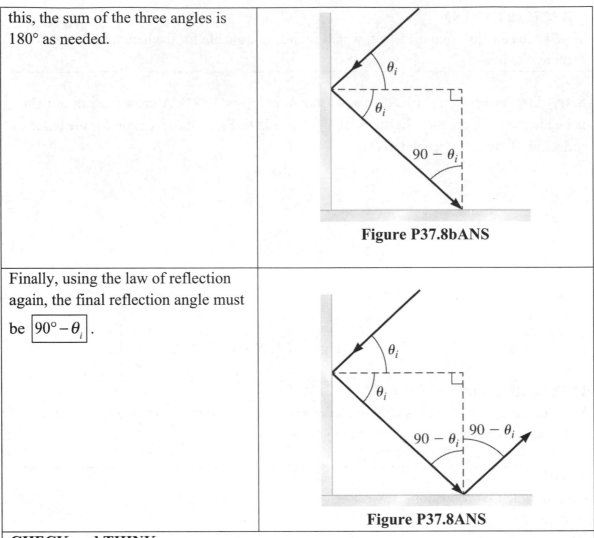

Figure P37.8bANS

Finally, using the law of reflection again, the final reflection angle must be $\boxed{90° - \theta_i}$.

Figure P37.8ANS

CHECK and THINK

Indeed, with the law of reflection and some geometry, we are able to relate the incident and final rays.

12. (N) Two mirrors make a 45.0° angle as shown in Figure P37.11. A narrow beam of light is incident on one mirror. The angle of incidence is 22.3°. Find the angle of reflection from the second mirror.

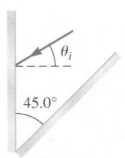

Figure P37.11

354

INTERPRET and ANTICIPATE	
We essentially have to complete Problem 11 on our way to answering this question. Since this problem concerns only reflections off of plane mirrors, we will use the law of reflection and geometry to determine the answer.	

SOLVE

We first sketch the situation.

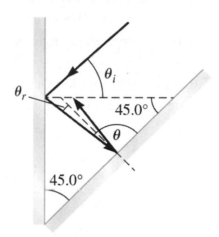

Figure P37.11ANS

Consider the reflection off of mirror 1. Note that the incident light reflects at $\theta_r = \theta_i$ according to the law of reflection (Eq. 37.3). Note also that triangle ABD is a right, isosceles triangle and in order for the interior angles to add to 180°, the angle at the vertex labeled D must be 45°. In triangle ACD, the interior angles must sum to 180°, so we can determine the angle labeled θ.	$\theta_{r,1} = \theta_{i,1}$ $\theta + \theta_{r,1} + 45.0° = \theta + \theta_{i,1} + 45.0° = 180.0°$ $\theta = 135.0° - \theta_{i,1}$
The angle of incidence on the second mirror is $90° - \theta$ (for instance, if $\theta = 90°$, it means that the ray would be aimed straight into the wall and the angle of incidence would be 0°). In addition, the angle of reflection off the second mirror equals the angle of incidence (law of reflection, $\theta_{r,2} = \theta_{i,2}$). This is the goal of Problem 11.	$\theta_{i,2} = \theta - 90.0° = \left(135.0° - \theta_{i,1}\right) - 90.0°$ $\theta_{i,2} = 45.0° - \theta_{i,1} \qquad \text{(mirror 2)}$ $\theta_{r,2} = \theta_{i,2} = 45.0° - \theta_{i,1}$

Chapter 37 – Reflection and Images Formed by Reflection

Now, we can apply this equation in the case that $\theta_{i,1} = 22.3°$.	$\theta_{r,2} = 45.0° - \theta_{i,1}$ $\theta_{r,2} = 45.0° - 22.3° = \boxed{22.7°}$

CHECK and THINK

Using geometry and the law of reflection, we can get pretty far, but we have to be a bit careful in keeping track of labels and angles. In this case, we relate the reflection off the second mirror to the incident angle on the first mirror.

15. (N) Light rays strike a plane mirror at an angle of 45.0° as shown in Figure P37.15. At what angle should a second mirror be placed so that the reflected rays are parallel to the first mirror?

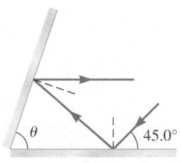

Figure P37.15

INTERPRET and ANTICIPATE

Use the law of reflection and geometry to solve the problem.

SOLVE We sketch a picture to walk through the solution, but the goal is use the geometry to relate angles and fill in the picture piece by piece.	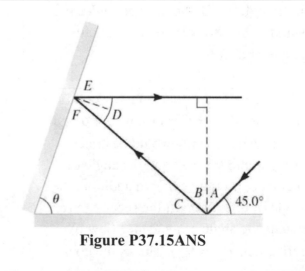 **Figure P37.15ANS**

356

© 2016 Cengage Learning. All Rights Reserved. May not be scanned, copied or duplicated, or posted to a publicly accessible website, in whole or in part.

Angle θ_A and the 45.0° angle form a right angle, so they must sum to 90°: Thus, θ_A = 45.0°.

Because of the law of reflection (Eq. 37.3), angle $\theta_A = \theta_B = 45.0°$ and $\theta_C = 45.0°$.

In the triangle including θ_B and θ_D, the interior angles must sum to 180 degrees: $\theta_B + \theta_D + 90.0° = 180.0°$, so $\theta_D = 180.0° - 90.0° - 45.0° = 45.0°$.

The law of reflection requires that $\theta_E = \theta_F$

$\theta_E + \theta_F + \theta_D = 180.0°$ because they are angles on one side of a straight line, so $2\theta_F = 180.0° - \theta_D = 135.0°$ so $\theta_F = 135.0°/2 = 67.5°$.

Finally, in the triangle including θ_F and θ_C, the interior angles add to 180° and so $\theta_F + \theta_C + \theta = 180.0°$. Therefore $\theta = 180.0° - 67.5° - 45.0° = \boxed{67.5°}$.

CHECK and THINK
This was a fair amount of work and there are a couple ways to go about organizing it, but if you are methodical in relating angles using geometry, it's possible to build relationships among the angles across a ray diagram such as this.

19. (N) You and your roommate share a plane mirror. She stands behind you so that her face is 18 in. behind your face. Your face is 3.5 ft from the mirror. How far does the image of your roommate's face appear to be from you? Give your answer in feet.

INTERPRET and ANTICIPATE
For a plane mirror, the magnitude of the image distance equals that of the object distance (Eq. 37.4). Since everything is in a straight line, we only need to worry about distances and don't need to deal with angles!

SOLVE
First, sketch the situation described.

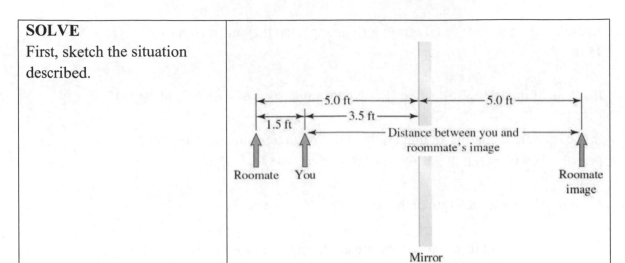

Figure P37.19ANS

The roommate is at a distance of 5.0 feet in front of the mirror (1.5 feet behind you plus 3.5 feet from you to the mirror), so her image is 5.0 feet behind the mirror (Eq. 37.4).

The total distance from you to the image of your roommate is the 3.5 feet that you are from the mirror plus the 5.0 feet from the mirror to the image location, or a total of 8.5 feet from you.

CHECK and THINK
We can also solve this problem by considering how far the light travels from your roommate to get to your eye. In this case, it travels 5 feet to the mirror and then 3.5 feet back to your eye, so we get the 8.5 feet answer again.

22. (A) An object moves toward a plane mirror with speed v_p at an angle θ with respect to the normal to the interface as shown in Figure P37.22. Determine the relative velocity between the object and the image.

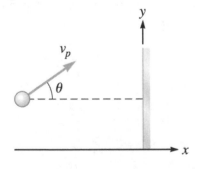

Figure P37.22

INTERPRET and ANTICIPATE

For a plane mirror, the magnitude of the image distance equals that of the object distance (Eq. 37.4), so the image is always equidistant on the other side of the mirror. We can express the velocity of the particle and then the reflected image in order to be able to calculate the relative velocity.

SOLVE	
First we write the velocity of the particle with respect to the mirror.	$\vec{v}_p = v\cos\theta\,\hat{i} + v\sin\theta\,\hat{j}$

The key is that the image is always at the same distance from the mirror as the particle (Eq. 37.4). So, if the image moves to the right towards the mirror $\left(+\hat{i}\right)$ the image will move to the left towards the mirror $\left(-\hat{i}\right)$ and vice versa. If the particle moves up or down though, the image will move along with. The consequence is that the x velocity of the image is the opposite of the particle's while the y velocity is the same.	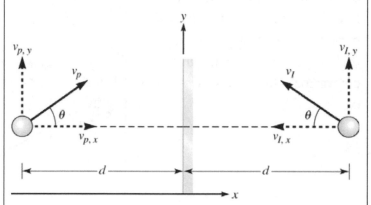 **Figure P37.22ANS** $\vec{v}_I = -v\cos\theta\,\hat{i} + v\sin\theta\,\hat{j}$

We can now find the relative velocity.	$\vec{v}_{pI} = \vec{v}_p - \vec{v}_I = \boxed{2v\cos\theta\,\hat{i}}$

CHECK and THINK

This is an example of a familiar phenomenon. As you walk near a mirror, if you move closer or further, your reflection moves closer or further from you accordingly.

23. (N) A rectangular room used for ballet practice is 12.0 m long and 6.20 m wide. The ballet teacher has mounted a 1.45-m-wide mirror on one of the 12.0-m-long walls of the room so that she can at all times observe the students lined at the handrail along the opposite wall. At one instant, the teacher is facing away from the students, a distance of 1.10 m from the mirror. What is the extent of the line of students that the teacher can see while looking forward?

Chapter 37 – Reflection and Images Formed by Reflection

The mirror is a plane mirror, so the image of the students is as far behind the mirror as the students are in front of the mirror. We could also consider the two edges of the mirror and use the law of reflection to determine which students are in view.

SOLVE
One way to view this problem is as shown. The image of the students is just as far behind the mirror as the students are in front of the mirror. Therefore, we can imagine the row of students 6.20 m behind the mirror and consider the two edges of the mirror to determine which ones the teacher can see.

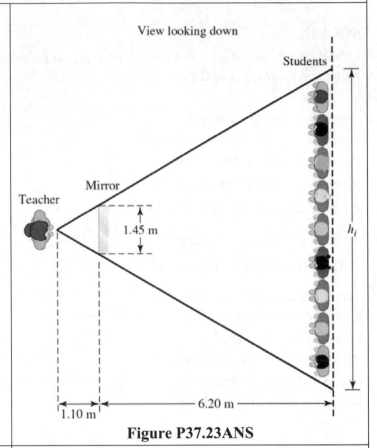

Figure P37.23ANS

The image of the students is then 1.10 m + 6.20 m = 7.30 m from the teacher. We can find the length of the line of students in view using similar triangles.

$$\frac{h_i}{1.45 \text{ m}} = \frac{7.30 \text{ m}}{1.10 \text{ m}}$$

Solve for the length of the image of students in view, h_i.

$$h_i = (1.45 \text{ m})\left(\frac{7.30 \text{ m}}{1.10 \text{ m}}\right) = \boxed{9.62 \text{ m}}$$

CHECK and THINK
You might also use the law of reflection, though you'll find that the method above is a little simpler. You may have experienced a similar effect when looking into a mirror in your bathroom, where you are able to see a wider field of view behind you—you can see things that are off to the side and behind you, even when looking forward.

30. (N) The magnitude of the radius of curvature of a convex spherical mirror is 22.0 cm.
a. What are the location and magnification of the image of an object placed 16.0 cm in front of the mirror? Is this image upright or inverted?
b. What are the location and magnification of the image of an object placed 44.0 cm in front of the mirror? Is this image upright or inverted?

INTERPRET and ANTICIPATE
In both cases, we can use the mirror equation and the formula for magnification. Given the object position and focal length, we can calculate the image position and magnification. We will also need to be careful about signs and make sure that we're consistent with Table 37.2.

SOLVE	
a. Apply the mirror equation (Eq. 37.8) with the definition of the focal length of a spherical mirror (Eq. 37.6, $f = R/2$, where R is the radius of curvature of the mirror).	$$\frac{1}{d_o} + \frac{1}{d_i} = \frac{1}{f} = \frac{2}{R}$$
Now consider the known quantities using the sign conventions (Table 37.2). The object is a real object, so it is positive. The mirror is convex, so the focal length is negative.	$$d_o = 16.0 \text{ cm}$$ $$f = \frac{R}{2} = -\frac{22.0 \text{ cm}}{2} = -11.0 \text{ cm}$$
Plug in values and solve for the image distance (location). Since it is negative, it is a virtual image, behind the mirror (Table 37.2).	$$\frac{1}{16.0 \text{ cm}} + \frac{1}{d_i} = \frac{1}{-11.0 \text{ cm}}$$ $$\frac{1}{d_i} = -\frac{1}{11.0 \text{ cm}} - \frac{1}{16.0 \text{ cm}} = -0.153 \text{ cm}^{-1}$$ $$d_i = \frac{1}{-0.153 \text{ cm}^{-1}} = \boxed{-6.52 \text{ cm}}$$
The magnification is given by Eq. 37.7. Since the value is smaller than one, the image is reduced compared to the object. Since it is positive, it is upright (oriented in the same direction as the object).	$$M = -\frac{d_i}{d_o} = -\frac{(-6.52 \text{ cm})}{16.0 \text{ cm}} = \boxed{0.407}$$

b. Repeat the steps above using d_o =44.0 cm. First, apply the mirror equation. We again find that the image is virtual (behind the mirror) since the sign is negative.	$\dfrac{1}{44.0 \text{ cm}} + \dfrac{1}{d_i} = \dfrac{1}{-11.0 \text{ cm}}$ $\dfrac{1}{d_i} = -\dfrac{1}{11.0 \text{ cm}} - \dfrac{1}{44.0 \text{ cm}} = -0.114 \text{ cm}^{-1}$ $d_i = \dfrac{1}{-0.114 \text{ cm}^{-1}} = \boxed{-8.80 \text{ cm}}$
Now calculate the magnification. We find that it is smaller than in part (a), but similarly the image is reduced ($\lvert M \rvert < 1$) and $\boxed{\text{upright}}$ ($M > 0$).	$M = -\dfrac{d_i}{d_o} = -\dfrac{(-8.80 \text{ cm})}{44.0 \text{ cm}} = \boxed{0.200}$

CHECK and THINK

With the mirror equation and equation for magnification, we can determine the location and properties of the image. The main issue is that we need to be careful to use the sign conventions when assigning or interpreting values in these equations.

---

33. (N) An object is placed 25.0 cm from the surface of a convex mirror and an image is formed. The same object is placed 20.0 cm in front of a plane mirror and, again, an image is formed. Suppose in each case the object is located at $x = 0$ and the mirrors lie along the positive x axis. If the images formed in the two mirrors lie at the same x coordinate, determine the radius of curvature of the convex mirror.

INTERPRET and ANTICIPATE

We need to first determine the image distance for the convex mirror using the information given about the plane mirror. Then we can use the mirror equation to obtain the radius of curvature.

SOLVE	
First, draw a sketch.	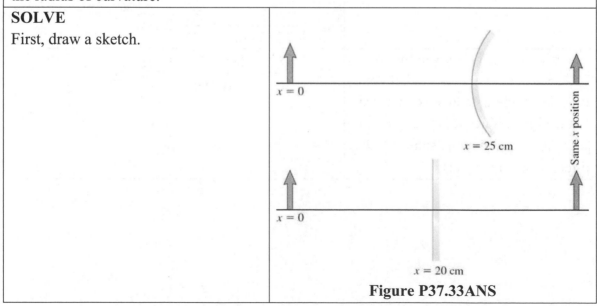

Figure P37.33ANS

For a plane mirror, the image distance equals the object distance (Eq. 37.4), so the position of the image relative to the object is twice the object distance, or 40.0 cm.	plane mirror: image position $x = 40.0$ cm
We are told this is also the location of the image for the convex mirror, therefore $d_o + \lvert d_i \rvert = 40.0$ cm. The object distance $d_o = +25.0$ cm (positive for a real object). This means that the image location is at a distance $\lvert d_i \rvert = 40.0$ cm $- d_o = 15.0$ cm. This is a virtual image, behind the mirror, so when applying the mirror equation, it is negative according to the sign convention (Table 37.2): $d_i = -15.0$ cm.	convex mirror: $d_o + \lvert d_i \rvert = 40.0$ cm $d_o = +25.0$ cm $\lvert d_i \rvert = 40.0$ cm $- d_o = 15.0$ cm $d_i = -15.0$ cm
We now apply the mirror equation (Eq. 37.8) for the convex mirror, where $d_o = 25.0$ cm, $d_i = -15.0$ cm, and f is unknown. Based on the sign conventions, we would expect that the focal length for a convex mirror will be negative, so this is encouraging.	$$\frac{1}{f} = \frac{1}{d_o} + \frac{1}{d_i}$$ $$\frac{1}{f} = \frac{1}{25.0 \text{ cm}} + \frac{1}{-15.0 \text{ cm}} = \frac{-2}{75.0 \text{ cm}}$$ $$f = -\frac{75.0 \text{ cm}}{2} = -37.5 \text{ cm}$$
According to Eq. 37.6, the radius of curvature is twice the focal length for a spherical mirror.	$$R = 2f = \boxed{-75.0 \text{ cm}}$$

CHECK and THINK

In this case, we applied the mirror equation and used the fact that a plane mirror creates an image at the same distance as the object. As usual, we need to be careful about signs, but we can learn a lot from these equations.

39. (N) The magnitude of the radius of curvature of a concave spherical mirror is 34.0 cm.

a. What are the location and magnification of the image of an object placed 60.0 cm in front of the mirror? Is this image real or virtual? Is it upright or inverted?

b. What are the location and magnification of the image of an object placed 34.0 cm in front of the mirror? Is this image real or virtual? Is it upright or inverted?

INTERPRET and ANTICIPATE We can use the mirror equation as well as the equation for magnification for both cases. An important consideration is to use the sign convention, both in assigning the sign of values used in the equation and in interpreting the results.	

SOLVE The mirror equation (Eq. 37.8) relates the object and image distance given the focal length.	$$\frac{1}{f} = \frac{1}{d_o} + \frac{1}{d_i}$$
For a spherical mirror, the focal length is half the radius of curvature (Eq. 37.6). Considering the sign conventions in Table 37.2, we also know that for a concave mirror, both R and f are positive.	$$f = \frac{R}{2} = +17.0 \text{ cm}$$
We can calculate the magnification in each case (Eq. 37.7). If the magnification is positive, the image is upright, and if it is negative, it is inverted. If the absolute value of the magnification is greater than one, the image is enlarged and if it's less than one, it is reduced in size.	$$M = -\frac{d_i}{d_o}$$
a. Apply these equations with $d_o = +60.0$ cm. Since this is a real object, the distance is positive. Solve for the image distance. We find that the distance is positive, so *the image is* boxed{*real*} and 23.7 cm in front of the mirror.	$$\frac{1}{d_i} = \frac{1}{f} - \frac{1}{d_o}$$ $$\frac{1}{d_i} = \frac{1}{17.0 \text{ cm}} - \frac{1}{60.0 \text{ cm}}$$ $$\frac{1}{d_i} = \frac{60.0 - 17.0}{(17.0)(60.0)} \text{ cm}^{-1} = 0.0422 \text{ cm}^{-1}$$ $$d_i = \frac{1}{0.0422 \text{ cm}^{-1}} = \boxed{23.7 \text{ cm}}$$

Now calculate the magnification, which we find is –0.395. Since $M < 0$ (i.e. negative), *the image is inverted* and about 40% the size of the original object.	$M = -\dfrac{d_i}{d_o} = -\dfrac{23.7 \text{ cm}}{60.0 \text{ cm}} = \boxed{-0.395}$
b. Now apply the equations with $d_0 = +34.0$ cm. Since this is a real object, the distance is positive. Solve for the image distance. We find that the distance is positive, so *the image is real* and 34.0 cm in front of the mirror.	$\dfrac{1}{d_i} = \dfrac{1}{f} - \dfrac{1}{d_o}$ $\dfrac{1}{d_i} = \dfrac{1}{17.0 \text{ cm}} - \dfrac{1}{34.0 \text{ cm}}$ $\dfrac{1}{d_i} = \dfrac{34.0 - 17.0}{(17.0)(34.0)} \text{ cm}^{-1} = \dfrac{1}{34.0} \text{ cm}^{-1}$ $d_i = \boxed{+34.0 \text{ cm}}$
Now calculate the magnification, which we find is –1.00. Since $M < 0$ (i.e. negative), *the image is inverted* and the same size as the original object.	$M = -\dfrac{d_i}{d_o} = -\dfrac{34.0 \text{ cm}}{34.0 \text{ cm}} = \boxed{-1.00}$

CHECK and THINK

Moving the object changes the location and magnification of the image, but we can calculate them easily as long as we keep track of the signs and do a little algebra!

42. (N) A concave mirror with a radius of curvature of 25.0 cm is used to form an image of an arrow that is 10.0 cm away from the mirror. If the arrow is 2.00 cm long and inverted (pointing below the optical axis), what is the height of the arrow's image?

INTERPRET and ANTICIPATE

We can make use of Eqs. 37.6, 37.7, and 37.8 to model the behavior of the concave mirror. The radius of curvature can be used to find the focal length. Then the focal length and object distance can be used to find the image distance. Lastly, this collected information can be used to find the image height. If the object is nearer to the mirror than the focal length, we expect a virtual image that is oriented in the same direction as the object. The object height must be negative because it is inverted.

SOLVE	
First, use Eq. 37.6 to express the focal length. A positive focal length is expected because the mirror is concave.	$f = \dfrac{r}{2} = \dfrac{25.0 \text{ cm}}{2}$

Now, use Eq. 37.8 to find the image distance. Note that the image distance is negative, which means that the image is virtual.	$$\frac{1}{d_o} + \frac{1}{d_i} = \frac{1}{f}$$ $$\frac{1}{d_i} = \frac{1}{f} - \frac{1}{d_o} = \frac{2}{25.0 \text{ cm}} - \frac{1}{10.0 \text{ cm}}$$ $$\frac{1}{d_i} = \frac{4}{50.0 \text{ cm}} - \frac{5}{50.0 \text{ cm}}$$ $$\frac{1}{d_i} = -\frac{1}{50.0 \text{ cm}}$$ $$d_i = -50.0 \text{ cm}$$
Then use Eq. 37.7 to find the image height. The image height is negative, but that is expected because the object height is negative, and the image orientation should remain unchanged since it is virtual.	$$\frac{h_i}{h_o} = -\frac{d_i}{d_o}$$ $$h_i = \left(-\frac{d_i}{d_o}\right) h_o$$ $$h_i = \left(-\frac{-50.0 \text{ cm}}{10.0 \text{ cm}}\right)(-2.00 \text{ cm})$$ $$h_i = \boxed{-10.0 \text{ cm}}$$

CHECK and THINK

We must be careful with the sign conventions for these three formulae. Virtual images will be the same orientation as their corresponding objects for a single mirror system. We can see here that the image is magnified compared to the object and is 5 times bigger than the object.

46. The upright image formed by a concave spherical mirror with a focal length of 14.0 cm is 2.50 times larger than the object.
a. (N) What is the distance of the object from the mirror?
b. (C) Is the image formed by the mirror real or virtual?
c. (G) Draw a ray diagram showing the locations of the object and the image by tracing at least three rays.

INTERPRET and ANTICIPATE

We can calculate the image distance with the focal length and object distance, being careful to observe sign conventions. Drawing a ray diagram in part (c) is a great way to confirm that our calculations are reasonable.

SOLVE a. We are told the magnification is 2.50. In addition, the fact that the image is upright (oriented in	$$M = -\frac{d_i}{d_o} = +2.50$$

the same direction as the object) means the magnification is positive, so $M = +2.50$ rather than -2.50. Using Eq. 37.7, we are able to relate the image distance to the object distance.	$d_i = -2.50d_o \qquad\qquad (1)$
Apply the mirror equation (Eq. 37.8).	$\dfrac{1}{f} = \dfrac{1}{d_o} + \dfrac{1}{d_i}$
Insert values. Since the mirror is concave, according to the sign convention in Table 37.2, $f > 0$. So, $f = +14.0$ cm. Use Equation (1) above as well and solve for the object distance.	$\dfrac{1}{14.0\text{ cm}} = \dfrac{1}{d_o} - \dfrac{1}{2.50d_o} = \dfrac{1.50}{2.50d_o}$ $d_o = (14.0\text{ cm})\left(\dfrac{1.50}{2.50}\right) = \boxed{8.40\text{ cm}}$
b. Using Equation (1) again, we see that the image distance is negative. Referring to the sign convention (Table 37.2), this indicates that the image is $\boxed{virtual.}$ That is, it is behind the mirror where we know light rays do not actually intersect to form the image.	$d_i = -2.50d_o = -2.50(8.40\text{ cm}) = -21.0\text{ cm}$
c. We draw a few rays from Table 37.1: Ray #2: drawn through the center of curvature of the lens, reflects back upon itself. Ray #3: parallel to the principal axis, reflects through the focus F. Ray #4: originates from F, reflects parallel to the principal axis.	 **Figure P37.46ANS**

Note that this agrees with our calculation: The three rays appear to come from a point behind the mirror (dashed lines) locating the virtual image. It is also upright and enlarged, corresponding with the facts that $M > 0$ and $|M| > 1$ respectively.	

CHECK and THINK

Not only can we calculate a lot of information with the mirror and magnification equations, but drawing a ray diagram is a quick way to make sure that everything looks right (and that we did not inadvertently make a sign mistake!). Table 37.2 (the sign conventions) is also very helpful in assigning and interpreting values.

---

49. (N) The focal length of a concave mirror is 30.0 cm. Find the two positions of the object in front of the mirror so that the image height is three times greater in magnitude than the object height.

INTERPRET and ANTICIPATE

A concave mirror can form either a real or a virtual image, depending on the location of the object. In either case, we'll write the mirror equation using the information given and seek two locations that satisfy the conditions.

SOLVE	
The two possibilities are that the magnification is –3 or +3. Let's start by considering the formula for magnification (Eq. 37.7) and assume that $M = -3$. This would indicate an inverted image. The fact that the image distance is positive (assuming the object is a real object and therefore positive), indicates that it is a real image.	$M = -\dfrac{d_i}{d_o} = -3 \qquad \rightarrow \qquad d_i = 3d_o$
Apply the mirror equation (Eq. 37.8) to determine the location of the object.	$\dfrac{1}{d_o} + \dfrac{1}{d_i} = \dfrac{1}{f}$ $\dfrac{1}{d_o} + \dfrac{1}{3d_o} = \dfrac{4}{3d_o} = \dfrac{1}{30.0 \text{ cm}}$ $d_o = \dfrac{4}{3}(30.0 \text{ cm}) = \boxed{40.0 \text{ cm}}$

Now consider the other possibility, $M = +3$, first by using the magnification equation. Since the magnification is positive, it is an upright image, and since the image distance is negative, it is a virtual image (behind the mirror).	$M = -\dfrac{d_i}{d_o} = 3 \quad \rightarrow \quad d_i = -3d_o$
Apply the mirror equation again to determine the location of the object in this case.	$\dfrac{1}{d_o} + \dfrac{1}{d_i} = \dfrac{1}{f}$ $\dfrac{1}{d_o} - \dfrac{1}{3d_o} = \dfrac{2}{3d_o} = \dfrac{1}{30.0 \text{ cm}}$ $d_o = \dfrac{2}{3}(30.0 \text{ cm}) = \boxed{20.0 \text{ cm}}$

CHECK and THINK

The two positions of the object in front of the mirror are 40.0 cm (forming a real, inverted image) and 20.0 cm (forming a virtual, upright image).

51. (N) An object in front of a concave mirror has a virtual image that is 11.0 cm from the mirror. The mirror's radius of curvature is 20.0 cm.
a. What is the object distance?
b. What is the magnification?

INTERPRET and ANTICIPATE
We will use the mirror equation and sign conventions to determine the object distance and from there we can calculate the magnification.

SOLVE **a.** Since the image is virtual, it is behind the mirror. The sign convention (Table 37.2) tells us that this is a negative value when we use the mirror equation.	$d_i = -11.0 \text{ cm}$
Find the focal length using Eq. 37.6. Again referring to Table 37.2, the focal length of a concave mirror is positive.	$f = +\dfrac{r}{2} = +\dfrac{20.0 \text{ cm}}{2} = +10.0 \text{ cm}$

Use the mirror equation (Eq. 37.8) to find the object distance: $\dfrac{1}{f}=\dfrac{1}{d_o}+\dfrac{1}{d_i}$	$\dfrac{1}{d_o}=\dfrac{1}{f}-\dfrac{1}{d_i}=\dfrac{1}{10.0\text{ cm}}-\dfrac{1}{-11.0\text{ cm}}$ $\dfrac{1}{d_o}=\dfrac{11.0+10.0}{(10.0)(11.0)}\text{ cm}^{-1}=0.191\text{ cm}^{-1}$ $d_o=\dfrac{1}{0.191\text{ cm}^{-1}}=\boxed{5.24\text{ cm}}$
b. Find the magnification using Eq. 37.7.	$M=-\dfrac{d_i}{d_o}=-\dfrac{-11.0\text{ cm}}{5.24\text{ cm}}=\boxed{2.10}$

CHECK and THINK

We've determined that the object is a real object, 5.24 cm in front of the mirror, and that the image is upright and about twice as large as the object.

53. (N) What must be the radius of curvature of a concave mirror to form an image of the Sun 2.0 cm in diameter?

INTERPRET and ANTICIPATE
The Sun is of course very far from the mirror, so the concave mirror will form the image of the Sun at its focal point. We can sketch the situation described and relate the focal length of the mirror to the image and object distances.

SOLVE	
If we consider the mirror equation with an object that is very far away, $1/d_o$ is practically zero, implying that the image will be formed at the focal length.	$\dfrac{1}{d_i}+\dfrac{1}{d_o}\approx\dfrac{1}{d_i}=\dfrac{1}{f}\qquad\rightarrow\qquad d_i=f$
Armed with this knowledge, we sketch the situation described.	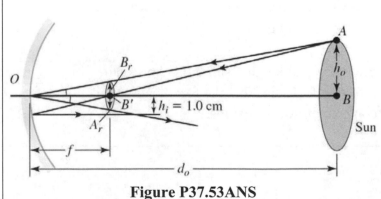 **Figure P37.53ANS**

Considering similar triangles, we can relate the ratio between the focal length f and the distance to the sun d_o to the ratio of the radius of the spot on the image $h_i = 1.0$ cm to the radius of the Sun h_o. (Note: It also works to use the diameter of each rather than the radius, since the ratio h_o/h_i will be the same in both cases.)	$$\frac{h_o}{h_i} = \frac{d_o}{f}$$
We look up the distance to the Sun and it's radius as 1.50×10^{11} m and 6.96×10^8 m. solve for the focal length and plug in values.	$$f = h_i \frac{d_o}{h_o}$$ $$f = (1.0 \text{ cm})\left(\frac{1.50 \times 10^{11} \text{ m}}{6.96 \times 10^8 \text{ m}}\right) = 216 \text{ cm}$$
For a spherical mirror, the radius of curvature is twice the focal length (Eq. 37.6).	$$R = 2f = 430 \text{ cm} = \boxed{4.3 \text{ m}}$$

CHECK and THINK

As with many of these problems, by drawing a sketch and relating the distances, we're able to deduce the focal length.

61. (A) An object is placed midway between two concave spherical mirrors as shown in Figure P37.61. The distance between the mirrors is D, and they have the same focal length. Determine the value(s) of D in terms of the focal length f for which only one image is formed in each mirror.

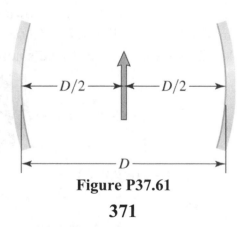

Figure P37.61

INTERPRET and ANTICIPATE The only way to form a single image is if the image is formed at the same location as the object. This way, the image that's formed by one mirror is located at the center, which is then imaged by the other mirror, which forms an image at the center, etc. An image formed elsewhere will serve as an object for the other mirror and produce an image in a new place.	

SOLVE Considering the mirror equation (Eq. 37.8) with the focal length given by Eq. 37.6 $f = r/2$, this implies that the image is formed at the center of curvature at a distance r.	$$\frac{1}{d_o} + \frac{1}{d_i} = \frac{1}{f} = \frac{2}{r}$$ Assuming $d_o = d_i$ implies $\frac{2}{d_i} = \frac{1}{f} = \frac{2}{r}$ and therefore $$d_i = r = 2f$$ That is, the object is at twice the focal length and the total width $D = 2d_i = \boxed{4f}$
The other possibility is that the first image is formed at infinity. This would happen if the focal length is equal to $D/2$. Solve for the focal length in this scenario.	$D = \boxed{2f}$

CHECK and THINK   We have successfully found a case where the image is located at the same place as the object, ensuring that there is only one image/object location.

63. (N) A critical characteristic of light rays that are reflected diffusely from an object such as your friend's nose is the extent to which adjacent rays diverge from each other. Imagine holding a gumball 2.0 cm in diameter in front of your friend's nose as shown in Figure P37.63. By what angle do the two tangential rays shown in the diagram diverge from each other when the center of the gumball is **a.** $D = 5.00$ cm, **b.** $D = 20.0$ cm, **c.** $D = 100.0$ cm, and **d.** $D = 100.0$ km from the tip of your friend's nose?

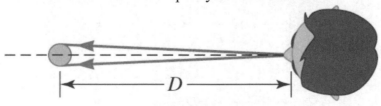

Figure P37.63

INTERPRET and ANTICIPATE

To get a handle on what we need to calculate, we first make a sketch. We can determine the angle with each value of D. As the distance D increases, the angle will become smaller (and as one consequence, the light rays from distant objects become essentially parallel).

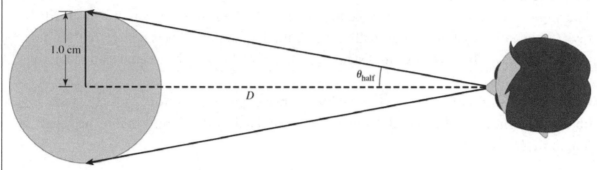

Figure P37.63ANS

SOLVE Use the triangle indicated, which includes a side of length D from the nose to the gumball, the 1.0 cm radius of the gumball, and the hypotenuse from the edge of the gumball to the nose. This defines an angle θ_{half} that is half the angle we desire, which we can determine from geometry.	$$\tan\theta_{half} = \frac{1.0 \text{ cm}}{D}$$
The total angle is twice the angle determined above. We can now apply this formula in each case.	$$\theta = 2\theta_{half} = 2\tan^{-1}\left(\frac{1.0 \text{ cm}}{D}\right)$$
a. $D = 5.00$ cm	$$\theta = 2\tan^{-1}\left(\frac{1.0 \text{ cm}}{5.00 \text{ cm}}\right) = \boxed{23°}$$
b. $D = 20.0$ cm	$$\theta = 2\tan^{-1}\left(\frac{1.0 \text{ cm}}{20.0 \text{ cm}}\right) = \boxed{5.7°}$$
c. $D = 100.0$ cm	$$\theta = 2\tan^{-1}\left(\frac{1.0 \text{ cm}}{100.0 \text{ cm}}\right) = \boxed{1.1°}$$

| **d.** $D = 100.0$ km | $$\theta = 2\tan^{-1}\left(\frac{1.0 \text{ cm}}{100.0 \text{ km}}\right)$$ $$\theta = 2\tan^{-1}\left(\frac{1.0 \times 10^{-2} \text{ m}}{100.0 \times 10^{3} \text{ m}}\right) = \boxed{\left(1.1 \times 10^{-5}\right)^{\circ}}$$ |

CHECK and THINK

The angle does indeed increase. Note that in part (d), the angle is so small that the two rays (from each edge of the gumball) are essentially parallel. This is generally true and useful to know—light rays from distant objects can essentially be treated as parallel rays.

66. (N) The height of an image formed by a convex spherical mirror is 40.0% of the object's height. The object and the image are separated by 48.0 cm. What is the focal length of the mirror?

INTERPRET and ANTICIPATE

We have another opportunity to test out the mirror and magnification equations. As usual, we will extract the relevant variables and follow the sign conventions.

SOLVE Convex mirrors only produce virtual images, which are behind the mirror. Therefore, the picture you should have in mind about the layout is an object, then the mirror, and finally the image. We are given the total distance (48.0 cm) and we know from the sign convention that the image distance will be negative (Table 37.2). The object is a real object and therefore d_o is a positive number.	$$48.0 \text{ cm} = \left	d_i\right	+ d_o = -d_i + d_o$$
We are also told that the image is 40.0% the size of the object. With the magnification equation (Eq. 37.8), this allows us to relate the ratio of the heights to the distances and then express the image distance in terms of the object distance.	$$M = -\frac{d_i}{d_o} = \frac{h_i}{h_o} = 0.400$$ $$d_i = -0.400 d_o$$
Insert this expression into the equation above and solve for the object distance.	$$48.0 \text{ cm} = -\left(-0.400 d_o\right) + d_o = 1.40 d_o$$ $$d_o = \frac{48.0 \text{ cm}}{1.40} = 34.3 \text{ cm}$$

Determine the image distance. It is negative, which we expect for a virtual image.	$d_i = -0.400d_o = -0.400(34.3 \text{ cm})$ $d_i = -13.7 \text{ cm}$
Finally, use the mirror equation (Eq. 37.8) to determine the focal length. The focal length is negative, which we expect for a convex mirror based on Table 37.2.	$\dfrac{1}{f} = \dfrac{1}{d_o} + \dfrac{1}{d_i}$ $\dfrac{1}{f} = \dfrac{1}{34.3 \text{ cm}} + \dfrac{1}{-13.7 \text{ cm}} = -0.0437 \text{ cm}^{-1}$ $f = \dfrac{1}{-0.0437 \text{ cm}^{-1}} = \boxed{-22.9 \text{ cm}}$

CHECK and THINK

As with many problems in this chapter, the mirror equation, magnification, and sign conventions allow us to determine a lot of information.

67. (E) Observe your reflection in the back of a spoon. From that observation, estimate the radius of curvature of the spoon. *Hint*: Model the spoon as a spherical mirror.

INTERPRET and ANTICIPATE
Specific answers will vary based on the spoon you use. If you imagine that the spoon was cut out of a spherical section, the radius of curvature refers to the radius of the sphere that would have been used. By looking at a spoon, you might guess that this is on the order of centimeters.

SOLVE	
You should actually try this yourself, but suppose that your face is about 20 cm long and when you hold the spoon 25 cm away, the image is upright and around 2 cm. We can first determine the magnification using Eq. 37.7. It is much less than one, as expected, since your image on the spoon is much smaller.	$M = -\dfrac{d_i}{d_o} = \dfrac{h_i}{h_o} \approx \dfrac{2 \text{ cm}}{20 \text{ cm}} = 0.1$
This also provides a relationship between the image and object distance.	$d_i = -0.1d_o$

Apply the mirror equation (Eq. 37.8) and solve for the focal length.	$\dfrac{1}{d_o} + \dfrac{1}{d_i} = \dfrac{1}{f}$
	$\dfrac{1}{d_o} - \dfrac{1}{0.1d_o} = -\dfrac{9}{d_o} = \dfrac{1}{f}$
	$f = -\dfrac{1}{9}d_o = -\dfrac{1}{9}(25\text{ cm}) = -2.8\text{ cm}$
Finally, using Equation 37.6, we can determine the radius of curvature, which is twice the focal length for a spherical mirror.	$r = 2f = \boxed{-6.0\text{ cm}}$

CHECK and THINK

The radius of curvature is 5 or 6 cm, which sounds reasonable. Easily observed quantities, such as the object position and size as well as the image size allow us to determine the focal length of the spoon.

Refraction and Images
Formed by Refraction

6. (N) A mirror is above a swimming pool and perpendicular to the surface of the water. A narrow beam is incident on the mirror (Fig. P38.5). If $\theta_i = 22.3°$, find the angle of refraction in the water.

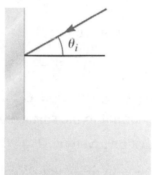

Figure P38.5

INTERPRET and ANTICIPATE

We can see that due to the law of reflection, the light ray will make an angle with respect to the water's surface equal to the angle of incidence with the mirror. This is because the ray reflects at the same angle of incidence from the mirror and the normal to the mirror is parallel to the surface of the water. Thus, for the purposes of the law of refraction, the angle of incidence on the water is actually $90° - \theta_i$. Then, we can apply Eq. 38.1 and solve for the angle of refraction. We expect the angle of refraction to be less than $90° - \theta_i$ since the index of refraction of the water is greater than that of air.

SOLVE	
Use Eq. 38.1 to find the angle of refraction in the water.	$n_{air} \sin(90° - \theta_i) = n_{water} \sin\theta_t$ $\theta_t = \sin^{-1}\left(\dfrac{1.0002926}{1.333} \sin(90° - 22.3°) \right) = \boxed{44.0°}$

CHECK and THINK

Note that the angle of refraction is less than the angle of incidence on the water, $90° - 22.3° = 67.7°$. With some geometry and the law of reflection, we were able to determine the angle of incidence on the water.

9. (N) A mirror is above a swimming pool and tilted toward the surface of the water. A narrow beam is parallel to the surface of the water and is incident on the mirror (Fig. P38.9). The angle of incidence on the mirror's surface is 22.3°. Find the angle of refraction in the water.

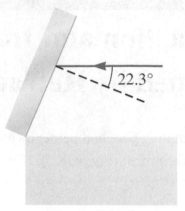

Figure P38.9

INTERPRET and ANTICIPATE

Using the law of reflection, we sketch the situation as shown in Figure P38.09ANS. By forming a right triangle, using the normal to the surface of the water, we can see that the angle of incidence on the water must be given by $\theta_i = 90° - 2(22.3°)$. Then, we can use Eq. 38.1 to find the angle of refraction in the water. We expect the angle of refraction to be less than $\theta_i = 90° - 2(22.3°)$, since the index of refraction of the water is greater than that of air.

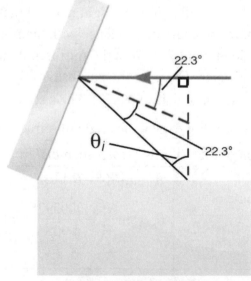

Figure P38.09ANS

SOLVE	$n_{air} \sin\left(90° - 2\left(22.3°\right)\right) = n_{water} \sin\theta_t$
Use Eq. 38.1 to find the angle of refraction in the water.	$\theta_t = \sin^{-1}\left(\dfrac{1.0002926}{1.333} \sin\left(90° - 2\left(22.3°\right)\right)\right) = \boxed{32.3°}$

CHECK and THINK

Note that the angle of refraction is less than the angle of incidence on the water,

$90° - 2\left(22.3°\right) = 45.4°$.

---

17. (N) A fish is 3.25 m below the surface of still water. Because of total internal reflection, it is hidden from the view of a fisher in a boat on the water as long as the boat is outside a circle of radius r. The center of the circle is directly above the fish (Figure P38.16). Find the minimum value of r.

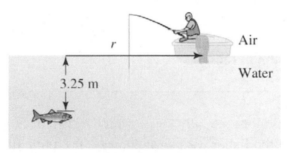

3.25 m

Air

Water

r

Figure P38.16

INTERPRET and ANTICIPATE

When a ray from the fish encounters the boundary at the critical angle, the light is totally reflected. If the boat is at the point where such a ray encounters the boundary, the fisher will not see the fish (Figure P38.17ANS). So, there is locus of such points that forms a circle above the fish, and if the fisher is farther laterally than this circle from the fish, then he or she won't see the fish. We can use Eq. 38.2 to find the critical angle for the light traveling from the water into the air. Then, we will use the geometry in Figure P38.17ANS to find the radius of the circle at the top of the water. This is the minimum value of r.

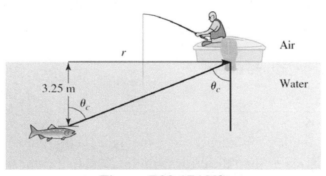

3.25 m

θ_c

θ_c

r

Air

Water

Figure P38.17ANS

Chapter 38 – Refraction and Images Formed by Refraction

<table>
<tr>
<td>

SOLVE

Using the indices of refraction for air and water, we find the critical angle from Eq. 38.2.

</td>
<td>

$$\sin\theta_c = \frac{n_t}{n_i}$$

$$\theta_c = \sin^{-1}\left(\frac{n_t}{n_i}\right) = \sin^{-1}\left(\frac{1.0002926}{1.333}\right)$$

$$\theta_c = 48.6°$$

</td>
</tr>
<tr>
<td>

We can see in our sketch that there is a right triangle formed with the depth, measured in the center of the circle, the radius of the circle at which the ray will be at the critical angle, and the line along the direction of the ray from the fish towards the circle. Use the tangent function to find the minimum radius, r.

</td>
<td>

$$\tan\theta_c = \frac{r}{3.25\text{ m}}$$

$$r = (3.25\text{ m})\tan\left[\sin^{-1}\left(\frac{1.0002926}{1.333}\right)\right] = \boxed{3.69\text{ m}}$$

</td>
</tr>
</table>

CHECK and THINK

As long as the fisher is 3.69 m laterally away from the fish, or more, he or she will not see the fish. Note that if the fisher were further away, the angle of incidence on the water's surface would be greater than the critical angle and the light would still be totally internally reflected, as we described when sketching the circle. The radius of 3.69 m is definitely the minimum distance.

---

19. (N) A beam of light travels through a fiber with an index of refraction of 1.55. The fiber is surrounded by a medium with an index of refraction of 1.36. At what minimum angle of incidence is all the light reflected back into the fiber?

INTERPRET and ANTICIPATE

The condition will be met if the light has an angle of incidence greater than or equal to the critical angle between the two media. Thus, use Eq. 38.2 to find the minimum angle of incidence for internal reflection, which is the critical angle. The incident medium has an index of refraction of 1.55.

<table>
<tr>
<td>

SOLVE

Use Eq. 38.2 to find the critical angle.

</td>
<td>

$$\theta_c = \sin^{-1}\frac{n_t}{n_i} = \sin^{-1}\frac{1.36}{1.55} = \boxed{61.3°}$$

</td>
</tr>
</table>

CHECK and THINK

This is the minimum angle of incidence since the light is still internally reflected if the angle of incidence is greater than the critical angle. Note that if the two materials were reversed, so that the inner material has an index of refraction of 1.36, there would be no internal reflection for light within that material surrounded by the $n = 1.55$ material. You could use Eq. 38.2 to show that there is no critical angle in that scenario!

22. (N) Consider a beam of 486-nm blue light within a slab of material. The beam is incident on the slab's interface with air. What is the critical angle for total internal reflection if the slab is made of **a.** ice ($n = 1.309$), **b.** fluorite ($n = 1.434$), and **c.** diamond ($n = 2.417$)?

INTERPRET and ANTICIPATE

In each case, we can use Eq. 38.2 to find the critical angle for light in that material, incident on air. We expect that the critical angle will decrease as the index of refraction of the incident medium increases.

SOLVE	
a. Apply Eq. 38.2 to find the critical angle for light traveling from ice into air.	$\theta_c = \sin^{-1} \dfrac{n_t}{n_i} = \sin^{-1} \dfrac{1.0002926}{1.309} = \boxed{49.83°}$
b. Apply Eq. 38.2 to find the critical angle for light traveling from fluorite into air.	$\theta_c = \sin^{-1} \dfrac{n_t}{n_i} = \sin^{-1} \dfrac{1.0002926}{1.434} = \boxed{44.23°}$
c. Apply Eq. 38.2 to find the critical angle for light traveling from diamond into air.	$\theta_c = \sin^{-1} \dfrac{n_t}{n_i} = \sin^{-1} \dfrac{1.0002926}{2.417} = \boxed{24.45°}$

CHECK and THINK

We did indeed find that the critical angle decreases as the index of refraction of the incident medium increases. This all assumes the transmitted medium is the same in each case (air in this problem).

26. (N) Dispersion occurs when the index of refraction of a medium depends on the wavelength of incident light, resulting in different angles of refraction for each wavelength. A beam of white light is incident on a block of zinc crown glass with $\theta_i = 34.0°$. The index of refraction of this glass for 589-nm yellow light is 1.528, while the index of refraction for 486-nm blue light is 1.517. **a.** What is the refraction angle θ_t for yellow light in the glass? **b.** What is the refraction angle θ_t for blue light in the glass?

INTERPRET and ANTICIPATE

In each case, we can use Eq. 38.1 to find the angle of refraction for each color of light. We expect that the angle of refraction of the yellow light will be less than that of the blue light because the index of refraction of the glass for the yellow light is greater than that for the blue light.

SOLVE	
a. Use Eq. 38.1 for the yellow light.	$\theta_{yellow} = \sin^{-1}\left[\left(\dfrac{n_{air}}{n_{yellow}}\right)\sin\theta_i\right]$ $\theta_{yellow} = \sin^{-1}\left[\left(\dfrac{1.0002926}{1.528}\right)\sin(34.0°)\right]$ $\theta_{yellow} = \boxed{21.5°}$
b. Use Eq. 38.1 for the blue light.	$\theta_{blue} = \sin^{-1}\left[\left(\dfrac{n_{air}}{n_{blue}}\right)\sin\theta_i\right]$ $\theta_{blue} = \sin^{-1}\left[\left(\dfrac{1.0002926}{1.517}\right)\sin(34.0°)\right]$ $\theta_{blue} = \boxed{21.6°}$

CHECK and THINK

The angle of refraction of the yellow light is less than that of the blue light. Though the difference in angle is very small, if the light rays are allowed to travel far enough, we will see the spread in the color. What we call "white light" is light that consists of all colors and can be separated by allowing the light to pass through a prism, or a piece of glass such as that described here.

35. (N) A paperweight is made from a transparent material with an index of refraction of 1.75. The paperweight is a hemisphere of radius R. At its base is a flattened ladybug. Where is the image of the ladybug, and what is its magnification? Assume that the index of refraction for air is 1.00.

INTERPRET and ANTICIPATE

Use Eq. 38.3 to find the image distance, since the exposed part of the paperweight can be thought of as part of a spherical surface. Note that the radius of curvature, here, is negative because from the perspective of the object (the ladybug) the surface is concave. Also, the ladybug's object distance is the same as the magnitude of the radius of curvature, since the ladybug is at the center of the sphere defined by the surface. When finding the magnification, using Eq. 38.4, we expect that the ladybug's image is magnified. This is because when viewed from above, the light should have traveled through the glass and then refracted away from the normal as it entered the air. This would make the image look bigger.

SOLVE

First, use Eq. 38.3 with the object distance equal to the radius of curvature and the radius of the surface equal to the negative of the radius of curvature, as described above. Solve for the image distance.

$$\frac{n_i}{d_o} + \frac{n_t}{d_i} = \frac{(n_t - n_i)}{r}$$

$$\frac{n_i}{R} + \frac{n_t}{d_i} = \frac{(n_t - n_i)}{-R}$$

$$\frac{n_i}{R} + \frac{n_t}{d_i} = -\frac{n_t}{R} + \frac{n_i}{R}$$

$$\frac{n_t}{d_i} = -\frac{n_t}{R}$$

$$d_i = \boxed{-R}$$

Use Eq. 38.4 to find the magnification of the ladybug.

$$M = -\frac{n_i}{n_t}\frac{d_i}{d_o}$$

$$M = -\frac{1.75}{1.00}\left(\frac{-R}{R}\right)$$

$$M = \boxed{1.75}$$

CHECK and THINK

The ladybug does appear to be magnified, as expected. Note that the image distance indicates that the image is located exactly where the object (ladybug) is located! The sign difference can be understood when considering the sign convention for real and virtual images.

38. (N) A Lucite slab ($n = 1.485$) 5.00 cm in thickness forms the bottom of an ornamental fish pond that is 40.0 cm deep. If the pond is completely filled with water, what is the apparent thickness of the Lucite plate when viewed from directly above the pond?

INTERPRET and ANTICIPATE

We need to find the apparent depth of the bottom of the slab and the top of the slab separately in order to determine the apparent thickness. This can be done using Eq. 38.4 with a magnification of $M = 1$, as it would be when viewed from above with all surfaces being flat. We have to be careful though! The top surface of the slab is viewed through the water only, so we need only apply Eq. 38.4 once to find the apparent depth of the top of the slab. The bottom of the slab, however, is viewed through the water and the slab itself. Thus, we must find the apparent depth of the bottom of the slab, as viewed through the slab first, and then find the apparent depth of this new bottom, as viewed through the water. Then, we can subtract the final apparent depths of the top and bottom of the slab to find the apparent thickness.

SOLVE Use Eq. 38.4 with $M = 1$ to express the apparent depth of the top of the slab, as viewed through the water.	$d_{i,\text{top}} = -\dfrac{n_t}{n_i} d_{o,\text{top}} = -\left(\dfrac{1.0002926}{1.333}\right)(40.0 \text{ cm})$
Now, use Eq. 38.4 with $M = 1$ to find the apparent depth of the bottom of the slab, as viewed through the slab itself, with the transmitted medium being the water and the incident medium being the Lucite with a thickness of 5.00 cm.	$d_{i,\text{bottom1}} = -\left(\dfrac{1.333}{1.485}\right)(5.00 \text{ cm})$
This image is the object for the observer in air above, in locating the bottom of the Lucite slab. We now need the apparent depth of this bottom. The depth of this new object would be the difference of the depth of the water and the apparent depth in the Lucite we just found.	$d_{o,\text{bottom2}} = 40.0 \text{ cm} - \left[-\left(\dfrac{1.333}{1.485}\right)(5.00 \text{ cm})\right]$ $d_{o,\text{bottom2}} = 40.0 \text{ cm} + \left(\dfrac{1.333}{1.485}\right)(5.00 \text{ cm})$
Now use Eq. 38.4 again, as before, to find the apparent depth of the bottom as viewed from the air.	$d_{i,\text{bottom2}} = -\left(\dfrac{1.0002926}{1.333}\right)\left(40.0 \text{ cm} + \left(\dfrac{1.333}{1.485}\right)(5.00 \text{ cm})\right)$

Now take the difference between the apparent depth of the bottom and the top as viewed from the air to find the apparent thickness.

$$\Delta d_i = d_{i,\text{top}} - d_{i,\text{bottom2}}$$

$$\Delta d_i = -\left(\frac{1.0002926}{1.333}\right)(40.0 \text{ cm}) - \left[-\left(\frac{1.0002926}{1.333}\right)\left(40.0 \text{ cm} + \left(\frac{1.333}{1.485}\right)(5.00 \text{ cm})\right)\right]$$

$$\Delta d_i = \boxed{3.37 \text{ cm}}$$

CHECK and THINK
The slab appears to be thinner than it actually is. We must be careful in determining the various materials that light must pass through before it reaches the observer. The difficulty in determining the apparent depth of the bottom of the slab, as viewed by a person in the air, in this problem is one example of this.

45. (N) A converging lens has a focal length of 30.0 cm and is made of glass ($n = 1.50$). If $|r_1| = |r_2| \equiv |r|$, what is $|r|$?

INTERPRET and ANTICIPATE
Use Eq. 38.7 to find the radius of curvature. Note that since the lens is a converging lens, the first surface has a positive radius of curvature and the second surface will have a negative radius of curvature. Since we have taken care of the signs for the radii of curvature, we should get a positive value when we solve for the radius of curvature.

SOLVE
Use Eq. 38.7 with the information we determined above and solve for $|r|$.

$$\frac{1}{f} = (n-1)\left[\frac{1}{r_1} - \frac{1}{r_2}\right]$$

$$\frac{1}{f} = (n-1)\left[\frac{1}{|r|} - \frac{1}{-|r|}\right] = \frac{2(n-1)}{|r|}$$

$$f = \frac{|r|}{2(n-1)}$$

$$|r| = f[2(n-1)] = (30.0 \text{ cm})[2(1.50-1)] = \boxed{30.0 \text{ cm}}$$

CHECK and THINK
We found that there is a special case, among others, where the magnitude of radius of curvature for each lens surface will be equal to the focal length of the lens. That is, of course, not categorically true. Note that is we been asked the same question, but with a diverging lens. The result would be the same, but negative, as the focal length of the diverging lens must also be negative, by convention.

48. (N) The radius of curvature of the left-hand face of a flint glass biconvex lens ($n = 1.60$) has a magnitude of 8.00 cm, and the radius of curvature of the right-hand face has a magnitude of 11.0 cm. The incident surface of a biconvex lens is convex regardless of which side is the incident side. What is the focal length of the lens if light is incident on the lens from the left?

INTERPRET and ANTICIPATE
Use Eq. 38.7 to find the focal length. Note that the lens is biconvex, meaning that both lens surfaces appear to be convex when viewed head-on. The meaning of this is that for a light beam traveling from left to right, the first surface will behave as a convex surface,

but when the light reaches the second surface, it will behave as a concave surface. We must express the second surface's radius of curvature as a negative number. Given the shape of this lens, we expect it to be a converging lens, and so the focal length should be positive.

SOLVE	
Use Eq. 38.7, with the radius of curvature of the first surface as +8.00 cm and the radius of curvature of the second surface as -11.0 cm. Solve for the focal length	$\frac{1}{f} = (n-1)\left[\frac{1}{r_1} - \frac{1}{r_2}\right] = (1.60 - 1.00)\left[\frac{1}{8.00 \text{ cm}} - \frac{1}{(-11.0 \text{ cm})}\right]$ $f = \dfrac{1}{(1.60-1.00)\left[\dfrac{1}{8.00 \text{ cm}} - \dfrac{1}{(-11.0 \text{ cm})}\right]} = \boxed{7.72 \text{ cm}}$

CHECK and THINK

The focal length is positive, indicating, we were correct in thinking this should be a converging lens. Note the shape of this lens – it is fatter in the middle than it is at its ends. This is actually a feature of converging lenses. Can you think about what happened in using Eq. 38.7 and explain why a lens that is "fatter in the middle than at its ends" must certainly be a converging lens? Think about the radii of curvature and how they are used to determine the focal length.

51. (N) A diverging lens with focal length $|f| = 15.5 \text{ cm}$ produces an image with a magnification of +0.750. What are the object and image distances?

INTERPRET and ANTICIPATE
Because we know the focal length and the magnification, we can use Eq. 38.5 and Eq. 38.6 to find the object and image distances. If we construct those two equations with the given information, to model the behavior of the lens, we will find that we have two unique equations with the same two unknowns. This should be a solvable system of equations and it would be a matter of algebra to solve for those two unknowns – the object and image distances. Because this is a diverging lens, we must express the focal length as a negative number. Also, the image distance should be negative, assuming the object is a real object. This is because the diverging lens will create a virtual image of a real object, regardless of its position (on the left side of the lens for light traveling from left to right through the lens).

SOLVE	
Express Eq. 38.6 and solve for image distance in terms of the object distance.	$M = -\dfrac{d_i}{d_o}$ $d_i = -Md_o$

Now, express Eq. 38.5 and substitute the result from Eq. 38.6 for the image distance. Then, this equation can be simplified in an effort to solve for the object distance. Here, find a common denominator for the two fractions on the left side of the equation.	$$\frac{1}{d_i} + \frac{1}{d_o} = \frac{1}{f}$$ $$\frac{1}{-Md_o} + \frac{1}{d_o} = \frac{1}{f}$$ $$\frac{1-M}{-Md_o} = \frac{1}{f}$$
Now, solve for the object distance and use the given information to find its value.	$$\frac{1}{d_o} = -\frac{M}{f(1-M)}$$ $$d_o = -\frac{f(1-M)}{M} = -\frac{(-15.5 \text{ cm})(1-0.750)}{0.750}$$ $$d_o = \boxed{5.17 \text{ cm}}$$
Go back and solve for the image distance from Eq. 38.6.	$$d_i = -(0.750)(5.17 \text{ cm}) = \boxed{-3.88 \text{ cm}}$$

CHECK and THINK

The image is virtual (because the image distance is negative), upright (because the magnification is positive), and diminished (because the magnification is less than 1).

57. (N) An object is placed a distance of $4.00f$ from a converging lens, where f is the lens's focal length. **a.** What is the location of the image formed by the lens? **b.** Is the image real or virtual? **c.** What is the magnification of the image? **d.** Is the image upright or inverted?

INTERPRET and ANTICIPATE	
We will use Eq. 38.5 and Eq. 38.6 to model the behavior of this lens. Given that the object distance is expressed in terms of f, it is likely that the image distance will be as well. Since the object is beyond the focal point and this is a converging lens, the image should be real and inverted. This means we expect the image distance to be positive, the magnification to be negative.	
SOLVE **a.** Use Eq. 38.5 to find the image distance. The result depends on f, as expected.	$$\frac{1}{d_o} + \frac{1}{d_i} = \frac{1}{f}$$ $$\frac{1}{4.00f} + \frac{1}{d_i} = \frac{1}{f}$$ $$\frac{1}{d_i} = \frac{1}{f} - \frac{1}{4.00f} = \frac{3.00}{4.00f}$$ $$d_i = \frac{4.00}{3.00}f = \boxed{1.33f}$$

b. The image is real because the lens is a converging lens. This means the focal length is positive. Thus, given our answer to part (a), the image distance will also be positive.

c. Use Eq. 38.6 to find the magnification.

$$M = -\frac{d_i}{d_o} = -\frac{\frac{4.00}{3.00}f}{4.00f} = -\frac{1}{3.00} = \boxed{-0.333}$$

d. Since the magnification is negative, the image is inverted, relative to the orientation to the object.

CHECK and THINK

Our expectations were met regarding the orientation and type of image we would find created in this scenario. It is helpful in many problems to think through these kinds of questions before applying the equations in a model. In this way, you are prepared to evaluate the results and will hopefully know quickly if something went wrong.

69. (N) CASE STUDY: Susan wears corrective lenses. The prescription for her right eye is −7.75 diopters, and the prescription for her left eye is −8.00 diopters. Is she nearsighted or farsighted? If she is nearsighted, what is the far point for each of her eyes? If she is farsighted, what is the near point for each of her eyes? CHECK and THINK: In which eye does she have better vision?

INTERPRET and ANTICIPATE

Because the power of each lens is negative, this means the focal length of each lens will be negative. Thus, she is wearing diverging lenses. Diverging lenses are used to treat myopia. Thus, she is myopic, or nearsighted. We can use Eq. 38.8 with the power of each lens to find the focal length of each lens. Then, we note that an object at infinity would make an image at her far point. Thus, we use Eq. 38.5 to find the image location of an object at infinity for each eye to determine the far point in each eye. The eye with the greater far point has the better vision.

SOLVE
Use Eq. 38.8 to find the focal length of the lens used with the left eye.

$$f_{left} = \frac{1}{D} = \frac{1}{-8.00 \text{ m}^{-1}} = -0.125 \text{ m}$$

Now, use Eq. 38.5 with an object at infinity to find the image distance, which is the far point of her left eye.	$$\frac{1}{d_o} + \frac{1}{d_i} = \frac{1}{f}$$ $$\frac{1}{\infty} + \frac{1}{d_{far,left}} = \frac{1}{-0.125 \text{ m}}$$ $$d_{far,left} = -0.125 \text{ m}$$
The negative sign is just an indication that a virtual image is created, so the left eye far point is $\boxed{12.5 \text{ cm}}$.	
Use Eq. 38.8 to find the focal length of the lens used with the right eye.	$$f_{right} = \frac{1}{D} = \frac{1}{-7.75 \text{ m}^{-1}} = -0.129 \text{ m}$$
Now, use Eq. 38.5 with an object at infinity to find the image distance, which is the far point of her right eye.	$$\frac{1}{d_o} + \frac{1}{d_i} = \frac{1}{f}$$ $$\frac{1}{\infty} + \frac{1}{d_{far,right}} = \frac{1}{-0.129 \text{ m}}$$ $$d_{far,right} = -0.129 \text{ m}$$
The negative sign is just an indication that a virtual image is created, so the right eye far point is $\boxed{12.9 \text{ cm}}$. This means she has slightly better vision in her left eye.	

CHECK and THINK

An optometrist must diagnose each eye individually as the lenses needed may be slightly different for each eye. The goal is to have the final image created by each eye created at the same distance for the most relaxed viewing. The lens needs to put the image at the far point for each eye, hence the method of solution we have chosen here.

75. (N) An object 2.50 cm tall is 15.0 cm in front of a thin lens with a focal length of 5.00 cm. A thin lens with a focal length of –12.0 cm is placed 2.50 cm beyond this converging lens as shown in Figure P38.75. Determine the final image height and the position of the final image relative to the second lens.

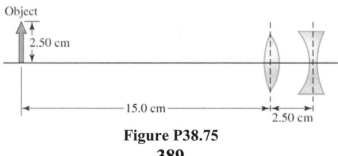

Figure P38.75

INTERPRET and ANTICIPATE

In order to model the behavior of each lens, we utilize Eq. 38.5 and Eq. 38.6. However, we must model the lenses in the order through which the light passes through them. This is because the second lens will "see" the image from the first lens as the object, not the original object. We must first find the location and height of the image from the first lens. Then, we can consider the distance between the lenses and the location of the first image to determine the object distance and height for the second lens. Because the second lens has a negative focal length, it is a diverging lens, which means that if the object for the second lens turns out to be a real object (it is to the left of the diverging lens with the light traveling from left to right), the final image should be virtual. However, if the object for the second lens turns out to be a virtual object (it is to the right of the diverging lens with the light traveling from left to right), the final image should be real!

SOLVE First, find the image distance for the first lens, using Eq. 38.5.	$\dfrac{1}{d_o} + \dfrac{1}{d_i} = \dfrac{1}{f}$ $\dfrac{1}{15.0 \text{ cm}} + \dfrac{1}{d_{i1}} = \dfrac{1}{5.00 \text{ cm}}$ $\dfrac{1}{d_{i1}} = -\dfrac{1}{15.0 \text{ cm}} + \dfrac{1}{5.00 \text{ cm}}$ $d_{i1} = 7.50 \text{ cm}$
The image from the first lens is a real image and is actually behind the second lens. Given the lenses are separated by 2.50 cm, we determine the object distance for the second lens. The object distance is negative so this is a virtual object for the second lens.	$d_{o2} = L - d_{i1} = 2.50 \text{ cm} - 7.50 \text{ cm}$ $d_{o2} = -5.00 \text{ cm}$
Use Eq. 38.5 to find the image distance for the second lens. The final image distance is positive, so it is to the right of the second lens (with light traveling from left to right), and the image is real.	$\dfrac{1}{d_o} + \dfrac{1}{d_i} = \dfrac{1}{f}$ $\dfrac{1}{-5.00 \text{ cm}} + \dfrac{1}{d_i} = -\dfrac{1}{12.0 \text{ cm}}$ $\dfrac{1}{d_i} = -\dfrac{1}{12.0 \text{ cm}} + \dfrac{1}{5.00 \text{ cm}} = -\dfrac{5}{60.0 \text{ cm}} + \dfrac{12}{60.0 \text{ cm}}$ $d_i = 60.0 \text{ cm}/7 = \boxed{8.57 \text{ cm}}$

Now, use Eq. 38.6 to find the magnification for the first lens.	$M_1 = -\dfrac{d_{i1}}{d_{o1}}$ $M_1 = -\dfrac{7.50 \text{ cm}}{15.0 \text{ cm}}$
Now, use Eq. 38.6 to find the magnification for the second lens.	$M_2 = -\dfrac{d_{i2}}{d_{o2}}$ $M_2 = -\dfrac{60.0 \text{ cm}/7}{-5.00 \text{ cm}} = \dfrac{60.0 \text{ cm}/7}{5.00 \text{ cm}}$
The total magnification is the product of the magnification from each lens.	$M = M_1 M_2 = \left(-\dfrac{7.50 \text{ cm}}{15.0 \text{ cm}}\right)\left(\dfrac{60.0 \text{ cm}/7}{5.00 \text{ cm}}\right)$
Now, use Eq. 38.6, with this total magnification to find the final image height.	$M = \dfrac{h_i}{h_o}$ $h_i = M h_o = \left(-\dfrac{7.50 \text{ cm}}{15.0 \text{ cm}}\right)\left(\dfrac{60.0 \text{ cm}/7}{5.00 \text{ cm}}\right)(2.50 \text{ cm})$ $h_i = \boxed{-2.14 \text{ cm}}$

CHECK and THINK

Because the final image height is negative, by convention, the image is upside down, or inverted. It is also slightly smaller than the object height, so the image is diminished in size by comparison. If we change the distance between the lenses, many of these aspects of the final image can change.

80. CASE STUDY: A group of students is given two converging lenses. Lens A has a focal length of 12.5 cm, and lens B has a focal length of 50.0 cm. The diameter of each lens is 6.50 cm. The students are asked to construct a telescope from these lenses if possible, and they have this discussion:

Avi: To make a telescope, we pick lens B to be the objective and lens A to be the eyepiece. Lens B has the greater focal length, so it has to be the objective.

Cameron: Both lenses are the same focal length—6.50 cm. It doesn't matter which is the objective.

Shannon: It does matter, because the magnification depends on their relative focal lengths. We still want to get the best magnification.

a. (C) What do you think?

b. (N) If a telescope can be constructed from these lenses, describe its design. What are its LGP and angular magnification? Compare the LGP to the value for your fully open pupil?

INTERPRET and ANTICIPATE **a.** In order to diagnose the dialogue, let's consider who is correct in his or her statement. Shannon and Avi are correct. The magnification depends on which lens is the eyepiece and which is the objective. To get the greatest magnification $m_{\text{tele}} = -f_o/f_e$ (Eq. 38.21), we would want to use lens B as the objective and lens A as the eyepiece. However, the LGP does not depend on this choice. To express the LGP and angular magnification, we can use Eq. 38.21 and Eq. 38.22. We also consider the design of the telescope as described below and compare the LGP and angular magnification to that for a human pupil.	

SOLVE **b.** Choose A to be the eyepiece and B to be the objective. Separate them by the sum of their focal lengths (about 62.5 cm). The LGP depends on the diameter of the objective. We take the diameter of the human pupil to be about 8 mm. Thus, using Eq. 38.22, we can create the ratio of each LGP.	$$\frac{(LGP)_{\text{tscope}}}{(LGP)_{\text{eye}}} = \left(\frac{65.0 \text{ mm}}{8 \text{ mm}}\right)^2 = 66$$
Thus, the telescope's light gathering power is 66 times better than your eye.	
The angular magnification is given by Eq. 38.21. Use this to find the angular magnification of the telescope.	$$m_{\text{tele}} = -\frac{f_o}{f_e} = -\frac{50.0 \text{ cm}}{12.5 \text{ cm}} = -4$$
Usually the negative sign is dropped, and we say that the telescope's magnification is 4×. The angular magnification of the eye is 1×, so 4× is the ratio of the telescopes angular magnification to that for the pupil.	

CHECK and THINK

If a greater magnification is desired in the case of the telescope, the curvature of the lenses must be adjusted to create more appropriate focal lengths. Also, the size of the lenses could be increased so that more light is captured by the lenses. This may aid in seeing fainter objects.

88. (N) A thick glass container ($n = 1.523$) with inner and outer vertical walls has its interior filled with water ($n = 1.333$) and has a quarter lying on the bottom. A ray of light with wavelength 486 nm has reflected from the quarter and strikes the interface between the water and the inner wall of the glass with an angle of incidence of 15.0°. **a.** What is the frequency of the light while in the water? **b.** What is the light ray's transmitted angle when it finally leaves the glass, passing into the air?

INTERPRET and ANTICIPATE

To find the frequency of the light while in the water, we can Eq. 36.8 to find the speed of the light while in the water, and then use Eq. 17.8 to find the frequency of the light, assuming the given wavelength is the wavelength when the light is in the water. We can apply Snell's law (Eq. 38.1) at each interface (water-glass and glass-air) to find the transmitted angle when the light leaves the glass. Since the air has a lower index of refraction than the water, we expect the transmitted angle to be larger than the angle of incidence in the water, assuming the walls of the glass are parallel (a fair assumption).

SOLVE **a.** Use Eq. 36.8 to find the speed of light in water.	$v = \dfrac{c}{n} = \dfrac{3.00 \times 10^8 \text{ m/s}}{1.333}$
Use Eq. 17.8 to find the frequency of the light while in water.	$f = \dfrac{v}{\lambda} = \dfrac{3.00 \times 10^8 \text{ m/s}/1.333}{486 \times 10^{-9} \text{ m}} = \boxed{4.63 \times 10^{14} \text{ Hz}}$
b. Apply Eq. 38.1 to the first interface between the water and the glass, and then again at the second interface between the glass and the air. Note that we assume the sides of the glass are parallel, which means that the ray that enters the glass from the water with an angle of refraction, θ_{glass}, would have an equal angle of incidence, θ_{glass}, at the second interface between the glass and the air.	$n_{\text{water}} \sin \theta_{\text{water}} = n_{\text{glass}} \sin \theta_{\text{glass}}$ $n_{\text{glass}} \sin \theta_{\text{glass}} = n_{\text{air}} \sin \theta_{\text{air}}$

Now, we can set the left side of the top equation equal to the right side of the bottom equation to find the transmitted angle into the air.	$n_{air} \sin\theta_{air} = n_{water} \sin\theta_{water}$ $$\theta_{air} = \sin^{-1}\left[\frac{n_{water} \sin\theta_{water}}{n_{air}}\right]$$ $$\theta_{air} = \sin^{-1}\left[\frac{(1.333)\sin(15.0°)}{1.0002926}\right]$$ $$\theta_{air} = \boxed{20.2°}$$

CHECK and THINK

If the walls of the glass are not assumed to be parallel, we would need to know how they were oriented to be able to model the scenario. There would also likely be some geometry required to relate angles as the light traveled across each boundary.

94. (N) Light in water is incident on diamond, with $\theta_i = 34.5°$. **a.** What is the angle of reflection? **b.** What is the angle of refraction in the diamond?

INTERPRET and ANTICIPATE

The law of reflection states that the angle of incidence is equal to the angle of reflection regardless of the medium. We can use Eq. 37.3 to find the angle of reflection. Eq. 38.1 can be used to find the angle of refraction of the light in the diamond. The angle of refraction should be less than the angle of incidence because the diamond has a greater index of refraction than the surrounding water.

SOLVE **a.** Use Eq. 37.3 to find the angle of reflection.	$\theta_r = \theta_i = \boxed{34.5°}$
b. Use Eq. 38.1 to find the angle of refraction.	$$\theta_t = \sin^{-1}\left[\left(\frac{n_{water}}{n_{diamond}}\right)\sin\theta_i\right]$$ $$\theta_t = \sin^{-1}\left[\left(\frac{1.333}{2.417}\right)\sin(34.5°)\right]$$ $$\theta_t = \boxed{18.2°}$$

CHECK and THINK

As expected, the angle of refraction was less than the angle of incidence. When an electromagnetic wave strikes a boundary between to media, usually, some of the wave will pass through to the next medium and refract and some of the wave will reflect.

107. (N) A block is constructed from layers of cubic zirconia ($n = 2.14$), flint glass ($n = 1.80$), and quartz ($n = 1.54$) as shown in Figure P38.13. The block is surrounded by air. For what values of the incident angle θ_i does total internal reflection occur at the quartz-air interface?

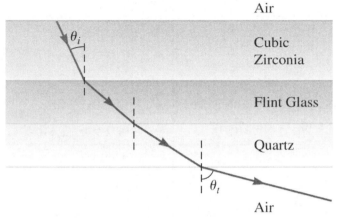

Figure P38.13

INTERPRET and ANTICIPATE

Note that because the layers and interfaces are all parallel, the angle of refraction in the top part of a layer will be equal to the angle of incidence in the bottom part of a layer. If we look at Figure P38.13ANS, we can visualize the application of Snell's law (Eq. 38.1) at each boundary. For the condition of total internal reflection, we can find the minimum angle of incidence necessary on the quartz-air interface by using Eq. 38.2. Once we find the minimum angle that causes the critical angle at the quartz-air interface, we can say any angle greater than this minimum angle will also cause internal reflection.

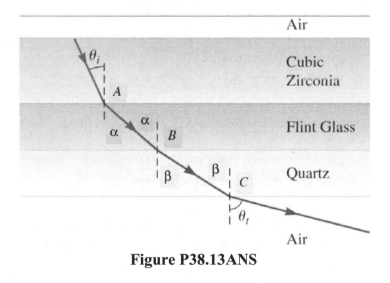

Figure P38.13ANS

SOLVE Apply Eq. 38.2 to the quartz-air interface to define the minimum angle of incidence on that interface.	$\theta_c = \sin^{-1}\left(\dfrac{n_{air}}{n_{quartz}}\right) = \sin^{-1}\left(\dfrac{1.0002926}{1.54}\right)$
Now, write Eq. 38.1 at each interface.	$n_{cz} \sin \theta_i = n_{flint} \sin \alpha$ $n_{flint} \sin \alpha = n_{quartz} \sin \beta$
We would like to solve these relationships for the critical angle and the quartz-air interface, or $\beta = \theta_c$.	$n_{cz} \sin \theta_i = n_{flint} \sin \alpha$ $n_{flint} \sin \alpha = n_{quartz} \sin \theta_c$
Because of the relationships we can equate the two equations and solve for the angle of incidence.	$n_{cz} \sin \theta_i = n_{quartz} \sin \beta = n_{quartz} \sin \theta_c$ $\sin \theta_i = \dfrac{n_{quartz} \sin \theta_c}{n_{cz}}$ $\sin \theta_i = \dfrac{(1.54)\sin\left(\sin^{-1}\left(\dfrac{1.0002926}{1.54}\right)\right)}{2.14}$ $\theta_i = \sin^{-1}\left[\dfrac{(1.54)\sin\left(\sin^{-1}\left(\dfrac{1.0002926}{1.54}\right)\right)}{2.14}\right]$ $\theta_i = 27.9°$
Any angle that is bigger than this initially will cause total internal reflection in the quartz, so $\boxed{\theta_i \geq 27.9°}$.	
CHECK and THINK Light incident on the cubic zirconia-flint glass interface with an angle greater than or equal to the 27.9° will not exit the quartz at the bottom.	

114. (N) An object is 20.0 cm to the left of a converging lens with a focal length of 30.0 cm. A second lens is placed 15.0 cm to the right of the converging lens. The second lens is a diverging lens with focal length of −20.0 cm. **a.** If the object has a height of 17.0 cm and is upright, find the magnification due to each lens and the magnification of the final image. **b.** What is the final image height?

INTERPRET and ANTICIPATE

In order to model the behavior of each lens, we utilize Eq. 38.5 and Eq. 38.6. However, we must model the lenses in the order through which the light passes through them. This is because the second lens will "see" the image from the first lens as the object, not the original object. We must first find the location and height of the image from the first lens. Then, we can consider the distance between the lenses and the location of the first image to determine the object distance and height for the second lens. Through successive application of these equations for each lens, we can determine the magnification of each lens and the final image, as well as the final image height.

SOLVE	
a. Use Eq. 38.5 to find the image distance for the first lens.	$\dfrac{1}{d_{o1}} + \dfrac{1}{d_{i1}} = \dfrac{1}{f_1}$ $\dfrac{1}{d_{i1}} = \dfrac{1}{f_1} - \dfrac{1}{d_{o1}} = \dfrac{1}{30.0 \text{ cm}} - \dfrac{1}{20.0 \text{ cm}} = \dfrac{-1}{60.0 \text{ cm}}$ $d_{i1} = -60.0 \text{ cm}$
Then, calculate the object distance for the second lens using the distance between the lenses. The first image is a virtual image and is in front of the first lens, further away from the second lens than the first lens itself!	$d_{o2} = w - d_{i1} = 15.0 \text{ cm} - (-60.0 \text{ cm})$ $d_{o2} = 75.0 \text{ cm}$
Now, use Eq. 38.5 to find the image distance for the second lens.	$\dfrac{1}{d_{o2}} + \dfrac{1}{d_{i2}} = \dfrac{1}{f_2}$ $\dfrac{1}{d_{i2}} = \dfrac{1}{f_2} - \dfrac{1}{d_{o2}}$ $\dfrac{1}{d_{i2}} = \dfrac{1}{-20.0 \text{ cm}} - \dfrac{1}{75.0 \text{ cm}} = \dfrac{-19}{3.00 \times 10^2 \text{ cm}}$ $d_{i2} = -15.8 \text{ cm}$
With the relevant object and image distances for each lens, we can now express the magnification due to each lens by using Eq. 38.6.	$M_1 = -\dfrac{d_{i1}}{d_{o1}} = -\dfrac{-60.0 \text{ cm}}{20.0 \text{ cm}} = \dfrac{60.0 \text{ cm}}{20.0 \text{ cm}} = \boxed{3.00}$ $M_2 = -\dfrac{d_{i2}}{d_{o2}} = -\dfrac{-15.8 \text{ cm}}{75.0 \text{ cm}} = \dfrac{15.8 \text{ cm}}{75.0 \text{ cm}} = \boxed{0.211}$
The total magnification is the product of the individual magnifications.	$M = M_1 M_2 = \left(\dfrac{60.0 \text{ cm}}{20.0 \text{ cm}} \right)\left(\dfrac{15.8 \text{ cm}}{75.0 \text{ cm}} \right) = \boxed{0.632}$

b. Use Eq. 38.6 to find the image height.	$h_i = Mh_o = (0.632)(17.0 \text{ cm}) = \boxed{10.7 \text{ cm}}$

CHECK and THINK

He magnification of this object would change if the distance between the lenses is altered. Currently, the final image is diminished and upright, or in the same orientation as the object. It is possible, by playing with the distance between the lenses, that we might cause the image to be magnified or inverted.

120. (N) Consider a converging lens with a 16.0-cm focal length. How far is the object from the lens if a virtual image is formed **a.** 22.0 cm in front of the lens and **b.** 42.0 cm in front of the lens?

INTERPRET and ANTICIPATE

In both cases, we can use Eq. 38.5 to find the object distance. Also, in both cases the image is in front of the lens, so they are virtual images. We must make the image distances negative. Since the lens is a converging lens, recall that a virtual image may be created by a real object if the object is closer than the focal length to the lens. We expect that the object distance, in both cases, is less than 16.0 cm.

SOLVE **a.** Use Eq. 38.5 to find the object distance for this first case.	$\dfrac{1}{d_o} + \dfrac{1}{d_i} = \dfrac{1}{f}$ $\dfrac{1}{d_o} + \dfrac{1}{-22.0 \text{ cm}} = \dfrac{1}{16.0 \text{ cm}}$ $\dfrac{1}{d_o} = \dfrac{1}{16.0 \text{ cm}} + \dfrac{1}{22.0 \text{ cm}}$ $d_o = \boxed{9.26 \text{ cm}}$
b. Use Eq. 38.5 to find the object distance for this second case.	$\dfrac{1}{d_o} + \dfrac{1}{d_i} = \dfrac{1}{f}$ $\dfrac{1}{d_o} + \dfrac{1}{-42.0 \text{ cm}} = \dfrac{1}{16.0 \text{ cm}}$ $\dfrac{1}{d_o} = \dfrac{1}{16.0 \text{ cm}} + \dfrac{1}{42.0 \text{ cm}}$ $d_o = \boxed{11.6 \text{ cm}}$

CHECK and THINK

If the object distance were greater than the focal length of the converging lens, then the images would be real and the image distances would be positive.

39

Relativity

4. (N) In an airport terminal there are two fast-moving sidewalks (9.0 km/h); one carries its passengers south, and the other carries its passengers north. Each sidewalk is 0.50 km long. At the instant a woman steps onto the north end of the southbound sidewalk, a man steps onto the south end of the northbound sidewalk. He stands still with respect to the sidewalk, while she walks south at 5.0 km/h. **a.** How long after stepping onto the sidewalk do they pass each other? (Report your answer to the nearest second.) **b.** How far does each person travel in that time? (Report your answer in kilometers.)

INTERPRET and ANTICIPATE We will need to create expressions for the positions of the man and woman as functions of time. We will assume he begins at $x = 0$ and that the woman begins at the location $x = 0.50 \times 10^3$ m, given that they start on opposite ends of the sidewalk. We will treat north as the positive direction and south as the negative direction. Once we have expressions for their positions as a function of time in this coordinate system we have proposed (the man beginning at the origin), we can equate the expressions and solve for the time when they are at the same position. This is when they must cross paths. Because she is walking south as the sidewalk moves south, we expect she is moving towards the origin faster than he is moving away from it. Thus, we expect they will cross paths less than half the original separation distance. Once we know the time, we can use this with either of their position formulas to determine the location where they cross. We can then express how far they have each traveled.	

SOLVE	
a. First, we create an expression for the position of the man as a function of time, assuming he begins at $x = 0$.	$x_{man} = x_i + v_{man}\Delta t = 0 + \left(9.0 \times 10^3 \text{ m/h}\right)\Delta t$ $x_{man} = \left(9.0 \times 10^3 \text{ m/h}\right)\Delta t$
Next, we use Eq. 39.11 to express the woman's velocity relative to the ground. Note that she is traveling towards the origin, or in the negative x direction.	$v_{woman} = v' + v_{rel} = -5.0 \times 10^3 \text{ m/h} - 9.0 \times 10^3 \text{ m/h}$ $v_{woman} = -14.0 \times 10^3 \text{ m/h}$

We use her velocity to create an expression for her position as a function of time.	$x_{woman} = x_i + v_{woman} \Delta t$ $x_{woman} = 0.50 \times 10^3 \text{ m} - (14.0 \times 10^3 \text{ m/h}) \Delta t$
They will pass each other when their positions are the same, so equate their functions of position and solve for Δt.	$x_{man} = x_{woman}$ $(9.0 \times 10^3 \text{ m/h}) \Delta t = 0.50 \times 10^3 \text{ m} - (14.0 \times 10^3 \text{ m/h}) \Delta t$ $(23.0 \times 10^3 \text{ m/h}) \Delta t = 0.50 \times 10^3 \text{ m}$ $\Delta t = \dfrac{0.50 \times 10^3 \text{ m}}{23.0 \times 10^3 \text{ m/h}} = 0.022 \text{ h} = \boxed{78 \text{ s}}$
b. Use the equation we created for the position of the man as a function to find the distance he travels in that time (since he started at the origin, $x = 0$).	$x_{man} = (9.0 \times 10^3 \text{ m/h}) \Delta t$ $x_{man} = (9.0 \times 10^3 \text{ m/h}) \left(\dfrac{0.50 \times 10^3 \text{ m}}{23.0 \times 10^3 \text{ m/h}} \right)$ $x_{man} = 2.0 \times 10^2 \text{ m} = \boxed{0.20 \text{ km}}$
Then, since the woman traveled from the other end, starting at 0.50 km, we subtract the 0.20 km to find the distance she traveled.	$d_{woman} = 0.50 \text{ km} - 0.20 \text{ km} = \boxed{0.30 \text{ km}}$

CHECK and THINK
In this problem, we had a coordinate system that was not moving, and could determine the relative velocity of both people relative to this coordinate system. We could then track their motion in a similar fashion to how we began this book. What we find in this chapter is that if the people are moving near the speed of light, though, they may not agree on how far they have each traveled, or how much time had passed in doing so. And yet, they may both be correct!

6. (N) Jason is driving north down a highway at 60.0 mph when he sees a car ahead. Kevin is driving that car north at 53.0 mph. **a.** What are the magnitude and direction of Kevin's velocity according to Jason? **b.** How far does Kevin travel in Jason's frame of reference in the time it takes Jason to travel 3.0 mi?

INTERPRET and ANTICIPATE
We can use Eq. 39.11 to express the relative velocity of Kevin, as determined by Jason. We can then use Eq. 39.1 to find the time it takes Jason to travel 3.0 mi, and then Eq. 39.10 to find the distance traveled by Kevin in Jason's frame of reference. Because they

are traveling in the same direction, we expect that the distance traveled in Jason's moving frame will be less than the distance traveled, according to an observer on the side of the road. In this problem, we will treat north as the positive direction and south as the negative direction.

SOLVE	$v_x = v_x' + v_{rel}$
a. We use Eq. 39.11 to find the velocity of Kevin according to Jason.	$53.0 \text{ mph} = v_x' + 60.0 \text{ mph}$
	$v_x' = -7.0 \text{ mph}$
	Or, we could say $\boxed{7.0 \text{ mph to the south}}$.
b. Use Eq. 39.1 to find the time it takes Jason to travel 3.0 mi.	$\Delta t = \dfrac{3.0 \text{ mi}}{60.0 \text{ mph}} = 0.050 \text{ h}$
Now, use Eq. 39.10 to find the distance traveled by Kevin in Jason's frame of reference.	$\Delta x = \Delta x' + v_{rel}\Delta t$
	$\Delta x = 0 + (-7.0 \text{ mi/h})(0.050 \text{ h}) = -0.35 \text{ mi}$
	$\lvert \Delta x \rvert = \boxed{0.35 \text{ mi}}$

CHECK and THINK

As expected, Jason does not perceive that Kevin has traveled that far in his frame of reference. If they were traveling in opposite directions, we would expect to find that Jason perceived Kevin to travel farther than the 3.0 mi, due to Jason's moving frame of reference.

13. (N) The primed frame has speed $0.75c$ in the x direction according to an observer in the laboratory frame. At $t = t' = 0$, the origins of the two frames coincide. At $t' = 16.0$ s, an observer in the primed frame sees a balloon hovering at $(x', y') = (6.0 \text{ m}, 12.0 \text{ m})$.
a. What can you say about the position of the balloon as seen in the laboratory frame?
b. How much time has elapsed in the laboratory frame when the observer in the primed frame makes this observation?

INTERPRET and ANTICIPATE

First, we recognize that it is only the x position of the balloon about which observers in the laboratory and primed frames might disagree. The y position of the balloon should appear to be the same in both frames of reference, based on Eq. 39.13. We can use Eq. 39.15 to find the x position in the laboratory frame. We can also use Eq. 39.19 to find the time in the laboratory frame when the 16.0 s have passed in the primed frame. Given that the measurements are being made on the object in the primed frame, we expect the position in the laboratory frame to be much greater and the time required to travel that distance greater, as well, than those values as found by the observer in the primed frame.

SOLVE	
a. Both observers agree that the y position of the balloon is 12.0 m, since this will be invariant, based on Eq. 39. 13.	$y = y' = \boxed{12.0 \text{ m}}$
We can use Eq. 39.15 to find the x position at that time, as measured in the laboratory frame.	$x = \gamma\left(x' + v_{rel}t'\right) = \dfrac{1}{\sqrt{1-\left(v_{rel}/c\right)^2}}\left(x' + v_{rel}t'\right)$ $x = \dfrac{1}{\sqrt{1-\left(0.75c/c\right)^2}}\left(6.0 \text{ m} + (0.75)\left(3.00\times10^8 \text{ m/s}\right)(16.0 \text{ s})\right)$ $x = \boxed{5.4\times10^9 \text{ m}}$
b. Use Eq. 39.19 to find the time in the laboratory frame.	$t = \gamma\left(t' + \dfrac{v_{rel}x'}{c^2}\right)$ $t = \dfrac{1}{\sqrt{1-\left(0.75c/c\right)^2}}\left(16.0 \text{ s} + \dfrac{(0.75)\left(3.00\times10^8 \text{ m/s}\right)(6.0 \text{ m})}{\left(3.00\times10^8 \text{ m/s}\right)^2}\right)$ $t = \boxed{24 \text{ s}}$

CHECK and THINK

Events occurring in the primed frame take a longer time to occur, as viewed by the observer in the laboratory frame. The reverse is also true! However, if the primed frame ever stops moving and returns to the laboratory frame, the observer in the primed frame is now in a frame of reference where more time has passed than his or her own previously. Effectively, that observer has aged less than the observer in the laboratory frame. This is often called "The Twin Paradox." Look it up in this chapter!

20. (N) What is the speed with which a ruler 12.0 in (30.48 cm) long, as measured in its own reference frame, must be moving if its apparent length is 20.0 cm?

INTERPRET and ANTICIPATE

Because we know the apparent length and the length of the ruler measured in its own frame of reference, we can use Eq. 39.25 to find the necessary speed to cause this to occur. Certainly, a relativistic speed, say at least 10% the speed of light, is expected.

SOLVE Use Eq. 39.25 to find the speed of the ruler.	$\dfrac{1}{\gamma} = \dfrac{L}{L'} = \dfrac{20.0 \text{ cm}}{30.48 \text{ cm}}$ $\sqrt{1-\left(v_{rel}/c\right)^2} = \dfrac{20.0 \text{ cm}}{30.48 \text{ cm}}$ $1-\left(v_{rel}/c\right)^2 = \left(\dfrac{20.0 \text{ cm}}{30.48 \text{ cm}}\right)^2$ $\left(v_{rel}/c\right)^2 = 1-\left(\dfrac{20.0 \text{ cm}}{30.48 \text{ cm}}\right)^2$ $v_{rel}/c = \sqrt{1-\left(\dfrac{20.0 \text{ cm}}{30.48 \text{ cm}}\right)^2}$ $v_{rel} = \sqrt{1-\left(\dfrac{20.0 \text{ cm}}{30.48 \text{ cm}}\right)^2}\, c = \boxed{0.755c}$

CHECK and THINK

As an object's speed becomes comparable to the speed of light, its dimension, along the direction it is moving, will appear contracted to an observer not moving with the object (in what we usually call the laboratory frame).

21. (N) A meterstick in the primed frame makes a 45.0° angle with the x' axis. The meterstick is at rest in the primed frame. As seen by an observer in the laboratory frame, the primed frame's relative speed in the x direction is $0.975c$. What is the length of the meterstick as observed in the laboratory frame?

INTERPRET and ANTICIPATE

Note that the meterstick can be thought of as having two component lengths – one along the x' direction and the other along the y' direction. This is because of the 45.0° angle with the x' axis. Because the primed frame has a relative speed in the x direction in the laboratory frame, it is only the component along the x' direction that will experience length contraction, as viewed by an observer in the laboratory frame. We can find the contracted x component length using Eq. 39.25, and understand that the y component will be exactly the same in both frames of reference. We can then use the Pythagorean theorem to find the length of the meterstick in the laboratory frame. Given the x component should be contracted, we expect the length of the meterstick to be less than 1 m. Note that the length of the meter stick will not limit significant figures since it is supposedly a representation of unity (1 m exactly).

SOLVE First, we find the length of the y component of the	$y = y' = \left(1 \text{ m}\right)\sin\left(45°\right)$

meterstick, which will be the same in each frame of reference because there is no relative motion in that direction.	
Use Eq. 39.25 to find the x-component, as viewed in the laboratory frame.	$$x = \frac{x'}{\gamma} = \frac{x'}{\dfrac{1}{\sqrt{1-\left(v_{rel}/c\right)^2}}}$$ $$x = \frac{\left(1\text{ m}\right)\cos\left(45°\right)}{\dfrac{1}{\sqrt{1-\left(0.975c/c\right)^2}}}$$ $$x = \left(1\text{ m}\right)\cos\left(45°\right)\left(\sqrt{1-\left(0.975\right)^2}\right)$$
Then, the length of the meter stick in the laboratory frame can be found using the Pythagorean theorem with these two components.	$$\ell = \sqrt{x^2 + y^2}$$ $$\ell = \sqrt{\left[\left(1\text{ m}\right)\cos\left(45°\right)\left(\sqrt{1-\left(0.975\right)^2}\right)\right]^2 + \left[\left(1\text{ m}\right)\sin\left(45°\right)\right]^2}$$ $$\ell = \boxed{0.724\text{ m}}$$

CHECK and THINK

The meterstick is contracted even though its entire length does not lie parallel to the direction its frame of reference is traveling. Only dimensions parallel to the relative velocity between two frames of reference will appear to be contracted.

28. (N) A neutron outside of a nucleus is an unstable particle. It decays in roughly 10.0 minutes. How fast would a neutron need to travel in order to be stable for 11.0 min as seen by an observer in the laboratory frame? How far would the neutron travel in that time according to the laboratory observer?

INTERPRET and ANTICIPATE

Given that we have the decay times in both frames of reference, we can find the relative speed between the two frames by using Eq. 39.26. Once we know the speed with which the neutron is moving, according to an observer in the laboratory frame, we can multiply the speed by the time observed in the laboratory frame to find the distance traveled in that frame ($d = v\Delta t$).

SOLVE Use Eq. 39.26 to find the speed.	$\Delta t = \gamma \Delta t'$ $\gamma = \dfrac{\Delta t}{\Delta t'}$ $\dfrac{1}{\sqrt{1-\left(v_{rel}/c\right)^2}} = \dfrac{\Delta t}{\Delta t'}$ $\sqrt{1-\left(v_{rel}/c\right)^2} = \dfrac{\Delta t'}{\Delta t}$ $1-\left(v_{rel}/c\right)^2 = \left(\dfrac{\Delta t'}{\Delta t}\right)^2$ $v_{rel} = \sqrt{1-\left(\dfrac{\Delta t'}{\Delta t}\right)^2}\,c = \sqrt{1-\left(\dfrac{10.0\ \text{min}}{11.0\ \text{min}}\right)^2}\,c = \boxed{0.417c}$
Now, use kinematics to find the distance traveled by the neutron in that 11.0 min according to the laboratory observer.	$d = v\Delta t = \left(0.417c\right)\left(11.0\ \text{min}\right)$ $d = \left(0.417\right)\left(3.00\times10^8\ \text{m/s}\right)\left(11.0\ \text{min}\right)\left(\dfrac{60\ \text{s}}{1\ \text{min}}\right)$ $d = \boxed{8.26\times10^{10}\ \text{m}}$

CHECK and THINK
The neutron must be traveling incredibly fast for the time dilation in this scenario, and thus, the neutron would travel quite far in that amount of time, according to the laboratory observer.

32. (N) It takes 1.0 h to bake a casserole in a frame moving at a speed of 2.0×10^7 m/s. How long is the baking time as measured by a cook in the laboratory frame?

INTERPRET and ANTICIPATE
Given the time to bake in the casserole's frame of reference, we can use Eq. 39.26 to find the time, as measured by an observer in the laboratory frame. We expect that the time will be greater than the 1 h measured in the casserole's frame.

SOLVE Use Eq. 39.26 to find the time in the laboratory frame.	$\Delta t = \gamma \Delta t' = \dfrac{1}{\sqrt{1-\left(v_{rel}/c\right)^2}}\Delta t'$ $\Delta t = \dfrac{1}{\sqrt{1-\left(\left(2.0\times10^8\ \text{m/s}\right)/\left(3.00\times10^8\ \text{m/s}\right)\right)^2}}\left(1.0\ \text{h}\right)$ $\Delta t = \boxed{1.3\ \text{h}}$

Chapter 39 – Relativity

37. (N) A source of hydrogen gas gives off Hα (Fig. 36.19A, page 1166). The line is emitted with a wavelength of 656 nm and observed with a wavelength of 652 nm. What is the constant speed of the source relative to the observer? Is the source moving toward or away from the observer?

Hα
656 nm

Hβ
486 nm

Hγ
434 nm

Hδ
410 nm

Figure P36.19A

INTERPRET and ANTICIPATE

We can actually answer the second question immediately. The source is moving towards the observer because the observed wavelength is shorter than the emitted wavelength. We can use Eq. 39.27 to find β and then use Eq. 39.18 to find the relative speed.

SOLVE	
Express Eq. 39.27 and solve for β.	$\lambda = \sqrt{\dfrac{1-\beta}{1+\beta}}\,\lambda'$
	$652 \text{ nm} = \sqrt{\dfrac{1-\beta}{1+\beta}}\,(656 \text{ nm})$
	$\sqrt{\dfrac{1-\beta}{1+\beta}} = \dfrac{652 \text{ nm}}{656 \text{ nm}}$
	$\dfrac{1-\beta}{1+\beta} = \left(\dfrac{652 \text{ nm}}{656 \text{ nm}}\right)^2$
	$1-\beta = \left(\dfrac{652 \text{ nm}}{656 \text{ nm}}\right)^2 + \left(\dfrac{652 \text{ nm}}{656 \text{ nm}}\right)^2 \beta$
	$\beta\left[1+\left(\dfrac{652 \text{ nm}}{656 \text{ nm}}\right)^2\right] = 1-\left(\dfrac{652 \text{ nm}}{656 \text{ nm}}\right)^2$
	$\beta = \dfrac{1-\left(\dfrac{652 \text{ nm}}{656 \text{ nm}}\right)^2}{1+\left(\dfrac{652 \text{ nm}}{656 \text{ nm}}\right)^2}$

Now, use Eq. 39.18 to find the relative speed.	$v_{rel} = \beta c = \dfrac{1 - \left(\dfrac{652 \text{ nm}}{656 \text{ nm}}\right)^2}{1 + \left(\dfrac{652 \text{ nm}}{656 \text{ nm}}\right)^2} c = \boxed{1.83 \times 10^6 \text{ m/s}}$

CHECK and THINK

While not greater than 10% of the speed of light, this is still a high value of speed, and the difference in length observed is only a few nm. The length contraction may be small, but it is measurable compared to the size of the wave.

44. (N) In a laboratory, two particles are fired sequentially. Particle 1 has speed $0.80c$ and is traveling to the east. A moment later, particle 2 is launched with speed $0.70c$.
a. What is the speed of particle 1 as measured in the reference frame of particle 2, if particle 2 is also traveling to the east?
b. What is the speed of particle 1 as measured in the reference frame of particle 2, if particle 2 is traveling to the west?

INTERPRET and ANTICIPATE

We can use Eq. 39.33 to answer both parts of this question, as long as we identify properly which particle is in the primed frame of reference and which is in the unprimed frame of reference, and consider the algebraic sign of the relative velocity given that the measured particle (particle 1) is always traveling to the east. In this problem, particle 2 is in the primed frame of reference in each case, measuring the speed of particle 1 in the unprimed frame of reference (since Eq. 39.33 is solved for the speed of an object in the unprimed frame of reference as measured in the primed reference frame). In part (a), both particles are traveling to the east, so the relative velocity is $+0.70c$. In part (b), particle 2 is traveling to the west, so the relative velocity is $-0.70c$.

SOLVE

a. Use Eq. 39.33 with the relative velocity of $0.70c$.	$v_x' = \dfrac{v_x - v_{rel}}{1 - \left(v_{rel}/c^2\right)v_x} = \dfrac{0.80c - 0.70c}{1 - \left(0.70c/c^2\right)(0.80c)}$ $v_x' = \boxed{0.23c}$ Or we could say, $0.23c$ to the east.
b. Use Eq. 39.33 with the relative velocity of $-0.70c$.	$v_x' = \dfrac{v_x - v_{rel}}{1 - \left(v_{rel}/c^2\right)v_x} = \dfrac{0.80c + 0.70c}{1 - \left(-0.70c/c^2\right)(0.80c)}$ $v_x' = \boxed{0.96c}$ Or we could say, $0.96c$ to the east.

CHECK and THINK

As we would expect, particle 1 appears to be moving slower when measured in a frame of reference moving in the same direction as particle 1, and faster when measured in a frame of reference moving in the opposite direction of particle 1. However, this is not a simple Galilean transformation, as in nonrelativistic mechanics.

45. (N) CASE STUDY: In a fictional spaceport, a fast-moving sidewalk moves at about $0.90c$. The sidewalk is 0.50 ly long. A man gets on the sidewalk, and a woman standing near the sidewalk watches him. In each case, find his speed in her frame in terms of c. **a.** He is at rest with respect to the sidewalk. **b.** He walks at 2.2 m/s with respect to the sidewalk. **c.** He uses fictional roller blades so that his speed relative to the sidewalk, and in the same direction, is $0.50c$.

INTERPRET and ANTICIPATE

In part (a), we expect that she should see him moving at the same speed as the sidewalk. There is no relative velocity to consider. The same is really true in part (b) because his speed relative to the sidewalk is so small compared to the speed of the sidewalk relative to the stationary observer. Finally, in part (c), we can use Eq. 39.31 to find his speed as measured in the by the woman (in the unprimed frame of reference for Eq. 39.31). Both the sidewalk and the man are moving in the same direction, so the relative velocity is positive in Eq. 39.31. We expect that his measured speed is greater because of the relative motion.

SOLVE

a. If he is at rest, he appears to move at the rate of the sidewalk, so $\boxed{0.90c}$.

b. Because he is moving so slow compared to the speed of the sidewalk, there is little reason to go through the process of transforming the velocity as viewed by the woman. He will appear to move with a speed of $\boxed{0.90c}$.

c. We can use the velocity transformation (Eq. 39.31) to find the velocity according to the woman.	$$v_x = \frac{v_x' + v_{rel}}{1+\left(v_{rel}/c^2\right)v_x'} = \frac{0.50c+0.90c}{1+\left(0.90c/c^2\right)(0.50c)} = \boxed{0.97c}$$

CHECK and THINK

Unless the velocities of objects, or the relative velocity between objects are relativistic, there is no need to use velocity transformations like that seen in Eq. 39.31. A simple Galilean transformation would suffice.

48. (N) The rest mass of a baseball is 0.145 kg. Find the momentum observed if the ball is moving at **a.** 90.0 mph (40.2 m/s)—a fast pitch, **b.** 250 km/s—the speed of the Sun around the center of the Milky Way galaxy, and **c.** $0.85c$ with respect to the laboratory.

<table>
<tr><td colspan="2">INTERPRET and ANTICIPATE
The issue to consider here is the when the classical calculation of momentum (Eq. 10.1) will suffice, and when the relativistic calculation of momentum (Eq. 39.40) will be required. In the first two parts of the problem, the velocities are not relativistic, and so Eq. 10.1 will suffice. However, in the final part of the problem, we will need to use Eq. 39.40 to find the momentum. The momentum should increase as we solve each part of the problem in succession.</td></tr>
<tr><td colspan="2">SOLVE</td></tr>
<tr><td>a. The ball is moving at a nonrelativistic velocity, so its momentum is essentially just its classical momentum (Eq. 10.1).</td><td>$p = mv = (0.145 \text{ kg})(40.2 \text{ m/s}) = \boxed{5.83 \text{ kg} \cdot \text{m/s}}$</td></tr>
<tr><td>b. Again, this velocity is not relativistic, so the momentum will be its classical momentum (Eq. 10.1).</td><td>$p = mv = (0.145 \text{ kg})(250.0 \times 10^3 \text{ m/s})$
$p = \boxed{3.63 \times 10^4 \text{ kg} \cdot \text{m/s}}$</td></tr>
<tr><td>c. Use Eq. 39.40 to find the momentum.</td><td>$p = \gamma m_{\text{rest}} v = \dfrac{m_{\text{rest}} v}{\sqrt{1-(v/c)^2}} = \dfrac{(0.145 \text{ kg})(0.850c)}{\sqrt{1-(0.850c/c)^2}}$
$p = \boxed{7.02 \times 10^7 \text{ kg} \cdot \text{m/s}}$</td></tr>
</table>

CHECK and THINK
While Eq. 39.40 is valid for each of these cases, it is appropriate to use Eq. 10.1 and treat the object classically as long as the velocity is less than 10% of the speed of light. It is then that the correction factor for relativity (the denominator of Eq. 39.40) is nearly equal to 1 and does not change the result of the calculation very much.

51. (N) In transmission electron microscopy (TEM), electrons are accelerated to have kinetic energies of hundreds of thousands of electron-volts (1 eV = 1.602×10^{-19} J). Suppose an electron has a kinetic energy of 300.0 keV. **a.** What is the rest mass energy of the electron? **b.** What is the total energy of the electron? **c.** What is the speed of the electron? Answer using three significant figures.

INTERPRET and ANTICIPATE
We can use Eq. 39.44 to express the rest mass energy of the electron. Then, the total energy of the electron is found using Eq. 39.45. Lastly, we can use Eq. 39.43 to find the relativistic mass of the moving electron with the kinetic energy given in the problem statement. This result can be used in Eq. 39.42 to find the speed of the electron under these conditions. We expect this speed to be relativistic, given the large amount of kinetic energy (for an object of its small mass).

SOLVE	
a. Use Eq. 39.44 with the rest mass to find the rest mass energy of the electron.	$E_{rest} = m_{rest}c^2 = (9.109 \times 10^{-31} \text{ kg})(3.00 \times 10^8 \text{ m/s})^2$ $E_{rest} = \boxed{8.20 \times 10^{-14} \text{ J or 512 keV}}$

b. The total energy can be found using Eq. 39.45.

$$E = K + m_{rest}c^2$$

$$E = (300.0 \times 10^3 \text{ eV})\left(\frac{1.602 \times 10^{-19} \text{ J}}{1 \text{ eV}}\right) + (9.109 \times 10^{-31} \text{ kg})(3.00 \times 10^8 \text{ m/s})^2$$

$$E = \boxed{1.30 \times 10^{-13} \text{ J or 812 keV}}$$

c. Use Eq. 39.43 to solve for the relativistic mass of the electron according to a lab observer.

$$K = mc^2 - m_{rest}c^2$$

$$m = \frac{K + m_{rest}c^2}{c^2}$$

$$m = \frac{(300.0 \times 10^3 \text{ eV})\left(\frac{1.602 \times 10^{-19} \text{ J}}{1 \text{ eV}}\right) + (9.109 \times 10^{-31} \text{ kg})(3.00 \times 10^8 \text{ m/s})^2}{(3.00 \times 10^8 \text{ m/s})^2}$$

$$m = 1.445 \times 10^{-30} \text{ kg}$$

| Find the speed of the particle according to the lab observer, using Eq. 39.42. | $m = \gamma m_{rest} = \dfrac{1}{\sqrt{1 - (v/c)^2}} m_{rest}$

 $\sqrt{1 - (v/c)^2} = \dfrac{m_{rest}}{m}$

 $v/c = \sqrt{1 - \left(\dfrac{m_{rest}}{m}\right)^2}$

 $v = \sqrt{1 - \left(\dfrac{m_{rest}}{m}\right)^2}\, c = \sqrt{1 - \left(\dfrac{9.109 \times 10^{-31} \text{ kg}}{1.445 \times 10^{-30} \text{ kg}}\right)^2}\, c$

 $v = \boxed{0.7762c}$ |

CHECK and THINK

The total energy of the electron is the sum of its rest mass energy and its kinetic energy. As the electron moves faster, its relativistic mass will increase and approach infinity as the speed approaches the speed of light.

56. (E) Assume the Sun's luminosity (the same as its power) is constant over its entire 10-billion-year life. Estimate the amount of mass it must consume in nuclear fusion to put out the required amount of energy. Give your answer in kilograms and in solar masses.

INTERPRET and ANTICIPATE	
We must take the Sun's power and multiply by the time scale given to find the total amount of energy that the Sun would radiate. Then, we can use Eq. 39.44 to find the amount of mass conversion this would represent. We will use the power of the Sun given in Example 39.10. At the end, when we want to convert kilograms to solar masses, we will use the fact that the mass of the Sun is about 2×10^{30} kg.	
SOLVE Multiply the Sun's power by the 10-billion-year lifetime, expressed in seconds. This is the total energy emitted in that time.	$E = Pt$ $E = \left(3.84 \times 10^{26} \text{ W}\right)\left(10 \times 10^{9} \text{ yr}\right)\left(\dfrac{365 \text{ day}}{1 \text{ yr}}\right)\left(\dfrac{24 \text{ h}}{1 \text{ day}}\right)\left(\dfrac{3600 \text{ s}}{1 \text{ h}}\right)$ $E = 1 \times 10^{44} \text{ J}$
Use Eq. 39.44 to determine the amount of mass that was consumed.	$E = mc^2$ $m = E/c^2$ $m = 1 \times 10^{44} \text{ J}/\left(3.00 \times 10^{8} \text{ m/s}\right)^2 = \boxed{1 \times 10^{27} \text{ kg}}$
Convert the answer to be expressed in terms of solar masses.	$m = \left(1 \times 10^{27} \text{ kg}\right) \times \left(\dfrac{1 \text{ solar mass}}{2 \times 10^{30} \text{ kg}}\right)$ $m = \boxed{0.0005 \text{ solar masses}}$
CHECK and THINK In that time, the Sun would have consumed 0.05% of what we consider a solar mass.	

60. (N) Hα ($f_{\text{emt}} = 4.57 \times 10^{14}$ Hz) is emitted from the surface of a neutron star. Its radius is 6.96×10^{4} m and its mass is 2 solar masses. What is the frequency of the light observed far from the neutron star?

INTERPRET and ANTICIPATE
We can use Eq. 39.46 to find the observed frequency. This equation considers the effect of the mass and size of the star on the frequency of the electromagnetic waves emitted by the star. We expect that the frequency will decrease due to the effects described by general relativity.

SOLVE Use Eq. 39.46 to find the observed frequency.	$\dfrac{f_{obs}}{f_{emt}} = \left(1 - \dfrac{2GM}{c^2 r}\right)^{\frac{1}{2}}$
	$f_{obs} = \left(1 - \dfrac{2\left(6.67\times10^{-11}\ \text{N}\cdot\text{m}^2/\text{kg}\right)(2)\left(1.99\times10^{30}\ \text{kg}\right)}{\left(3.00\times10^8\ \text{m/s}\right)^2\left(6.96\times10^4\ \text{m}\right)}\right)^{\frac{1}{2}} 4.57\times10^{14}\ \text{Hz}$
	$f_{obs} = \boxed{4.37\times10^{14}\ \text{Hz}}$

CHECK and THINK

The star's mass and its extent cause a drag-like effect on the electromagnetic waves that are emitted, resulting in a decrease in frequency of the waves, according to observers away from the star.

63. (N) The Milky Way galaxy has a supermassive black hole at its center. Its mass is 3.6 million solar masses. What is its Schwarzschild radius? Give your answer in kilometers and in Earth radii.

INTERPRET and ANTICIPATE

We can use Eq. 39.51 to find the radius. This Schwarzschild radius is a required radius of a sphere for a given amount of mass, such that the escape velocity at the surface of the sphere would be the speed of light.

SOLVE Use Eq. 39.51 to find the radius.	$R_{sch} \approx 3\dfrac{M}{M_\odot}\ \text{km} = 3\dfrac{\left(3.6\times10^6\right)M_\odot}{M_\odot}\ \text{km}$
	$R_{sch} = \boxed{1.1\times10^7\ \text{km} = 1700 R_\oplus}$

CHECK and THINK

In principle, we could find the radius of such a sphere with the mass of an apple, but the sphere would be vanishingly small! Go ahead and try it and consider your result.

68. (N) The energy released in the first peacetime nuclear weapon test at Bikini Atoll in 1946 was approximately 6.3×10^{16} J. What mass of plutonium-239 would have the same rest mass energy as the amount of energy released in this explosion?

INTERPRET and ANTICIPATE

Here, we can use Eq. 39.44 to find an amount of rest mass that would correspond to the same amount of rest-mass energy. Fission events release energy which can be related to a difference in rest mass energy between the initial particles and the products. The result of our calculation should be comparable to the amount of plutonium consumed in this explosion.

SOLVE	$E_{rest} = m_{rest}c^2$
Use Eq. 39.44 to find an amount of rest mass that would correspond to the same amount of rest-mass energy.	$m_{rest} = E_{rest}/c^2 = \left(6.3 \times 10^{16} \text{ J}\right)/\left(3.00 \times 10^8 \text{ m/s}\right)^2$
	$m_{rest} = \boxed{0.70 \text{ kg}}$

CHECK and THINK
Finding enriched, fissionable, plutonium in nature is actually quite rare. Much of the world's stockpile of this material has been created in the laboratory. Such a seemingly small amount of mass can release quite a lot of energy!

73. (N) On its way back to the Earth, an interplanetary spaceship with a proper length of 180.0 m is observed by flight controllers to pass a stationary beacon on the Moon in 0.550 μs. What is the speed of the spacecraft in the Earth's reference frame?

INTERPRET and ANTICIPATE
We can express the length of the ship, as observed by the beacon (Earth's reference frame), in terms of the speed of the ship and the time it took to pass the beacon, $L = v\Delta t$. Then, we can use Eq. 39.25 to find the speed of the ship in Earth's reference frame because we also know the proper length of the ship (its length in its own frame of reference).

SOLVE	
We begin by relating the length of the spacecraft according to the flight controllers to the time they measure and the speed of the craft, v.	$L = v\Delta t$
Then, we can use Eq. 39.25 to express the length.	$L = \dfrac{L'}{\gamma} = v\Delta t$ $L'\sqrt{1-\left(v/c\right)^2} = v\Delta t$

Now, solve for v.	$$L'\sqrt{1-(v/c)^2} = v\Delta t$$ $$1-(v/c)^2 = \left(\frac{v\Delta t}{L'}\right)^2$$ $$\left(\frac{v\Delta t}{L'}\right)^2 + (v/c)^2 = 1$$ $$v^2\left[\left(\frac{\Delta t}{L'}\right)^2 + \left(\frac{1}{c}\right)^2\right] = 1$$ $$v = \sqrt{\frac{1}{\left[\left(\frac{\Delta t}{L'}\right)^2 + \left(\frac{1}{c}\right)^2\right]}}$$ $$v = \sqrt{\frac{1}{\left[\left(\frac{0.550\times10^{-6}\text{ s}}{180.0\text{ m}}\right)^2 + \left(\frac{1}{3.00\times10^8\text{ m/s}}\right)^2\right]}}$$ $$v = 2.21\times10^8\text{ m/s} = \boxed{0.737c}$$

CHECK and THINK

As the ship moves faster, the time it takes to pass the beacon will be less and this is in part due to the fact that its perceived length is shortened – not just because it is moving faster!

74. (N) What is the speed of a proton that has a momentum 4.00 times greater than its classical momentum?

INTERPRET and ANTICIPATE Here, we should express the classical momentum of a proton moving with a speed v and the relativistic momentum of a proton moving with a speed v. Then equate the relativistic momentum to 4.00 times the classical momentum, and solve for v. We expect the speed to be relativistic if these two expressions are not equal to each other (as opposed to one being 4.00 times the other).	
SOLVE Eq. 10.1 can be used to express the classical momentum, where the mass is the rest mass of the proton.	$$p_{cl} = m_{rest}v$$
Use Eq. 39.40 to express the relativistic momentum of the proton.	$$p_{rel} = \gamma m_{rest}v$$

Now, express that the relativistic momentum is 4.00 times the classical momentum.	$p_{rel} = (4.00)\, p_{cl}$
Using these expressions, we can see that γ should be equal to 4.00.	$p_{rel} = (4.00)\, p_{cl} = \gamma m_{rest} v = \gamma p_{cl}$ $\gamma = 4.00$
Solve for v with this value of γ.	$\gamma = 4.00 = \dfrac{1}{\sqrt{1-(v/c)^2}}$ $\sqrt{1-(v/c)^2} = \dfrac{1}{4.00}$ $1-(v/c)^2 = \left(\dfrac{1}{4.00}\right)^2$ $v/c = \sqrt{1-\left(\dfrac{1}{4.00}\right)^2}$ $v = \sqrt{1-\left(\dfrac{1}{4.00}\right)^2}\, c = \boxed{0.968c}$

CHECK and THINK

The speed is certainly relativistic. The findings of classical physics must be corrected as objects begin to have speeds comparable to the speed of light. Here. The momentum of the proton is actually 4.00 times greater than what would be predicted by classical physics; a consequence of its great speed.

78. (N) In December 2012, researchers announced the discovery of ultramassive black holes, with masses up to 40 billion times the mass of the Sun (seen as the bright spot at the center of the galaxy near the center of Fig. P39.78).
a. What is the Schwarzschild radius of a black hole with a mass 40 billion times that of the Sun? **b.** Suppose this black hole is 1.3 billion ly from the Earth. What is the angular radius of a galaxy that is 1.7 billion ly behind it, as viewed from the Earth?

Figure P39.78

INTERPRET and ANTICIPATE
We can find the Schwarzschild radius by using Eq. 39.51, since we know the mass of the black hole. Then, we can use Eq. 39.49 to find to find the angular radius. Note that since the distances are in ly, we will need to convert the expression in parentheses in Eq. 39.49 to have units of m so that each quantity will be in metric units.

SOLVE	
a. Use Eq. 39.51 to find the Schwarzschild radius.	$R_{sch} \approx 3\dfrac{M}{M_\odot} = 3\dfrac{\left(40 \times 10^9\right) M_\odot}{M_\odot} = \boxed{1.2 \times 10^{11} \text{ km}}$

b. Use Eq. 39.49 to find the angular radius. Note that with the distances in light years, the factor inside the parentheses will have units of light years, so we multiple the denominator by a factor 9.46×10^{15} m/ly to convert to metric units.

$$\theta_r = \sqrt{\frac{4GM}{c^2}\left(\frac{d_{src} - d_{lens}}{d_{src}d_{lens}}\right)}$$

$$\theta_r = \sqrt{\frac{4\left(6.67 \times 10^{-11}\right)\left(40 \times 10^9\right)\left(1.9891 \times 10^{30}\right)}{\left(3.00 \times 10^8\right)^2}\left(\frac{(1.7 - 1.3) \times 10^9}{\left(1.7 \times 10^9\right)\left(1.3 \times 10^9\right)\left(9.46 \times 10^{15}\right)}\right)}$$

$$\theta_r = \boxed{2.1 \times 10^{-6} \text{ rad}}$$

CHECK and THINK
The angular radius is quite small because all of the mass must be packed into a small volume (resulting in a very high density). The radius of this volume is the Schwarzschild radius.

81. (N) How much work is required to increase the speed of a helium nucleus ($m_{He} = 6.64 \times 10^{-27}$ kg) from $0.750c$ to $0.990c$?

INTERPRET and ANTICIPATE	
The work required would be equal to the change in kinetic energy. We can express the kinetic energy at any moment using Eq. 39.43. Then, we must construct an equation for the work necessary, which would be the final kinetic energy minus the initial kinetic energy.	
SOLVE Eq. 39.43 is the relativistic kinetic energy of a particle.	$$K = mc^2 - m_{rest}c^2 = (\gamma - 1)m_{rest}c^2$$

Express the work as the difference between the final and initial kinetic energies of the helium nucleus and solve.

$$W = \Delta K$$

$$W = (\gamma_f - 1)m_{rest}c^2 - (\gamma_i - 1)m_{rest}c^2$$

$$W = (\gamma_f - \gamma_i)m_{rest}c^2$$

$$W = \left(\frac{1}{\sqrt{1-(v_f/c)^2}} - \frac{1}{\sqrt{1-(v_i/c)^2}} \right)mc^2$$

$$W = \left(\frac{1}{\sqrt{1-(0.990)^2}} - \frac{1}{\sqrt{1-(0.750)^2}} \right)(6.64 \times 10^{-27} \text{ kg})(3.00 \times 10^8 \text{ m/s})^2$$

$$W = \boxed{3.33 \times 10^{-9} \text{ J}}$$

CHECK and THINK

As the helium nucleus accelerates, we would find that the amount of energy needed to increase the speed by $0.10c$ would increase with each successive bump in speed. In other words, the energy needed to increase the speed from $0.750c$ to $0.760c$ would be less than the energy needed to increase the speed from $0.850c$ to $0.860c$. This is a consequence of the theory of special relativity.

40

The Origin of Quantum Physics

3. (N) A stove top burner operates at a temperature of 350.0°F (Fig. P40.3). Modeling the burner as a black body, what is the peak wavelength emitted by the burner?

Figure P40.3

INTERPRET and ANTICIPATE	
First, we will need to use Equations 19.1 and 19.2 to convert the given temperature to the Kelvin temperature scale. This will allow us to use Eq. 40.1 to determine the peak wavelength from the burner. We would expect that the result should indicate a peak that is in the red or infrared part of the spectrum.	
SOLVE First, use Equations 19.1 and 19.2 to convert the temperature to the Kelvin temperature scale.°	$T_{\text{K}} = \dfrac{5}{9}\left(T_{\text{F}} - 32°\right) + 273.15$ $T_{\text{K}} = \dfrac{5}{9}\left(350.0° - 32°\right) + 273.15$
Then, use Eq. 40.1 to find the peak wavelength.	$\lambda_{\text{max}} T = 2.898 \times 10^{-3} \text{ m} \cdot \text{K}$ $\lambda_{\text{max}} = \left(2.898 \times 10^{-3} \text{ m} \cdot \text{K}\right) \big/ T$ $\lambda_{\text{max}} = \left(2.898 \times 10^{-3} \text{ m} \cdot \text{K}\right) \Big/ \left(\dfrac{5}{9}\left(350.0° - 32°\right) + 273.15\right)$ $\lambda_{\text{max}} = \boxed{6.443 \times 10^{-6} \text{ m}}$

CHECK and THINK

The result here is equivalent to 6443 nm which is in the infrared part of the electromagnetic spectrum. Note that this is just the peak wavelength emitted when modeling the burner as a black body. There is a distribution of wavelengths being emitted, many of which will be in the red portion of the spectrum, giving the burner its red color when hot.

6. (E) Model yourself as a black body and estimate your peak wavelength.

INTERPRET and ANTICIPATE

To begin, you must first estimate your body's temperature in Kelvins. You can then use Eq. 40.1 to determine the peak wavelength emitted. If you think about night vision cameras and why they are useful for spotting people against an environmental background, we expect that the peak might be in the infrared region of the spectrum.

SOLVE	
A normal body temperature is about 37°C. Use Eq. 19.2 to find the temperature in Kelvins.	$T_K = T_C + 273.15$ $T_K = 37°C + 273.15 = 310 \text{ K}$
Then, use Eq. 40.1 to find the peak emitted wavelength.	$\lambda_{max} T = 2.898 \times 10^{-3} \text{ m} \cdot \text{K}$ $\lambda_{max} = \left(2.898 \times 10^{-3} \text{ m} \cdot \text{K}\right)/T$ $\lambda_{max} = \left(2.898 \times 10^{-3} \text{ m} \cdot \text{K}\right)/(310 \text{ K})$ $\lambda_{max} = \boxed{9.3 \times 10^{-6} \text{ m}}$

CHECK and THINK

As we expected, the peak wavelength is in the infrared portion of the electromagnetic spectrum. Everything with a temperature will emit some form of radiation in releasing its thermal energy. This is one of the means by which two systems can exchange thermal energy. Note, though, that Eq. 40.1 only determines the peak wavelength when modeling the object as a black body – an ideal system.

11. (N, C) A molecule vibrates with a minimum frequency of 9.50×10^{13} Hz. What is the minimum quantum of energy that this molecule can have? Can this molecule vibrate at 12.5×10^{13} Hz? If not, why not? If so, what is its energy when it vibrates at this frequency?

Chapter 40 – The Origin of Quantum Physics

> **INTERPRET and ANTICIPATE**
>
> We can use Eq. 40.7 to determine the minimum quantum of energy the molecule can possess by choosing $n = 1$ in that relationship. Greater frequencies, and thus energies, are allowed for the molecule, but they must be whole number multiples of the respective minimum values. Greater values could be found by changing n to 2, 3, 4…. In order to determine if the 12.5×10^{13} Hz is allowed, we can divide it by the minimum frequency and see if the result is a whole number. If it is, then it is allowed. If not, then it cannot vibrate at that frequency because greater values of both the energy and frequency must be whole number multiples of the minima.

SOLVE	
Use Eq. 40.7 with $n = 1$ to find the minimum energy.	$E = nhf = (1)\left(6.626\times10^{-34}\text{ J}\cdot\text{s}\right)\left(9.50\times10^{13}\text{ Hz}\right)$ $E = \boxed{6.29\times10^{-20}\text{ J}}$
Now, divide the given frequency by this minimum value to see if the result is a whole number. Because it is not, the frequency is not allowed.	$\dfrac{12.5\times10^{13}\text{ Hz}}{9.50\times10^{13}\text{ Hz}} = 1.32$

> **CHECK and THINK**
>
> Some examples of other allowed frequencies of vibration would be:
>
> $f_2 = 2\left(9.50\times10^{13}\text{ Hz}\right) = 1.90\times10^{14}\text{ Hz}$
>
> $f_3 = 3\left(9.50\times10^{13}\text{ Hz}\right) = 2.85\times10^{14}\text{ Hz}$
>
> $f_4 = 4\left(9.50\times10^{13}\text{ Hz}\right) = 3.80\times10^{14}\text{ Hz}$
>
> The corresponding energies could be found using Eq. 40.7 with $n = 2, 3, 4$….

16. A material has a work function of 2.3 eV.

a. (N) What frequency of light is necessary to eject electrons from its surface?

> **INTERPRET and ANTICIPATE**
>
> In order to determine the frequency of light necessary to eject electrons from the metal, we consider the scenario where the electrons are ejected with no kinetic energy. This means they received just enough energy to be freed and no more. This can be modeled using Eq. 40.10 with $K_{max} = 0$, where we find the energy of the photons required. Then, use Eq. 40.8 to find the frequency of the photons.

SOLVE	
First, using Eq. 40.10, with $K_{max} = 0$, we see that the necessary energy of a photon is the same as the work function of the metal.	$E_{light} = K_{max} + W_0 = W_0 = 2.3\text{ eV}$

Now use Eq. 40.8 to find the frequency of the photon.	$E = hf$ $f = E_{light}/h = (2.3 \text{ eV})/(4.136 \times 10^{-15} \text{ eV} \cdot \text{s})$ $f = \boxed{5.6 \times 10^{14} \text{ Hz}}$

CHECK and THINK

The work function is a measure of the necessary photon energy to eject an electron from the metal. If the work function is greater, then the necessary energy, and frequency, of the photon will also be greater.

b. (C) Assuming light of the frequency you found in part (a), does the number of ejected electrons increase if the intensity of the light is increased? How might you explain this in terms of the particle representation of light?

Yes, the number of electrons ejected would increase if the intensity is increased. This is because there are more photons per square meter if the intensity goes up. Thus, there are more incidents of photons striking the metal, and more electrons ejected per unit time.

c. (N) What is the stopping potential for this material if it is illuminated by light with a frequency of 6.50×10^{14} Hz?

INTERPRET and ANTICIPATE

Note that this frequency is greater than the frequency we found in part (a). That means that this photon will have a greater energy than that from part (a), and that an electron ejected by a photon with this frequency will have some kinetic energy. The stopping potential, required to ensure all electrons are stopped, can be found using Eq. 40.11. First, though, we must use Eq. 40.8 to find the energy of the light.

SOLVE Use Eq. 40.8 to express the energy of the light.	$E_{light} = hf = (4.136 \times 10^{-15} \text{ eV} \cdot \text{s})(6.50 \times 10^{14} \text{ Hz})$
Then, use Eq. 40.11 to find the stopping potential for this material illuminated by this light source.	$eV_0 = E_{light} - W_0$ $eV_0 = (4.136 \times 10^{-15} \text{ eV} \cdot \text{s})(6.50 \times 10^{14} \text{ Hz}) - 2.3 \text{ eV}$ $V_0 = \boxed{0.4 \text{ V}}$

CHECK and THINK

In part (a), the stopping potential would have been 0 because the electrons did not have any kinetic energy after being ejected. Because the energy of this light is greater than the work function, there is a nonzero maximum value of kinetic energy possible, and thus, now a nonzero stopping potential can be found.

d. (N, C) What is the maximum kinetic energy of ejected electrons if light with a frequency of 3.50×10^{14} Hz is incident on this material? Explain.

SOLVE

The energy of a photon with this frequency can be found using Eq. 40.8, but note that this frequency is less than the frequency found in part (a). This means the energy of the photon will be less than the work function of the metal. Thus, no electrons are ejected and the maximum kinetic energy is $\boxed{0}$.

21. (N) A strange metallic rock is found and is being tested. Suppose that light with a frequency of 7.50×10^{14} Hz is incident upon the rock and a stopping potential of 1.00 V is needed to reduce the electron current to zero in a photoelectric experiment.
a. What is the maximum kinetic energy of an electron ejected by this light from this material?
b. What is the work function of this material?

INTERPRET and ANTICIPATE

The maximum kinetic energy can be found using the given stopping potential and Eq. 40.9. We can then use Eq. 40.11 to find the work function of the metal because the energy of the light can be determined using the given frequency and Eq. 40.8. The work function should be a positive value and less than the energy of the light.

SOLVE	
a. Use Eq. 40.9 to find the maximum kinetic energy.	$K_{max} = eV_0 = e(1.00 \text{ V}) = \boxed{1.00 \text{ eV}}$
b. Now, use Eq. 40.8 to express the energy of the photons in the light.	$E_{light} = hf = \left(4.136 \times 10^{-15} \text{ eV} \cdot \text{s}\right)\left(7.50 \times 10^{14} \text{ Hz}\right)$
Then, use Eq. 40.11 to find the work function of the metal.	$W_0 = E_{light} - K_{max}$ $W_0 = \left(4.136 \times 10^{-15} \text{ eV} \cdot \text{s}\right)\left(7.50 \times 10^{14} \text{ Hz}\right) - 1.00 \text{ eV}$ $W_0 = \boxed{2.10 \text{ eV}}$

CHECK and THINK

The work function is less than the energy of the light, and thus it turned out to be

> positive, meaning it is possible for the photons to eject electrons from this material as described. If the frequency were increased, the maximum kinetic energy of ejected electrons would increase and the stopping potential would also need to increase.

23. (N) Iron has a work function of 4.50 eV. What is the minimum frequency of light needed to eject electrons from iron?

INTERPRET and ANTICIPATE	
We must first express the minimum energy of light required to eject electrons from the iron. This can be found using Eq. 40.10 with $K_{max} = 0$, since we want just enough energy to free electrons. Then, Eq. 40.8 can be used to find the frequency of the light.	
SOLVE Use Eq. 40.10 to find the energy of the light required.	$E_{light} = K_{max} + W_0 = W_0 = 4.50$ eV
Now, use Eq. 40.8 to find the frequency of the light.	$E = hf$ $f = E_{light}/h = (4.50 \text{ eV})/(4.136 \times 10^{-15} \text{ eV} \cdot \text{s})$ $f = \boxed{1.09 \times 10^{15} \text{ Hz}}$

CHECK and THINK	
As we have seen in examples, and in similar problems, this is the required frequency to just have enough energy to overcome the work function and free an electron from the metal. Less frequency would not free any electrons and more would cause the electrons to have some kinetic energy after they are released.	

27. (N) Suppose that the ratio of the wavelengths of two photons is $\lambda_1/\lambda_2 = 4.00$. What is the ratio of their momenta, p_1/p_2 ?

INTERPRET and ANTICIPATE	
In order to find the ratio of the momenta, we must use Eq. 40.14 to express the momentum of each photon in terms of its wavelength. Then, we can divide those expressions to determine the ratio of the momenta in terms of the ratio of the wavelengths. This is useful because we know the value of the ratio of the wavelengths.	
SOLVE Begin by writing Eq. 40.14 for the first photon.	$p_1 = \dfrac{h}{\lambda_1}$

Now, write Eq. 40.14 for the second photon.	$p_2 = \dfrac{h}{\lambda_2}$
Express the ratio, p_1/p_2, and simplify.	$\dfrac{p_1}{p_2} = \dfrac{h/\lambda_1}{h/\lambda_2} = \dfrac{\lambda_2}{\lambda_1} = \dfrac{1}{4.00} = \boxed{0.250}$

CHECK and THINK

Planck's constant cancelled when we created the ratio. Because the momentum is inversely proportional to the wavelength, the ratio of the two momentum is dependent on the inverse of the ratio of the wavelengths.

30. (A) A photon undergoes a shift in wavelength due to a collision with an electron so that its wavelength changes by 10.0%. Find an expression for the scattering angle in terms of the initial wavelength of the photon.

INTERPRET and ANTICIPATE
When the photon scatters, its wavelength should increase or stay the same because it either loses energy, or has the same energy with which it began. In this case, we know that the final wavelength should be 1.10 times the initial wavelength (because the wavelength changes by 10%). We can use Eq. 40.15 to find an expression for the scattering angle at which this photon would be directed. The angle should depend on the initial wavelength and the Compton wavelength, since the final wavelength can be expressed in terms of the initial wavelength.

SOLVE	
First, we should express the final wavelength in terms of the initial wavelength.	$\lambda_f = (1.10)\lambda_i$
Now, write Eq. 40.15 and solve for the scattering angle.	$\lambda_f - \lambda_i = \lambda_C(1-\cos\varphi)$ $1.10\lambda_i - \lambda_i = \lambda_C(1-\cos\varphi)$ $\dfrac{0.10\lambda_i}{\lambda_C} = 1 - \cos\varphi$ $\cos\varphi = 1 - \dfrac{0.10\lambda_i}{\lambda_C}$ $\boxed{\varphi = \cos^{-1}\left[1 - \dfrac{0.10\lambda_i}{\lambda_C}\right]}$

CHECK and THINK

Our expectations are met in the final scattering angle formula we have derived for this case. Note that the result depends on the initial wavelength! This means that even though this result models all cases in which the scattered wavelength is 10% greater than the initial, the actual angle depends on the initial wavelength. A "longer" photon that scatters with a 10% greater wavelength will scatter at a completely different angle than a "shorter" photon that scatters with a 10% greater wavelength.

36. (N) Suppose that you perform the Compton experiment with X-rays of wavelength 1.984×10^{-11} m. What is the frequency (in exaHertz) of the scattered X-rays observed at **a.** 0°, **b.** 45° and, **c.** 180°?

INTERPRET and ANTICIPATE

In each case, we know the initial wavelength and the scattering angle. This means we can use Eq. 40.15 to express the final wavelength in each case. As the scattering angle increases, with the same initial wavelength, we expect that the final wavelength of the scattered X-ray will increase. Once we have an expression for the final wavelength, we can use Eq. 34.20 to find the frequency of the scattered X-ray. Because the wavelength should increase with increasing scattering angle, the final frequencies of the scattered X-rays should decrease as the scattering angle increases.

SOLVE	
a. Write Eq. 40.15 for the first case described and solve for the final wavelength.	$\lambda_f - \lambda_i = \lambda_C (1 - \cos\varphi)$ $\lambda_f = \lambda_i + \lambda_C (1 - \cos\varphi)$ $\lambda_f = 1.984 \times 10^{-11}$ m $+ (2.426 \times 10^{-12}$ m$)(1 - \cos 0°)$ $\lambda_f = 1.984 \times 10^{-11}$ m
Now, use Eq. 34.20 to find the frequency. We express the result in exaHertz (EHz), as requested in the problem statement.	$c = f\lambda$ $f = c/\lambda = (3.00 \times 10^8$ m/s$)/(1.984 \times 10^{-11}$ m$)$ $f = \boxed{15.1 \text{ EHz}}$
b. Write Eq. 40.15 for the first case described and solve for the final wavelength.	$\lambda_f = \lambda_i + \lambda_C (1 - \cos\varphi)$ $\lambda_f = 1.984 \times 10^{-11}$ m $+ (2.426 \times 10^{-12}$ m$)(1 - \cos 45°)$

Now, use Eq. 34.20 to find the frequency. We express the result in exaHertz (EHz), as requested in the problem statement.	$f = c/\lambda$ $$f = \frac{3.00 \times 10^8 \text{ m/s}}{\left(1.984 \times 10^{-11} \text{ m} + \left(2.426 \times 10^{-12} \text{ m}\right)\left(1 - \cos 45°\right)\right)}$$ $f = \boxed{14.6 \text{ EHz}}$
c. Write Eq. 40.15 for the first case described and solve for the final wavelength.	$\lambda_f = \lambda_i + \lambda_C\left(1 - \cos\varphi\right)$ $\lambda_f = 1.984 \times 10^{-11} \text{ m} + \left(2.426 \times 10^{-12} \text{ m}\right)\left(1 - \cos 180°\right)$
Now, use Eq. 34.20 to find the frequency. We express the result in exaHertz (EHz), as requested in the problem statement.	$f = c/\lambda$ $$f = \frac{3.00 \times 10^8 \text{ m/s}}{\left(1.984 \times 10^{-11} \text{ m} + \left(2.426 \times 10^{-12} \text{ m}\right)\left(1 - \cos 180°\right)\right)}$$ $f = \boxed{12.1 \text{ EHz}}$

CHECK and THINK

The frequency did indeed decrease as the scattering angle was increased. Note that the first case (a) is an example of the X-ray being forward scattered. Its wavelength, frequency and energy are not changed. What does this mean about the energy, or behavior of the electron? In order to think about this question, consider the conservation of energy for this situation.

38. (N) An X-ray with a frequency of 8.7700×10^{16} Hz undergoes a Compton scattering process and is detected at an angle of 45.000° relative to its original velocity.
a. What is the wavelength of the photon after it has scattered?
b. What is the kinetic energy of the electron after its collision with the X-ray?

INTERPRET and ANTICIPATE
We can use Eq. 40.15 to find the wavelength of the scattered photon, though we must first find the initial wavelength by using Eq. 34.20. Once we have done this, we can determine the energy of both the initial and scattered photon (Eq. 40.17). It is then a matter of using conservation of energy to determine the kinetic energy of the electron involved in the scattering process. Note that the energy of the scattered photon must be less than the energy of the initial photon for the electron to have gained energy.

SOLVE	
a. First, use Eq. 34.20 to express the initial wavelength.	$\lambda = c/f = \left(3.00 \times 10^8 \text{ m/s}\right)/\left(8.7700 \times 10^{16} \text{ Hz}\right)$

Now, use Eq. 40.15 to find the wavelength of the scattered photon.	$\lambda_f - \lambda_i = \lambda_C(1 - \cos\varphi)$ $\lambda_f = \lambda_i + \lambda_C(1 - \cos\varphi)$ $\lambda_f = (3.00\times10^8 \text{ m/s})/(8.7700\times10^{16} \text{ Hz})$ $\qquad\qquad + (2.426\times10^{-12} \text{ m})(1 - \cos 45.000°)$ $\lambda_f = \boxed{3.4215\times10^{-9} \text{ m}}$
b. Use Eq. 40.17 to express the energy of the initial photon.	$E_i = \dfrac{hc}{\lambda_i}$ $E_i = \dfrac{(4.136\times10^{-15} \text{ eV}\cdot\text{s})(3.00\times10^8 \text{ m/s})}{(3.00\times10^8 \text{ m/s})/(8.7700\times10^{16} \text{ Hz})}$ $E_i = 362.73 \text{ eV}$
Now, use Eq. 40.17 to express the energy of the scattered photon.	$E_i = \dfrac{hc}{\lambda_i}$ $E_i = \dfrac{(4.136\times10^{-15} \text{ eV}\cdot\text{s})(3.00\times10^8 \text{ m/s})}{3.4215\times10^{-9} \text{ m}}$ $E_i = 362.65 \text{ eV}$
Using conservation of energy, we can now find the kinetic energy of the electron.	$E_i = E_f + E_e$ $E_e = E_i - E_f = 362.73 \text{ eV} - 362.65 \text{ eV}$ $E_e = \boxed{0.08 \text{ eV}}$

CHECK and THINK
During the scattering process, the initial energy is carried entirely by the photon, while after the collision that energy is shared between the scattered photon and the electron. The angle at which the photon is scattered will affect its wavelength, and thus, how much energy it and the electron have after the collision.

45. (N) Consider a photon and a proton carrying equal momenta of 2.13×10^{-25} kg · m/s.
a. What is the wavelength of a photon with this momentum?
b. What is the speed of a proton with this momentum?

Chapter 40 – The Origin of Quantum Physics

INTERPRET and ANTICIPATE

For part (a), we can use Eq. 40.14 to find the wavelength of the photon. In part (b) we can use Eq. 10.1 to find the speed of the proton. Once we find the speed of the proton, we should check to make sure the speed is not very near the speed of light (say, within 10% of c). If it were, then it would not be appropriate to use Eq. 10.1, and we instead should use an equation for relativistic momentum like Eq. 39.40.

SOLVE	
a. Use Eq. 40.14 to find the wavelength of the photon.	$\lambda = h/p$ $\lambda = \left(6.626\times10^{-34}\ \text{J}\cdot\text{s}\right)/\left(2.13\times10^{-25}\ \text{kg}\cdot\text{m/s}\right)$ $\lambda = \boxed{3.11\times10^{-9}\ \text{m}}$
b. Use Eq. 10.1 to find the speed of the proton.	$v = p/m$ $v = \left(2.13\times10^{-25}\ \text{kg}\cdot\text{m/s}\right)/\left(1.6726\times10^{-27}\ \text{kg}\right)$ $v = \boxed{127\ \text{m/s}}$

CHECK and THINK

Given the low value of the speed, there is no reason to worry about relativistic effects. Though we did not use Eq. 40.18 in this problem, it is worth noting the similarity between Eq. 40.18 and Eq. 40.14. One describes the momentum of a photon in terms of its wavelength and the other describes the momentum of a particle in terms of its wavelength. Though they appear identical, they are saying different things, while pointing to the wave-particle duality that exists in nature.

49. (N) If an electron has a kinetic energy 3.40 eV, what is the de Broglie wavelength of the electron?

INTERPRET and ANTICIPATE

We want to make use of a relationship between the kinetic energy and the momentum of a particle. If we can find the momentum in this way, we can then use Eq. 40.18 to express the de Broglie wavelength of the electron. The relationship we need is $K = p^2/(2m)$.

Using the mass of the electron, and the given kinetic energy, we can find the momentum this way. Note that we must convert the energy to Joules before proceeding, or do so as part of the calculation. Also, note that we should really check to see if the speed of the electron is such that relativistic concerns should be considered. We can do that here:

428

© 2016 Cengage Learning. All Rights Reserved. May not be scanned, copied or duplicated, or posted to a publicly accessible website, in whole or in part.

$$K = \frac{1}{2}mv^2$$

$$v = \sqrt{\frac{2K}{m}} = \sqrt{\frac{2(3.40 \text{ eV})(1.6 \times 10^{-19} \text{ J}/1 \text{ eV})}{9.109 \times 10^{-31} \text{ kg}}} = 1.09 \times 10^6 \text{ m/s}$$

While this is a high speed, it is not near 10% of the speed of light, and therefore we do not need to consider special relativity in our solution.

SOLVE Use the relationship $K = p^2/(2m)$, to express the momentum.	$K = \frac{p^2}{2m}$ $p = \sqrt{2m_e K}$ $p = \sqrt{2(9.109 \times 10^{-31} \text{ kg})(3.40 \text{ eV})((1.6 \times 10^{-19} \text{ J})/(1 \text{ eV}))}$
Use Eq. 40.18 to find the wavelength of the electron.	$\lambda = h/p$ $\lambda = \dfrac{(6.626 \times 10^{-34} \text{ J} \cdot \text{s})}{\sqrt{2(9.109 \times 10^{-31} \text{ kg})(3.40 \text{ cV})((1.6 \times 10^{-19} \text{ J})/(1 \text{ eV}))}}$ $\lambda = \boxed{6.66 \times 10^{-10} \text{ m}}$

CHECK and THINK

If the kinetic energy of the electron is increased, the momentum would also increase, and thus the wavelength would decrease. Controlling the energy of electrons in an electron microscope is an important key in controlling the wavelength and the imaging capabilities of the electrons.

54. (N) An alpha particle (a helium nucleus) moves with a speed of 2.46×10^5 m/s. What is the wavelength of this particle?

INTERPRET and ANTICIPATE

The speed is not relativistic so we can use Eq. 10.1 to express the momentum of the particle. We will need to calculate its mass, as part of this equation. The alpha particle is a helium nucleus, so we can model the mass as the sum of the masses of two protons and two neutrons. Once we have the momentum, we can use Eq. 40.18 to find the wavelength.

SOLVE	
Use Eq. 10.1 to express the momentum of the alpha particle. Note that we express the mass as two times the sum of the masses of a proton and neutron.	$p = mv$ $p = 2(1.6726\times10^{-27} \text{ kg} + 1.6749\times10^{-27} \text{ kg})(2.46\times10^{5} \text{ m/s})$
Now use Eq. 40.18 to find the wavelength.	$\lambda = h/p$ $\lambda = \dfrac{(6.626\times10^{-34} \text{ J}\cdot\text{s})}{\left[2(1.6726\times10^{-27} \text{ kg} + 1.6749\times10^{-27} \text{ kg})(2.46\times10^{5} \text{ m/s})\right]}$ $\lambda = \boxed{4.02\times10^{-13} \text{ m}}$

CHECK and THINK

The wavelength of an alpha particle can be comparable to that of an electron in an electron microscope without having to move as fast. This is due to its mass being so much larger than that of an electron.

57. (N) Suppose that an electron and a proton are traveling at the same speed. What is the ratio λ_p / λ_e of their wavelengths?

INTERPRET and ANTICIPATE	
In this problem, we need only express the wavelength of each particle using Eq. 40.18 and then create the ratio λ_p / λ_e. Assuming that the speed of the particles is not relativistic, we can use Eq. 10.1 to express their momenta and use the masses of the particles to solve. Because the proton will have much more momentum than the electron moving at the same speed (by virtue of its greater mass), and because the wavelength is inversely proportional to momentum, we expect that ratio will be less than 1.	
SOLVE Use Eq. 40.18 and Eq. 10.1 to express the wavelength of the proton.	$\lambda_p = \dfrac{h}{p_p} = \dfrac{h}{m_p v}$
Now, do the same to express the wavelength of the electron.	$\lambda_e = \dfrac{h}{p_e} = \dfrac{h}{m_e v}$

Create the ratio from the problem statement and simplify to find the result.	$\dfrac{\lambda_p}{\lambda_e} = \dfrac{h/m_e v}{h/m_p v} = \dfrac{m_e}{m_p}$ $\dfrac{\lambda_p}{\lambda_e} = \dfrac{9.109 \times 10^{-31} \text{ kg}}{1.6726 \times 10^{-27} \text{ kg}} = \boxed{5.446 \times 10^{-4}}$

CHECK and THINK

The ratio is less than 1 as expected. The greater momentum of the proton makes its wavelength much smaller than that of the electron moving at the same speed.

60. (N) Find **a.** the energy of a photon with a wavelength of 3.56 fm, and **b.** the kinetic energy of an electron with a de Broglie wavelength of 3.56 fm. **c.** Compare your results by finding the ratio of your answers (part (a)/part(b)).

INTERPRET and ANTICIPATE

In order to solve part (a), we utilize Eq. 40.17 with the wavelength of the photon. For part (b), however, we must determine the kinetic energy of a particle, which can be achieved by finding the momentum, using Eq. 40.18, and the kinetic energy-momentum relationship $K = p^2/(2m)$. Lastly, we can find the ratio using our results from (a) and (b).

SOLVE **a.** Use Eq. 40.17 to find the energy of the photon.	$E = \dfrac{hc}{\lambda}$ $E = \dfrac{\left(6.626 \times 10^{-34} \text{ J} \cdot \text{s}\right)\left(3.00 \times 10^{8} \text{ m/s}\right)}{3.56 \times 10^{-15} \text{ m}}$ $E = \boxed{5.58 \times 10^{-11} \text{ J}}$
b. Now, first express the momentum of the electron using Eq. 40.18.	$p = h/\lambda$ $p = \left(6.626 \times 10^{-34} \text{ J} \cdot \text{s}\right) / \left(3.56 \times 10^{-15} \text{ m}\right)$
Use $K = p^2/(2m)$ to find the kinetic energy of the electron.	$K = \dfrac{p^2}{2m}$ $K = \dfrac{\left[\left(6.626 \times 10^{-34} \text{ J} \cdot \text{s}\right) / \left(3.56 \times 10^{-15} \text{ m}\right)\right]^2}{2\left(9.109 \times 10^{-31} \text{ kg}\right)}$ $K = \boxed{1.90 \times 10^{-8} \text{ J}}$

c. Find the ratio using the results from parts (a) and (b).	$\dfrac{E}{K} = \dfrac{5.58 \times 10^{-11} \text{ J}}{1.90 \times 10^{-8} \text{ J}} = \boxed{2.94 \times 10^{-3}}$

CHECK and THINK

The ratio turned out to be less than 1, meaning that the energy of the photon was less than that of the electron. It takes a lot of energy to get even a small piece of matter, like the electron, to move fast enough so that it's wavelength is comparable to light with the same wavelength!

62. (N) An electron moves with a speed of $0.60c$.
a. Find the wavelength of the electron semi-classically by using $p = mv$ for the momentum of the electron.
b. Find the wavelength of the electron using special relativity to express the momentum of the electron.
c. Calculate the percent difference between your answers for parts (a) and (b).

INTERPRET and ANTICIPATE

Though this electron should be treated using special relativity, the problem asks us to find the momentum classically in part (a). We can then use Eq. 40.18 to find the wavelength. Then, in part (b), we can use Eq. 39.40 and Eq. 39.41 to find the momentum of the electron using special relativity. Again, we will find the wavelength using Eq. 40.18, and can then compare the two results in part (c) by calculating the percent difference. Because of our work in Chapter 39, we expect the classical momentum to be less than that found using special relativity, and thus the wavelength in part (b) should be less than that found in part (a).

SOLVE **a.** Use the given equation for the electron to express the momentum.	$p = mv = m_e(0.60c)$
Then, use Eq. 40.18 to find the wavelength.	$\lambda = \dfrac{h}{p} = \dfrac{6.626 \times 10^{-34} \text{ J} \cdot \text{s}}{m_e(0.60c)}$ $\lambda = \dfrac{6.626 \times 10^{-34} \text{ J} \cdot \text{s}}{(9.109 \times 10^{-31} \text{ kg})(0.60)(3.00 \times 10^8 \text{ m/s})}$ $\lambda = \boxed{4.0 \times 10^{-12} \text{ m}}$
b. Use Eq. 39.40 and Eq. 39.41 to express the momentum of the electron.	$p = \gamma mv = \dfrac{mv}{\sqrt{1-(v/c)^2}} = \dfrac{m_e(0.60c)}{\sqrt{1-(0.60)^2}}$

Then, use Eq. 40.18 to find the wavelength.	$\lambda = \dfrac{h}{p} = \dfrac{6.626 \times 10^{-34}\ \text{J} \cdot \text{s}}{\left(m_e \left(0.60c \right) \middle/ \left(\sqrt{1-\left(0.60\right)^2} \right) \right)}$ $\lambda = \dfrac{\left(6.626 \times 10^{-34}\ \text{J} \cdot \text{s}\right)\left(\sqrt{1-\left(0.60\right)^2}\right)}{\left(9.109 \times 10^{-31}\ \text{kg}\right)\left(0.60\right)\left(3.00 \times 10^8\ \text{m/s}\right)}$ $\lambda = \boxed{3.2 \times 10^{-12}\ \text{m}}$
c. Calculate the percent difference between parts (a) and (b).	$\%\text{diff} = \dfrac{\left| 4.0 \times 10^{-12}\ \text{m} - 3.2 \times 10^{-12}\ \text{m} \right|}{\left[\left(4.0 \times 10^{-12}\ \text{m} + 3.2 \times 10^{-12}\ \text{m} \right)\middle/ 2 \right]} \times 100\%$ $\%\text{diff} = \boxed{22\%}$

CHECK and THINK

Though the values appear to be very near one another, the corrections that arise from special relativity cause there to be a measureable difference. As expected, the wavelength is less in part (b) because the corrections from special relativity predict the correct, greater momentum.

68. (N) A cup containing 5.67 kg of water is placed in a microwave oven. The oven creates microwaves with a frequency of 2.45 GHz. The initial temperature of the water is 20.0°C and the specific heat capacity of water is 4186 J/kg · K.
a. How much energy is necessary to raise the water to its boiling point?
b. How many microwave photons are necessary to bring the water to its boiling point, assuming all of the energy from each photon is absorbed by the water?

INTERPRET and ANTICIPATE

Part (a) is a thermodynamics question that can be modeled using Eq. 21.5, in order to determine the amount of thermal energy required. Once we know the total energy needed, we can use Eq. 40.17 to find the energy contained in one microwave photon. We then need only to divide the total energy needed by the energy supplied by each photon, in order to find the number of photons required.

SOLVE	
a. Use Eq. 21.5 to find the thermal energy needed to change the temperature of the water.	$Q = mc\Delta T = \left(5.67\ \text{kg}\right)\left(4186\ \text{J/}\left(\text{kg} \cdot \text{K}\right)\right)\left(80.0\ \text{K}\right)$ $Q = \boxed{1.90 \times 10^6\ \text{J}}$

b. Use Eq. 40.17 to find the energy of one microwave photon.	$E = hf = \left(6.626 \times 10^{-34} \text{ J} \cdot \text{s}\right)\left(2.45 \times 10^{9} \text{ Hz}\right)$
Then, the number of photons, n, would be equal to the energy required from part (a) divided by the energy of one photon.	$n = \dfrac{Q}{E} = \dfrac{\left(5.67 \text{ kg}\right)\left(4186 \text{ J/(kg} \cdot \text{K)}\right)\left(80.0 \text{ K}\right)}{\left(6.626 \times 10^{-34} \text{ J} \cdot \text{s}\right)\left(2.45 \times 10^{9} \text{ Hz}\right)}$ $n = \boxed{1.17 \times 10^{30}}$

CHECK and THINK

Of course, in reality more photons should be required. Some energy will not stay within the water, as it loses energy through radiation and conduction. Still, this kind of back-of-the-envelope calculation gives us an estimate of what might be required, and a starting point for a more complicated model that could be developed.

---

71. (N) CASE STUDY: In Example 40.10 (page 1331), we found that the kinetic energy of the electrons in an electron microscope is 4 keV. What potential difference must be provided by the electron gun (Fig. 40.17)?

INTERPRET and ANTICIPATE

We can use a modified version of Eq. 40.9, where instead of a stopping potential, we are finding the electric potential through which the electron must move to acquire the final kinetic energy. After all, if the electron is to acquire that kinetic energy from an electrical potential difference, the model of a stopping potential is analogous assuming the electron started from rest. This is a fair assumption here.

SOLVE Use Eq. 40.9 such that the kinetic energy is derived from the electron accelerating through a potential difference V, assuming it started from rest.	$eV = K$ $V = \dfrac{K}{e} = \dfrac{4 \text{ keV}}{e} = \boxed{4 \text{ kV}} = 4 \times 10^{3} \text{ V}$

CHECK and THINK

When you divide the unit eV by the fundamental electric charge e, the result would be the unit of a V, almost as if the e cancels the e in eV. In our problem, we had keV, so the result has the units kV. Every volt through which an electron is accelerated will add (or subtract, depending on the direction of travel) 1eV of kinetic energy to that electron

---

75. (N) If 5.0% of the power radiated by a 100.0-W bulb is emitted as visible light at 550 nm, determine the number of photons that are emitted per second at this wavelength. Answer with two significant figures.

INTERPRET and ANTICIPATE	
First, we must determine the energy of each photon emitted by the bulb by using Eq. 40.17. We can then recognize that the unit of power, W, is actually Joules per second (J/s). If we take 5.0% of the power and divide it by the energy of one photon in Joules, the result will be the number of photons emitted each second, n.	
SOLVE Use Eq. 40.17 to find the energy of one photon of 550 nm light.	$E = \dfrac{hc}{\lambda} = \dfrac{\left(6.626 \times 10^{-34}\ \text{J} \cdot \text{s}\right)\left(3.00 \times 10^8\ \text{m/s}\right)}{550 \times 10^{-9}\ \text{m}}$
Now, express 5.0% of the power in terms of J/s. This is the power radiated, which we could call P_{rad}	$P_{\text{rad}} = (0.050)(100.0\ \text{W}) = (0.050)(100.0\ \text{J/s})$ $P_{\text{rad}} = 5.0\ \text{J/s}$
The number of photons emitted each second is then this radiated power, divided by the energy of one photon.	$n = \dfrac{P_{\text{rad}}}{E}$ $n = \dfrac{5.0\ \text{J/s}}{\left(6.626 \times 10^{-34}\ \text{J} \cdot \text{s}\right)\left(3.00 \times 10^8\ \text{m/s}\right) / \left(550 \times 10^{-9}\ \text{m}\right)}$ $n = \boxed{1.4 \times 10^{19}\ \text{s}^{-1}}$

CHECK and THINK

As we saw in our examination of black body radiation, the bulb will emit energy at many wavelengths. Here, we have examined a case of a subset of the energy emitted at a particular wavelength, and found the corresponding number of photons emitted each second.

76. (E) Estimate the number of photons that are emitted by the Sun per second.

INTERPRET and ANTICIPATE
In order to do this, we must estimate the amount of energy emitted by the Sun each second. If we then choose an appropriate wavelength that would represent the most abundant type of photon emitted, we can use the energy of that type of photon (Eq. 40.17) to derive an estimate for the number emitted each second. The Sun has a power of about $4 \times 10^{26}\ \text{W}$ or $4 \times 10^{26}\ \text{J/s}$. This means the Sun radiates $4 \times 10^{26}\ \text{J}$ each second. The peak wavelength of the light from the Sun is about 500 nm, and so we choose to use this value for the purposes of this solution. If you choose slightly different estimates, your answer will of course vary a little from the result to follow.

SOLVE Use Eq. 40.1 to find the energy of a 500-nm photon of light.	$E = \dfrac{hc}{\lambda} = \dfrac{\left(6.626\times10^{-34}\ \text{J}\cdot\text{s}\right)\left(3.00\times10^{8}\ \text{m/s}\right)}{500\times10^{-9}\ \text{m}}$
Divide the estimate for the Sun's emitted energy each second, by the energy of this type of photon to get the number emitted each second.	$n = \dfrac{4\times10^{26}\ \text{J}}{\left(6.626\times10^{-34}\ \text{J}\cdot\text{s}\right)\left(3.00\times10^{8}\ \text{m/s}\right)/\left(500\times10^{-9}\ \text{m}\right)}$ $n = \boxed{1\times10^{45}}$

CHECK and THINK

Of course there are plenty of photons at other wavelengths emitted by the Sun each second, but we get a decent estimate of the total number by choosing a wavelength near, or at the peak wavelength emitted by the Sun.

41

Schrödinger's Equation

3. (N) Suppose that a hockey puck's mass is 0.160 kg, and it is in a box of length 6.35 m (Example 41.1). What excited state would give an energy of 8.52 J?

INTERPRET and ANTICIPATE
Given the information in the problem statement, we can use Eq. 41.7 to find the excited state, n. Given the macroscopic nature of this problem, and knowing energy quantization is a quantum effect that tends to manifest on smaller size scales, we expect n to be quite large.

SOLVE	
Use Eq. 41.7 to find n, the quantum number of the excited state.	$$E_n = n^2 \frac{h^2}{8mL^2}$$ $$8.52 \text{ J} = n^2 \frac{\left(6.626 \times 10^{-34} \text{ J} \cdot \text{s}\right)^2}{8(0.160 \text{ kg})(6.35 \text{ m})^2}$$ $$n = \sqrt{\frac{8(8.52 \text{ J})(0.160 \text{ kg})(6.35 \text{ m})^2}{\left(6.626 \times 10^{-34} \text{ J} \cdot \text{s}\right)^2}} = \boxed{3.16 \times 10^{34}}$$

CHECK and THINK
The number is very large, but again, this is a macroscopic problem with many Joules of energy. As we see in other problems, the energy separation between subsequent states tends to be on the order of electron-volts, or eV, not Joules.

6. (N) Suppose we wanted to try to observe the quantization of energy for a 1.00-g peanut trapped in a box. If our detector has a resolution such that it is able to measure the energy difference between the ground state and first excited state, as long as the difference is greater than 0.00100 eV, what is the maximum width L of the box? Note that we, of course, cannot have a box this small. Thus, we do not observe the quantization of peanuts.

INTERPRET and ANTICIPATE
This problem seeks to reinforce why we will not observe energy quantization on a macroscopic scale, by having us find the required width of a box for a peanut, such that

we might be able to measure the energy separation between states. Here, we use Eq. 41.7 to express the energy of both the ground state and first excited state. We can then take the difference between these and set the result equal to the given energy difference to find L. We expect L will turn out to be very small, such that it isn't conceivable to create such a box and observe the energy quantization of the peanut (otherwise, that would be an awesome lab and we would totally do that!).

SOLVE Use Eq. 41.7 to express the energy of the ground state, $n = 1$.	$E_n = n^2 \dfrac{h^2}{8mL^2}$ $E_1 = (1)^2 \dfrac{\left(6.626 \times 10^{-34} \text{ J} \cdot \text{s}\right)^2}{8\left(1.00 \times 10^{-3} \text{ kg}\right)L^2}$
Now use Eq. 41.7 to express the energy of the first excited state, $n = 2$.	$E_n = n^2 \dfrac{h^2}{8mL^2}$ $E_2 = (2)^2 \dfrac{\left(6.626 \times 10^{-34} \text{ J} \cdot \text{s}\right)^2}{8\left(1.00 \times 10^{-3} \text{ kg}\right)L^2}$
Convert the energy difference to Joules.	$\Delta E = E_2 - E_1$ $\Delta E = \left(0.00100 \text{ eV}\right)\left(1.60 \times 10^{-19} \text{ J/1 eV}\right)$ $\Delta E = 1.60 \times 10^{-22} \text{ J}$
Finally, subtract the energy of the ground state from the first excited state, using the expressions we had created, and equate to this value of the energy difference. Then solve for L.	$(2)^2 \dfrac{\left(6.626 \times 10^{-34} \text{ J} \cdot \text{s}\right)^2}{8\left(1.00 \times 10^{-3} \text{ kg}\right)L^2} - (1)^2 \dfrac{\left(6.626 \times 10^{-34} \text{ J} \cdot \text{s}\right)^2}{8\left(1.00 \times 10^{-3} \text{ kg}\right)L^2} = 1.60 \times 10^{-22} \text{ J}$ $(3)\dfrac{\left(6.626 \times 10^{-34} \text{ J} \cdot \text{s}\right)^2}{8\left(1.00 \times 10^{-3} \text{ kg}\right)L^2} = 1.60 \times 10^{-22} \text{ J}$ $L^2 = \dfrac{(3)\left(6.626 \times 10^{-34} \text{ J} \cdot \text{s}\right)^2}{(8)\left(1.00 \times 10^{-3} \text{ kg}\right)\left(1.60 \times 10^{-22} \text{ J}\right)}$ $L = \sqrt{\dfrac{(3)\left(6.626 \times 10^{-34} \text{ J} \cdot \text{s}\right)^2}{(8)\left(1.00 \times 10^{-3} \text{ kg}\right)\left(1.60 \times 10^{-22} \text{ J}\right)}} = \boxed{1.01 \times 10^{-21} \text{ m}}$

CHECK and THINK

The width is far less than the width of a hydrogen atom, or even a hydrogen nucleus. There is no way we could create a box to hold this peanut such that the energy quantization would be observable, given the measureable energy difference in the problem statement.

9. Consider again the throwing of the dice in Example 41.2. In quantum mechanics, we will often speak of the current state of the system given a set of possible states. The sum of the faces of the two dice that are face-up after being thrown (or rolled) can be thought of as representing the state of the dice. This means the possible total, or state of the dice, could be 2, 3, 4, and so on, with 12 being the maximum value.

a. (N) How many possible ways are there for the state of the dice to equal 10?

b. (N) How many possible ways are there for the state of the dice to equal 6?

c. (C) Which of these two states is more likely to occur? Explain your answer.

INTERPRET and ANTICIPATE For each case, we just need to determine the possible ways the total could be found when the dice are thrown. Each time the dice are thrown, each die may turn up a number 1 through 6. In this problem we are only asked to examine the possible ways the state of the dice might equal 10, and 6. The one with the more possible ways the state could be achieved is the more likely state to occur. This is because when we throw the dice, there is no preferential value for each die (1, 2, 3, 4, 5, and 6 all occur with equal probability on each die). Because 10 is such a high total, it requires both dice to have a high value. Therefore, we expect that 6 is a more likely final state of the dice than 10.
SOLVE **a.** The dice could turn up as (6, 4), (4, 6), or (5, 5) and result in a total of 10. Thus, there are $\boxed{3}$ ways for the state of the dice to equal 10.
b. The dice could turn up as (1, 5), (5, 1), (2, 4), (4, 2), or (3, 3) and result in a total of 6. Thus, there are $\boxed{5}$ ways for the state of the dice to equal 6.
c. Because there are more possible ways for the 6 to occur than the 10, the state of $\boxed{6}$ is more likely to occur.
CHECK and THINK It turns out that the most likely final state of the dice is 7. Can you determine how many possible ways there are to roll a total of 7 on the dice? The result should be greater than 5, the number of ways to roll a total of 6.

15. (N) A proton is bound in an infinite square well. A photon with a frequency of 2.34×10^{15} Hz is emitted when the proton makes a transition from the $n = 6$ excited state to the ground state. What is the width of the square well?

INTERPRET and ANTICIPATE We know that the energy of the emitted photon should be equal to the energy difference between the two states, $n = 6$ and $n = 1$. We can use Eq. 40.8 to express the energy of the photon, and Eq. 41.7 to express the energy difference between the two states. Once this is done, the only unknown will be the width of the well, L.

Chapter 41 – Schröödinger's Equation

SOLVE	
Use Eq. 40.8 to express the energy of the photon.	$E = hf = \left(6.626 \times 10^{-34} \text{ J} \cdot \text{s}\right)\left(2.34 \times 10^{15} \text{ Hz}\right)$
Now use Eq. 41.7 to express the energy difference between the two states.	$E_6 - E_1 = \left(6\right)^2 \dfrac{h^2}{8mL^2} - \left(1\right)^2 \dfrac{h^2}{8mL^2} = \left(35\right)\dfrac{h^2}{8mL^2}$
Finally, equate the energy difference to the energy of the photon and solve for L.	$\left(6.626 \times 10^{-34} \text{ J} \cdot \text{s}\right)\left(2.34 \times 10^{15} \text{ Hz}\right) = \left(35\right)\dfrac{h^2}{8mL^2}$ $L = \sqrt{\dfrac{\left(35\right)h^2}{8m\left(6.626 \times 10^{-34} \text{ J} \cdot \text{s}\right)\left(2.34 \times 10^{15} \text{ Hz}\right)}}$ $L = \sqrt{\dfrac{\left(35\right)\left(6.626 \times 10^{-34} \text{ J} \cdot \text{s}\right)^2}{8\left(1.6726 \times 10^{-27} \text{ kg}\right)\left(6.626 \times 10^{-34} \text{ J} \cdot \text{s}\right)\left(2.34 \times 10^{15} \text{ Hz}\right)}}$ $L = \boxed{2.72 \times 10^{-11} \text{ m}}$

CHECK and THINK

The size is small, though the result is much larger than the size of a nucleus, Thus, we can attempt to understand the energy quantization of the proton by using the infinite square well model if we were to really measure the energies of photons emitted as a proton undergoes transitions from one quantum state to another.

18. (A) An infinite square well of width L is used to confine a particle. The well extends from $x = 0$ to $x = L$. Assume the particle is in the first excited state ($n = 2$).
a. What is the probability of detecting the particle between $x = 0$ to $x = L$?
b. What is the probability of detecting the particle between $x = 0$ to $x = 0.50L$?

INTERPRET and ANTICIPATE

We can use Eq. 41.18 to evaluate the probability of finding the particle in a certain range when it is in a state given by integrating over the given range. Looking at Eq. 41.18 and knowing we would like to integrate that function, we must make use of the following result, obtained from a table of integrals:

$$\int \sin^2 ax \, dx = \frac{x}{2} - \frac{\sin 2ax}{4a} + C$$

In the first case, we expect that the result should be 1, because the particle must be in the infinite square well. If we integrate over the known state using Eq. 41.8 from 0 to L. we should be guaranteed to find the particle. This is not true, however, when we consider the range from 0 to $0.50L$. We might guess that the result of the integral in this case should be 1/2, since we think we should have a 50% chance of finding the particle in half of the box.

SOLVE **a.** Express Eq. 41.8 for the first excited state.	$\dfrac{2}{L}\sin^2\left(k_2 x\right)$
Now integrate this result from 0 to L using the integral from above.	$P = \displaystyle\int_0^L \left|\psi_2\right|^2 dx = \int_0^L \dfrac{2}{L}\sin^2\left(k_2 x\right) dx$ $P = \dfrac{2}{L}\left[\dfrac{x}{2} - \dfrac{\sin 2k_2 x}{4k_2}\right]_0^L$ $P = \dfrac{2}{L}\left[\dfrac{L}{2} - \dfrac{\sin 2k_2\left(L\right)}{4k_2} - 0 + 0\right] = \dfrac{2}{L}\left[\dfrac{L}{2} - \dfrac{\sin 2k_2 L}{4k_2}\right]$
Now use Eq. 41.15 to express k_2, and solve for the probability P.	$P = \dfrac{2}{L}\left[\dfrac{L}{2} - \dfrac{\sin\left(2\pi/L\right)2L}{4\left(2\pi/L\right)}\right] = \dfrac{2}{L}\left[\dfrac{L}{2} - \dfrac{\sin 4\pi}{4\left(2\pi/L\right)}\right]$ $P = \dfrac{2}{L}\left[\dfrac{L}{2}\right]$ $P = \boxed{1}$
b. Again, express Eq. 41.18 for the first excited state.	$\dfrac{2}{L}\sin^2\left(k_2 x\right)$
Now integrate this result from 0 to $0.50L$ using the integral from above.	$P = \displaystyle\int_0^{0.50L} \left|\psi_2\right|^2 dx = \int_0^{0.50L} \dfrac{2}{L}\sin^2\left(k_2 x\right) dx$ $P = \dfrac{2}{L}\left[\dfrac{x}{2} - \dfrac{\sin 2k_2 x}{4k_2}\right]_0^{0.50L}$ $P = \dfrac{2}{L}\left[\dfrac{0.50L}{2} - \dfrac{\sin 2k_2\left(0.50L\right)}{4k_2}\right] = \dfrac{2}{L}\left[\dfrac{L}{4} - \dfrac{\sin k_2 L}{4k_2}\right]$

Now use Eq. 41.15 to express k_2, and solve for the probability P.	$P = \dfrac{2}{L}\left[\dfrac{L}{4} - \dfrac{\sin(2\pi/L)L}{4(2\pi/L)}\right] = \dfrac{2}{L}\left[\dfrac{L}{4} - \dfrac{\sin 2\pi}{4(2\pi/L)}\right]$ $P = \dfrac{2}{L}\left[\dfrac{L}{4}\right]$ $P = \boxed{\dfrac{1}{2}}$

CHECK and THINK

As expected, we had a 100% chance of finding the particle when integrating over the entire width of the well, and only a 50% chance when integrating over half the well. Note however, that the result here can depend on the state in which the particle currently resides. We also, could consider integrating over different ranges and observe that there are different probabilities, over the same range, depending on the current state of the particle.

21. (N) An electron is trapped in an infinite well of width 2.50 nm. Initially the system is in the $n = 3$ quantum state; it then jumps to the $n = 5$ quantum state. Determine the amount of energy lost or gained by the system. Give your answer in electron-volts.

INTERPRET and ANTICIPATE
Given that the electron is increasing the quantum number of its state, the system must be gaining energy, since we know the energy of the $n = 5$ state should be larger than the energy of the $n = 3$ state. We can use Eq. 41.16 to express the energy of each state, and then the energy difference. Lastly, we can convert the result to electron-volts.

SOLVE	
First, use Eq. 41.16 to express the energy of the $n = 3$ state.	$E_3 = (3)^2 \dfrac{h^2}{8mL^2}$
Now use Eq. 41.16 to express the energy of the $n = 5$ state.	$E_5 = (5)^2 \dfrac{h^2}{8mL^2}$
The energy difference is the energy for the $n = 5$ state minus the energy for the $n = 3$ state.	$\Delta E = E_f - E_i = E_5 - E_3 = (5)^2 \dfrac{h^2}{8mL^2} - (3)^2 \dfrac{h^2}{8mL^2}$ $\Delta E = (16)\dfrac{h^2}{8mL^2} = \dfrac{2h^2}{mL^2}$ $\Delta E = \dfrac{2(6.626\times10^{-34}\ \text{J}\cdot\text{s})^2}{(9.109\times10^{-31}\ \text{kg})(2.50\times10^{-9}\ \text{m})^2} = 1.54\times10^{-19}\ \text{J}$

Now, convert the result to electron-volts.	$\Delta E = \left(1.54 \times 10^{-19} \text{ J}\right)\left(1 \text{ eV} / 1.60 \times 10^{-19} \text{ J}\right) = \boxed{0.964 \text{ eV}}$

CHECK and THINK

When a confined particle is bound in a quantized energy state, it must gain energy in order to move to a higher energy state, and lose energy in order to relax to a lower energy state. We can find the amount of energy necessary, or emitted, in these transitions by making use of Eq. 41.16.

24. (E) A proton in the nucleus of an atom can be modeled crudely as a proton trapped in an infinite well with a width on the order of 10^{-14} m. What is the order of magnitude of its zero-point energy? Give your answer in joules and electron-volts.

INTERPRET and ANTICIPATE

We can use Eq. 41.16 to express find the order of magnitude of the zero-point energy. Once found in joules, we can then convert the answer to electron-volts.

SOLVE Use Eq. 41.16 to find the zero-point, or ground state, energy of the proton in the well. Note that we are only to answer with the order of magnitude.	$E_1 = (1)^2 \dfrac{h^2}{8mL^2}$ $E_1 = \dfrac{\left(6.626 \times 10^{-34} \text{ J} \cdot \text{s}\right)^2}{8\left(1.6726 \times 10^{-27} \text{ kg}\right)\left(10^{-14} \text{ m}\right)^2}$ $E_1 = 3.28 \times 10^{-13} \text{ J} = \boxed{10^{-13} \text{ J}}$
Now, convert the original calculation to electron-volts. Again, we answer only with the order of magnitude.	$E_1 = \left(3.28 \times 10^{-13} \text{ J}\right)\left(1 \text{ eV} / 1.60 \times 10^{-19} \text{ J}\right)$ $E_1 = \boxed{10^6 \text{ eV}}$

CHECK and THINK

Note that the bound particle in this case was a proton. If we were to bound an electron in a well of the same width (similar to the size of a nucleus), then the energy would be much greater. Both the mass of the bound particle and the width of the well are important factors in determining the quantized energy states.

33. (N) A proton is in a finite square well with $U_0 = 35.5$ MeV. It is found that $\Lambda_2 = 2.35 \times 10^{-15}$ m. What is E_2? Give your answer in joules and in electron-volts.

INTERPRET and ANTICIPATE

If we convert the potential barrier height to joules, we can use Eq. 41.21 to find the energy of the first excited state, $n = 2$. Then, we can convert our final answer to electron-

Chapter 41 – Schroödinger's Equation

volts. The bound state energy must be less than the barrier height, otherwise there would not be a penetration distance.

SOLVE First, express the barrier height in terms of joules.	$U_0 = (35.5 \times 10^6 \text{ eV})(1.60 \times 10^{-19} \text{ J/1 eV})$
Now, write Eq. 41.21 and solve for the energy of the first excited state.	$\Lambda_2 = \dfrac{h}{2\pi\sqrt{2m(U_0 - E_2)}}$ $\sqrt{2m(U_0 - E_2)} = \dfrac{h}{2\pi\Lambda_2}$ $2m(U_0 - E_2) = \dfrac{h^2}{4\pi^2\Lambda_2^2}$ $U_0 - E_2 = \dfrac{h^2}{8\pi^2 m\Lambda_2^2}$ $E_2 = U_0 - \dfrac{h^2}{8\pi^2 m\Lambda_2^2}$
Substitute the known quantities and solve for the energy.	$E_2 = (35.5 \times 10^6 \text{ eV})(1.60 \times 10^{-19} \text{ J/1 eV})$ $- \dfrac{(6.626 \times 10^{-34} \text{ J·s})^2}{8\pi^2 (1.6726 \times 10^{-27} \text{ kg})(2.35 \times 10^{-15} \text{ m})^2}$ $E_2 = \boxed{5.08 \times 10^{-12} \text{ J}}$

Now, convert the answer to electron-volts.

$$E_2 = \left[(35.5 \times 10^6 \text{ eV})(1.60 \times 10^{-19} \text{ J/1 eV}) - \frac{(6.626 \times 10^{-34} \text{ J·s})^2}{8\pi^2 (1.6726 \times 10^{-27} \text{ kg})(2.35 \times 10^{-15} \text{ m})^2} \right](1 \text{ eV}/1.60 \times 10^{-19} \text{ J})$$

$E_2 = \boxed{3.17 \times 10^7 \text{ eV}}$

CHECK and THINK
The bound state energy is less than the barrier height as required. Note that it is not possible to evaluate Eq. 41.21 for a penetration distance if we set $E_n > U_0$. This makes sense because if that were the case, the particle would have more energy than the barrier height and would pass over the barrier.

36. (E) CASE STUDY The probability of two protons tunneling in the Sun's core is $P_{\text{tunnel}} \sim 10^{-10}$. This means that out of each 10^{10} pairs of protons, one pair tunnels successfully. Make an estimate showing that this probability can account for the 10^{38} fusion reactions that take place each second in the Sun's core. *Hint:* Assume that the Sun's core contains about 40% of its mass.

INTERPRET and ANTICIPATE

We can use the mass of the Sun to determine the mass in the core. If can determine an estimate for how many pairs of protons are represented in this mass, we will have a value for how many fusion events between protons might be occurring at any one moment. This is because we can multiply the estimated number of proton pairs by the probability of tunneling, or probability of fusing, and gauge the number of fusion events. We can then compare our result to the 10^{38} fusion reactions in the core of the Sun.

SOLVE	
First, we express the mass of the core of the Sun after looking up the total mass of the Sun.	$M_{core} = 0.4\,M_\odot = (0.4)(1.99 \times 10^{30}\ \text{kg})$
Now, we determine the number of protons represented by this value.	$N_p = \dfrac{M_{core}}{m_p} = \dfrac{0.4\,M_\odot}{m_p} = \dfrac{(0.4)(1.99 \times 10^{30}\ \text{kg})}{1.6726 \times 10^{-27}\ \text{kg}} = 5 \times 10^{56}$
We recognize that some of the mass may not be protons, but may be neutrons, or may already be fused into helium nuclei. Thus, let's just use the order of magnitude of this number, or 1/5 of the number we calculated in proceeding to determine the number of fusion reactions each second.	$N_{p,\,\text{corrected}} = (1/5)5 \times 10^{56} = 1 \times 10^{56}$
It takes two protons to make a possible fusion pair, so at any moment, our number of pairs is half of our value.	$N_{p,\,\text{pairs}} = (1/2)1 \times 10^{56} = 5 \times 10^{55}$
Multiply the probability of a fusion event occurring by this number of pairs to determine how many events might occur each second, n.	$n = P_{\text{tunnel}} N_{p,\,\text{pairs}} = (1 \times 10^{-10})(5 \times 10^{55}) = \boxed{5 \times 10^{45}}$

CHECK and THINK

The number we found can certainly account for the 10^{38} fusion reactions, even if we are off by a few orders of magnitude.

39. (N) An electron with a kinetic energy of 45.34 eV is incident on a square barrier with $U_b = 54.43$ eV and $w = 2.400$ pm. What is the probability that the electron tunnels through the barrier?

INTERPRET and ANTICIPATE

With the given information, we can use Eq. 41.22 and Eq. 41.23 to find the probability that the electron tunnels through the barrier. The answer, as a decimal, must be less than 1, or we could say that the answer must be less than 100%, because the energy of the electron is less than the barrier height.

SOLVE

Use Eq. 41.23 to find the factor γ, for use in Eq. 41.22.

$$\gamma^2 = \frac{32\pi m w^2 \left(U_b - K_i\right)}{h^2}$$

$$\gamma^2 = \frac{32\pi \left(9.109\times10^{-31}\ \text{kg}\right)\left(2.400\times10^{-12}\ \text{m}\right)^2 \left(54.43\ \text{eV} - 45.34\ \text{eV}\right)\left(1.60\times10^{-19}\ \text{J}/1\ \text{eV}\right)}{\left(6.626\times10^{-34}\ \text{J}\cdot\text{s}\right)^2}$$

$$\gamma^2 = 1.747\times10^{-3}$$

$$\gamma = 4.180\times10^{-2}$$

Now, use Eq. 41.22 to find the probability of tunneling.	$P_{\text{tunnel}} = e^{-\gamma} = e^{-4.180\times10^{-2}} = \boxed{0.9591}$

CHECK and THINK

The probability is less than 100%, which makes sense given the electron's energy versus the barrier height. If the electron had more energy than the barrier height, we could not evaluate Eq. 41.22, because there would be no need to tunnel, the electron would just pass over the barrier.

43. (N) In a quantum wire, an electric current can be controlled by varying the height of a potential barrier. Suppose that the electrons each have a kinetic energy of 8.24 eV and that 1.33×10^4 electrons pass through the wire every second when there is no potential barrier. If the width of the barrier is 50.00 pm, what barrier energy height would cause the current to become 1.00% of its initial value?

INTERPRET and ANTICIPATE

First, we recognize that the current will become 1.00% of the initial value if only 1.00% of the electrons are able to tunnel through the barrier. Thus, the probability of tunneling must be 1.00%. We can use Eq. 41.22 to determine the value of γ, and then Eq. 41.23 to determine the barrier height. We expect the barrier height must be greater than the energy of the electrons. Otherwise, there would be no tunneling occurring.

SOLVE Use Eq. 41.22 to find the value of γ.	$P_{tunnel} = e^{-\gamma} = 0.0100$ $\gamma = -\ln(0.0100)$

Now use Eq. 41.23 to find the barrier height. Note that in the second to last step, we multiply the first term by $(1 \text{ eV}/1.60 \times 10^{-19} \text{ J})$ so that the two terms can be added together to give the final answer in electron-volts.

$$\gamma^2 = \frac{32\pi^2 m\omega^2 \left(U_b - K_i\right)}{h^2}$$

$$\left(U_b - K_i\right) = \frac{h^2\gamma^2}{32\pi^2 m\omega^2}$$

$$U_b = \frac{h^2\gamma^2}{32\pi^2 m\omega^2} + K_i$$

$$U_b = \frac{\left(6.626 \times 10^{-34} \text{ J}\cdot\text{s}\right)^2 \left(-\ln(0.0100)\right)^2}{32\pi^2 \left(9.109 \times 10^{-31} \text{ kg}\right)\left(50.00 \times 10^{-12} \text{ m}\right)^2}\left(1 \text{ eV}/1.60 \times 10^{-19} \text{ J}\right) + \left(8.24 \text{ eV}\right)$$

$$U_b = \boxed{89.2 \text{ eV}}$$

CHECK and THINK

The barrier energy height is much greater than the kinetic energy of the electron, which makes sense if the tunneling probability is to be 1.00%. If the electron's energy were increased, the probability of tunneling would go up.

45. (N) The energy difference between two adjacent energy levels in a quantum simple harmonic oscillator is 3.58 eV. What is the energy of the ground state?

INTERPRET and ANTICIPATE

Recall that subsequent energy levels in a quantum simple harmonic oscillator potential are equally spaced. We can use Eq. 41.26 to interpret the energy of the ground state, $n = 1$.

Chapter 41 – Schroödinger's Equation

SOLVE First, use Eq. 41.26 to express the energy of two subsequent energy levels, m and n, where $m > n$.	$E_m = \left(m - \dfrac{1}{2}\right)hf$ $E_n = \left(n - \dfrac{1}{2}\right)hf$
Now, the energy difference must be 3.58 eV.	$E_m - E_n = \left(m - \dfrac{1}{2}\right)hf - \left(n - \dfrac{1}{2}\right)hf = 3.58 \text{ eV}$ $(m - n)hf = 3.58 \text{ eV}$ $(1)hf = 3.58 \text{ eV}$ $hf = 3.58 \text{ eV}$
Use Eq. 41.26 to find the energy of the ground state.	$E_1 = \left(1 - \dfrac{1}{2}\right)hf = \left(\dfrac{1}{2}\right)hf = \left(\dfrac{1}{2}\right)(3.58 \text{ eV})$ $E_1 = \boxed{1.79 \text{ eV}}$

CHECK and THINK

While the energy difference between subsequent levels are identical in the quantum simple harmonic oscillator potential well, the energy of the ground state will always be half of that energy difference.

50. (A) Show that a free particle that is traveling in the negative x direction is equally likely to be found anywhere along the x axis.

INTERPRET and ANTICIPATE

We can use the second half of Eq. 41.31 to express the wave function for a free particle traveling in the negative x direction. If we were to try to compute the probability of finding the particle within a certain region, we would first need to compute the probability density, $|\psi(x)|^2 = \psi^*(x)\psi(x)$. We could then integrate over that certain region to determine the probability of finding the particle. However, if the particle is to be found anywhere equally, it must be the case that the integral would turn out the same no matter what region we chose. In other words, it should be the case that $|\psi(x)|^2 = \psi^*(x)\psi(x) = \psi^2$, or some constant, so that the integral for the probability turns out to be 1, or 100%, no matter over what region we would try to integrate. Note, to find the actual constant, we would need to normalize the wave function.

SOLVE First, we write Eq. 41.31, but only retain the second term, which would be the wave function for a free particle traveling in the negative x direction.	$\psi(x) = \psi_1 e^{-ikx}$
Now, we check to see if $\|\psi(x)\|^2 = \psi^2$, or a constant, which would be required for the particle to be equally likely to be found anywhere along the x axis.	$\|\psi(x)\|^2 = \psi^*(x)\psi(x) = \psi_1 e^{ikx}\psi_1 e^{-ikx} = \psi_1^2$

CHECK and THINK

The result of $\|\psi(x)\|^2$ is a constant, which means if we were to try and integrate over any region of space, we would get a probability of 1, after normalizing the wave function. There is no need to go further to prove that. Anytime the result of $\|\psi(x)\|^2$ is a constant, this will be the case, and is a feature we expect of the free particle.

54. (N) Review. An alpha particle is a helium nucleus consisting of two protons and two neutrons. It is moving with a speed of 1.10×10^7 m/s.
a. What is the momentum of the alpha particle?
b. If there is a 10.0% uncertainty in the momentum of this alpha particle, what is the minimum uncertainty in the position of the alpha particle?

INTERPRET and ANTICIPATE
Once we determine the mass of the alpha particle, we can use Eq. 10.1 to find the momentum of the alpha particle, since it is not quite at a relativistic speed. Then, assuming that momentum is the maximum uncertainty in the momentum of the particle, we can use Eq. 41.34 to find the minimum uncertainty in the position of the particle.

SOLVE **a.** The alpha particle is composed of two protons and two neutrons. Ignoring binding energies, we estimate the mass as being the sum of the masses of these four particles.	$m = 2m_p + 2m_n$ $m = 2(1.6726 \times 10^{-27} \text{ kg}) + 2(1.6749 \times 10^{-27} \text{ kg})$ $m = 2(1.6726 \times 10^{-27} \text{ kg} + 1.6749 \times 10^{-27} \text{ kg})$

Then, use Eq. 10.1 to find the momentum of the particle.	$p = mv$ $p = 2\left(1.6726\times10^{-27} \text{ kg} + 1.6749\times10^{-27} \text{ kg}\right)\left(1.10\times10^{7} \text{ m/s}\right)$ $p = \boxed{7.36\times10^{-20} \text{ kg}\cdot\text{m/s}}$
b. Use Eq. 41.34 to find the minimum uncertainty in the position of the particle, assuming the maximum uncertainty is given by the momentum of the particle.	$\Delta x \Delta p \geq \dfrac{h}{4\pi}$ $\Delta x \geq \dfrac{h}{4\pi\Delta p} = \dfrac{6.626\times10^{-34} \text{ J}\cdot\text{s}}{4\pi(0.100)\left(7.36\times10^{-20} \text{ kg}\cdot\text{m/s}\right)}$ $\Delta x \geq \boxed{7.16\times10^{-15} \text{ m}}$

CHECK and THINK

When considering Heisenberg's uncertainty principle, we can determine the minimum uncertainty of one quantity by using the maximum uncertainty of the other quantity. It is a common practice to assume the maximum uncertainty in a calculated quantity, like momentum, is at most equal to that calculated value, unless there is a reason that we could reduce the uncertainty (this might depend on how we measured the quantity).

59. (N) An infinite square well of width $L = 155$ pm is used to confine an electron. The well extends from $x = 0$ to $x = L$. Assume the electron is in the ground state.
a. What is the probability of detecting the electron between $x = 0$ to $x = 0.50L$?
b. What is the probability of detecting the electron between $x = 0$ to $x = 0.25L$?

INTERPRET and ANTICIPATE

In order to find the probability of finding the electron between 0 and 0.50L and in the ground state, we must integrate Eq. 41.18 for the ground state from 0 to 0.50L. In order to find the probability of finding the electron between 0 and 0.25L and in the ground state, we must integrate Eq. 41.18 for the ground state from 0 to 0.25L. In performing the integrals, we will find the following integral (from a table) to be useful:

$$\int \sin^2 ax \; dx = \frac{x}{2} - \frac{\sin 2ax}{4a} + C$$

The probability of finding the electron in either region depends on the shape of the wave function for the state being examined. Given that the shape of the ground state wave function is symmetric about the middle of the well, we would expect 1/2 in the first case, but we are not sure about the second case – the ground state wave function does not divide into equal parts when dividing the width of the well into four equal widths. That being said, the probability of finding the electron between 0 and 0.25L should be less than finding the particle between 0 and 0.50L.

SOLVE **a.** Express Eq. 41.8 for the ground state.	$\dfrac{2}{L}\sin^2\left(k_1 x\right)$
Use Eq. 41.18 to integrate this expression over the range from 0 to 0.50L. Then, solve for the probability. Note that we make use of Eq. 41.15 to express the value of k for the ground state. $k_1 = \left(\pi/L\right)$	$P = \displaystyle\int_0^{0.50L} \left|\psi_1\right|^2 dx = \int_0^{0.50L} \frac{2}{L}\sin^2\left(k_1 x\right) dx$ $P = \dfrac{2}{L}\left[\dfrac{x}{2} - \dfrac{\sin 2k_1 x}{4k_1}\right]_0^{0.50L}$ $P = \dfrac{2}{L}\left[\dfrac{0.50L}{2} - \dfrac{\sin 2k_1\left(0.50L\right)}{4k_1}\right] = \dfrac{2}{L}\left[\dfrac{L}{4} - \dfrac{\sin k_1 L}{4k_1}\right]$ $P = \dfrac{2}{L}\left[\dfrac{L}{4} - \dfrac{\sin\left(\pi/L\right)L}{4\left(\pi/L\right)}\right] = \dfrac{2}{L}\left[\dfrac{L}{4} - \dfrac{\sin\pi}{4\left(\pi/L\right)}\right] = \dfrac{2}{L}\left[\dfrac{L}{4}\right]$ $P = \dfrac{1}{2} = \boxed{0.50}$
b. Begin again, but this time integrate from 0 to 0.25L. Solve for the probability.	$P = \displaystyle\int_0^{0.25L} \left|\psi_1\right|^2 dx = \int_0^{0.25L} \frac{2}{L}\sin^2\left(k_1 x\right) dx$ $P = \dfrac{2}{L}\left[\dfrac{x}{2} - \dfrac{\sin 2k_1 x}{4k_1}\right]_0^{0.25L}$ $P = \dfrac{2}{L}\left[\dfrac{0.25L}{2} - \dfrac{\sin 2k_1\left(0.25L\right)}{4k_1}\right] = \dfrac{2}{L}\left[\dfrac{L}{8} - \dfrac{\sin\left(k_1 L/2\right)}{4k_1}\right]$ $P = \dfrac{2}{L}\left[\dfrac{L}{8} - \dfrac{\sin\left(\pi L/2L\right)}{4\left(\pi/L\right)}\right] = \dfrac{2}{L}\left[\dfrac{L}{8} - \dfrac{\sin\left(\pi/2\right)}{4\left(\pi/L\right)}\right]$ $P = \dfrac{2}{L}\left[\dfrac{L}{8} - \dfrac{L}{4\pi}\right]$ $P = \dfrac{1}{4} - \dfrac{1}{2\pi} = \dfrac{\pi}{4\pi} - \dfrac{2}{4\pi}$ $P = \boxed{\dfrac{\pi-2}{4\pi}}$

CHECK and THINK

The probability of finding the electron between 0 and 0.25L is indeed less than that of finding the electron between 0 and 0.50L, as we expected. Note that these probabilities may not be the same over the same ranges for a different bound state, like the second excited state, for example.

62. (N) A proton is trapped in an infinite well of length 0.350 nm. Find the energy of the first four excited states ($n = 2$ to $n = 5$). Give your answers in joules and electron-volts.

INTERPRET and ANTICIPATE We can use Eq. 41.16 to find the energy for each state when $n = 2$, 3, 4, and 5. The energy should increase as the quantum number n increases. In each case, we will perform the calculation to find the answer in joules, and then again, converting the result to electron-volts. In this way, we won't lose significant digits due to rounding, if we were just to convert our results in joules.	

SOLVE We first find the energy when $n = 2$.	$$E_2 = (2)^2 \frac{(6.626 \times 10^{-34} \text{ J} \cdot \text{s})^2}{8(1.6726 \times 10^{-27} \text{ kg})(0.350 \times 10^{-9})^2}$$ $$\boxed{E_2 = 1.07 \times 10^{-21} \text{ J}}$$ $$E_2 = (2)^2 \frac{(6.626 \times 10^{-34} \text{ J} \cdot \text{s})^2}{8(1.6726 \times 10^{-27} \text{ kg})(0.350 \times 10^{-9})^2} \cdot \left(\frac{1 \text{ eV}}{1.60 \times 10^{-19} \text{ J}}\right)$$ $$\boxed{E_2 = 6.70 \times 10^{-3} \text{ eV}}$$
Now find the energy when $n = 3$.	$$E_3 = (3)^2 \frac{(6.626 \times 10^{-34} \text{ J} \cdot \text{s})^2}{8(1.6726 \times 10^{-27} \text{ kg})(0.350 \times 10^{-9})^2}$$ $$\boxed{E_3 = 2.41 \times 10^{-21} \text{ J}}$$ $$E_3 = (3)^2 \frac{(6.626 \times 10^{-34} \text{ J} \cdot \text{s})^2}{8(1.6726 \times 10^{-27} \text{ kg})(0.350 \times 10^{-9})^2} \cdot \left(\frac{1 \text{ eV}}{1.60 \times 10^{-19} \text{ J}}\right)$$ $$\boxed{E_3 = 1.51 \times 10^{-2} \text{ eV}}$$
Now find the energy when $n = 4$.	$$E_4 = (4)^2 \frac{(6.626 \times 10^{-34} \text{ J} \cdot \text{s})^2}{8(1.6726 \times 10^{-27} \text{ kg})(0.350 \times 10^{-9})^2}$$ $$\boxed{E_4 = 4.29 \times 10^{-21} \text{ J}}$$ $$E_4 = (4)^2 \frac{(6.626 \times 10^{-34} \text{ J} \cdot \text{s})^2}{8(1.6726 \times 10^{-27} \text{ kg})(0.350 \times 10^{-9})^2} \cdot \left(\frac{1 \text{ eV}}{1.60 \times 10^{-19} \text{ J}}\right)$$ $$\boxed{E_4 = 2.68 \times 10^{-2} \text{ eV}}$$

Lastly, find the energy when $n = 5$.	$E_5 = (5)^2 \dfrac{\left(6.626 \times 10^{-34} \text{ J} \cdot \text{s}\right)^2}{8\left(1.6726 \times 10^{-27} \text{ kg}\right)\left(0.350 \times 10^{-9}\right)^2}$
	$E_5 = \boxed{6.70 \times 10^{-21} \text{ J}}$
	$E_5 = (5)^2 \dfrac{\left(6.626 \times 10^{-34} \text{ J} \cdot \text{s}\right)^2}{8\left(1.6726 \times 10^{-27} \text{ kg}\right)\left(0.350 \times 10^{-9}\right)^2} \cdot \left(\dfrac{1 \text{ eV}}{1.60 \times 10^{-19} \text{ J}}\right)$
	$E_5 = \boxed{4.19 \times 10^{-2} \text{ eV}}$

CHECK and THINK

As expected, the energy increases as the quantum number n increases. Note that in the infinite square well model, the difference in energy between subsequent levels increases as the energy of the levels increases.

63. (N) A proton is trapped in an infinite well of length 0.350 nm. Initially the system is in its ground state. It then absorbs a photon with an energy of 1.34×10^{-2} eV. What state is the system in as a result?

INTERPRET and ANTICIPATE

We can consider the results from the previous problem (P41.62) in determining what happens to the proton when it absorbs this energy. We must also calculate the ground state energy of the proton in this well. Given that the proton starts in the ground state and absorbs this energy, we can express the total energy of the proton by adding these two values together. We can then compare the total energy of the electron to those of the different bound states in P41.62. The state with the nearest value that is not greater than our proton's energy is the state the proton will occupy.

SOLVE Use Eq. 41.16 to calculate the ground state energy of the proton.	$E_n = n^2 \dfrac{h^2}{8mL^2}$
	$E_1 = (1)^2 \dfrac{\left(6.626 \times 10^{-34} \text{ J} \cdot \text{s}\right)^2}{8\left(1.6726 \times 10^{-27} \text{ kg}\right)\left(0.350 \times 10^{-9}\right)^2} \cdot \left(\dfrac{1 \text{ eV}}{1.60 \times 10^{-19} \text{ J}}\right)$
	$E_1 = 1.67 \times 10^{-3} \text{ eV}$
Now, add the ground state energy to the energy of the absorbed photon to get the total energy of the proton.	$E = 1.67 \times 10^{-3} \text{ eV} + 1.34 \times 10^{-2} \text{ eV} = 1.52 \times 10^{-2} \text{ eV}$

By comparison to the energy of the states in P41.62, we see that the state with the nearest energy that is not greater than the proton's energy is the $n = 3$ state.	$\boxed{n = 3}$

CHECK and THINK

Note that the energy of the photon was such that there is some excess energy. Once the proton settles into the $n = 3$ state, that other energy must be emitted as a photon, or transferred to another system in some fashion.

---

68. (C) A particle is in an infinite square well which has its width halved. Is the percentage change in the third excited state energy (E_4) the same as the percentage change in the ground state energy E_1? Explain your answer.

INTERPRET and ANTICIPATE
We can answer this question by examining Eq. 41.16 and what happens to the energy of a particular state when the well width is halved. We can then find the percentage change between the initial and final states and consider if the result would depend on the quantum number n.

SOLVE Express Eq. 41.16 for the initial width, L_i.	$E_{n_i} = n^2 \dfrac{h^2}{8mL_i^2}$
Now express the energy of the nth state after the width is halved.	$E_{n_f} = n^2 \dfrac{h^2}{8mL_f^2} = n^2 \dfrac{h^2}{8m\left(\frac{1}{2}L_i\right)^2}$ $E_{n_f} = n^2 \dfrac{h^2}{8\left(\frac{1}{4}\right)mL_i^2} = 4n^2 \dfrac{h^2}{8mL_i^2}$ $E_{n_f} = 4E_{n_i}$
Note that the result is independent of the quantum number, n. This means that the new energy for a particular state is 4 times greater than the previous energy for that state no matter which state we examine. Calculate the percent change.	$\%\text{change} = \dfrac{E_{n_f} - E_{n_i}}{E_{n_i}} \times 100\%$ $\%\text{change} = \dfrac{4E_{n_i} - E_{n_i}}{E_{n_i}} \times 100\% = \boxed{300\%}$

CHECK and THINK

Our result means that as the width of the well is adjusted, all of the energy levels experience the same percentage change based on their initial values. This does not mean that the energy difference between subsequent levels is unchanged, however! Consider problem 41.69 and investigate.

70. (A, C) The width of a potential barrier decreases at a constant rate (dw/dt) until the width becomes zero. Derive an approximate expression for the rate of change of the probability (dP_{tunnel}/dt) that a particle with mass m and kinetic energy K_i tunnels through the barrier with height U_b. Interpret the meaning of your answer.

INTERPRET and ANTICIPATE

We must take the derivative of P_{tunnel} in terms of time, realizing that γ depends on w, which depends on time, in Eq. 41.22. This means we will need to make use of Eq. 41.23 also. In the end, we will need to examine how P_{tunnel} depends on both dw/dt and w in order to interpret the effect.

SOLVE	
In Eq. 41.22, we wish to take the derivative with respect to time, knowing that γ depends on time.	$$\frac{d}{dt}P_{tunnel} = \frac{d}{dt}\left[e^{-\gamma}\right] = e^{-\gamma}\frac{d\gamma}{dt}$$
Solve Eq. 41.23 for γ.	$$\gamma^2 = \frac{32\pi^2 mw^2\left(U_b - K_i\right)}{h^2}$$ $$\gamma = \sqrt{\frac{32\pi^2 mw^2\left(U_b - K_i\right)}{h^2}}$$
Now we can take the derivative of this expression in terms of time, where w depends on time.	$$\frac{d\gamma}{dt} = \frac{d}{dt}\left[\sqrt{\frac{32\pi^2 mw^2\left(U_b - K_i\right)}{h^2}}\right]$$ $$\frac{d\gamma}{dt} = \frac{1}{2}\left(\frac{32\pi^2 mw^2\left(U_b - K_i\right)}{h^2}\right)^{-1/2}(2)\frac{32\pi^2 mw\left(U_b - K_i\right)}{h^2}\frac{dw}{dt}$$ $$\frac{d\gamma}{dt} = \sqrt{\frac{32\pi^2 m\left(U_b - K_i\right)}{h^2}}\frac{dw}{dt}$$
Finally, we can go back and build the expression for the derivative of P_{tunnel} with respect to time, since we now have all the pieces.	$$\frac{dP_{tunnel}}{dt} = e^{-\gamma}\frac{d\gamma}{dt}$$ $$\boxed{\frac{dP_{tunnel}}{dt} = \frac{dw}{dt}\sqrt{\frac{32\pi^2 m\left(U_b - K_i\right)}{h^2}}e^{-\sqrt{\frac{32\pi^2 mw^2\left(U_b - K_i\right)}{h^2}}}}$$

> ### CHECK and THINK
> As the width decreases, the exponential term increases, which means the probability of tunneling is going up. Also, the rate at which the decrease occurs does affect the probability of tunneling. They are related linearly.

77. (N) Consider the mix of stationary states described in Problem 76. If there is a 1 in 4, or 25%, chance of finding the particle in the first excited state ($n = 2$), what is the probability of finding the particle in the ground state?

> ### INTERPRET and ANTICIPATE
> The mix of stationary states in Problem 76 describes the possible states in which we might find the particle when we go to measure it. Here, there are two possibilities: Either the particle is in the $n = 1$ ground state, or the particle is in the $n = 2$ excited state. The particle must be in one of the two states when we measure. Note that the total probability of finding the particle must be 100%.
>
> ### SOLVE
> Since there is a 1 in 4, or 25%, chance of finding the particle in the $n = 2$ state, and the total probability of finding the particle must be 100%, then the probability of finding it in the ground state, p_1, can be found.
>
> $$p_1 + p_2 = 100\%$$
> $$p_1 + 25\% = 100\%$$
> $$p_1 = \boxed{75\%}$$
>
> ### CHECK and THINK
> Until the particle is measured, we might say that it is in either state, neither state, and both! The wave function describes the particle up until it is measured. It is one of the more puzzling moments where physics meets philosophy, as to how to interpret the wave function and what it represents. Is it merely a mathematical construct that is excellent at quantifying and predicting phenomenon, or does it really represent the existence of the particle?

42

Atoms

2. (E) Review Estimate the number of atoms that exist in a cubic centimeter of a solid material. This process will be aided by choosing an element and examining its density in the solid phase.

INTERPRET and ANTICIPATE	
If you try this problem yourself, you might choose a different material and try the calculation again. For the purposes of this solution, we choose a piece of graphite, which is made up of carbon atoms. Using the periodic table, we find that the atomic mass of a carbon atom is 12.011 u. Once we have a total mass for a cubic centimeter of graphite, we can divide that mass by the mass of one carbon atom to find an estimate for how many make up the material.	
SOLVE	
Using your library, or the internet, you can find a typical density for graphite that is given in g/cm^3. Here, we use the value of 2.26 g/cm^3.	$\rho = 2.26$ g/cm^3
To find the total mass for 1 cm^3 of graphite, we use Eq. 1.1. We then convert the result to kg.	$\rho = m/V$ $m = \rho V = \left(2.26 \text{ g/cm}^3\right)\left(1 \text{ cm}^3\right) = 2.26$ g $m = \left(2.26 \text{ g}\right) \times \left(\dfrac{1 \text{ kg}}{1000 \text{ g}}\right) = 0.00226$ kg
The mass of one carbon atom is 12.011 u, but we should convert that to kg before continuing.	$m_C = \left(12.011 \text{ u}\right) \times \left(\dfrac{1.66 \times 10^{-27} \text{ kg}}{1 \text{ u}}\right)$
Now we can find the number of carbon atoms in the volume by dividing the total mass by the mass of one carbon atom.	$N = \dfrac{m}{m_C} = \dfrac{0.00226 \text{ kg}}{\left(12.011 \text{ u}\right)\left(1.66 \times 10^{-27} \text{ kg/1 u}\right)}$ $N = 1.13 \times 10^{23} \approx \boxed{1 \times 10^{23}}$
CHECK and THINK	
That is a lot of atoms in a small space! A good rule of thumb for a linear mass density of solid materials is that there are between $10^7 - 10^8$ atoms per cm. This would be true in each dimension.	

9. (N) In the Bohr model of hydrogen, an electron makes a transition from the $n = 5$ excited state to the $n = 3$ excited state. What is the change in the orbital radius of the electron?

INTERPRET and ANTICIPATE We can use Eq. 42.7 to express the radius of the electron for each state, and then find the difference between the final radius and the initial radius. Because the initial excited state is a higher than the final excited state, the electron should be getting closer to the nucleus and the change in radius should be negative.	
SOLVE Use Eq. 42.7 to express the radius of the electron in each excited state.	$r_5 = (5)^2 r_B$ $r_3 = (3)^2 r_B$
Now, find the difference in radius between the initial and final state.	$\Delta r = r_f - r_i$ $\Delta r = (3)^2 r_B - (5)^2 r_B = -16 r_B$ $\Delta r = -16(5.29 \times 10^{-11} \text{ m})$ $\Delta r = \boxed{-8.46 \times 10^{-10} \text{ m}}$

CHECK and THINK   As expected, the change is negative. Note that even in a very high excited state, the electron is at most about 1×10^{-10} m away from the nucleus. This length is often used as an estimate for the size of the hydrogen atom.

13. (N) Consider photons incident on a hydrogen atom.
a. A transition from the $n = 3$ to the $n = 7$ excited state requires the absorption of a photon of what minimum energy?
b. A transition from the $n = 1$ ground state to the $n = 5$ excited state requires the absorption of a photon of what minimum energy?

INTERPRET and ANTICIPATE   In each case, since the electron is transitioning from a lower state to a higher one, energy must be absorbed. The minimum energy required is dictated by the magnitude of the difference in bound state energy between the two energy levels in the transition. We can use Eq. 42.10 to express the bound state energy of each level in either part of the problem. We then need only to find the magnitude of the difference between the energy levels that the electron occupies initially and after the transition. We expect answers that represent small amounts of energy, on the order of 1 eV, or 10 eV, since this is the typical order of the energy levels in hydrogen.

SOLVE **a.** First, use Eq. 42.10 to express the energy of the $n = 3$ state and the $n = 7$ state.	$E_3 = \dfrac{-13.6 \text{ eV}}{(3)^2}$ $E_7 = \dfrac{-13.6 \text{ eV}}{(7)^2}$
Then, find the magnitude of the difference between these energies. This is the minimum photon energy needed for the transition.	$E_{photon} = \left\| E_7 - E_3 \right\| = \left\| \dfrac{-13.6 \text{ eV}}{(7)^2} - \dfrac{-13.6 \text{ eV}}{(3)^2} \right\| = \boxed{1.23 \text{ eV}}$
b. Again, use Eq. 42.10 to express the energy of the $n = 1$ state and the $n = 5$ state.	$E_1 = \dfrac{-13.6 \text{ eV}}{(1)^2}$ $E_5 = \dfrac{-13.6 \text{ eV}}{(5)^2}$
Then, find the magnitude of the difference between these energies. This is the minimum photon energy needed for the transition.	$E_{photon} = \left\| E_5 - E_1 \right\| = \left\| \dfrac{-13.6 \text{ eV}}{(5)^2} - \dfrac{-13.6 \text{ eV}}{(1)^2} \right\| = \boxed{13.1 \text{ eV}}$

CHECK and THINK

We did find small amounts of energy on the order of 1 eV, or 10 eV. We could use these energies to find the wavelength, or frequency of the photons. This would allow us to determine the color of the photons.

17. (N) An atom in an excited state can, on average, exist in that state for 10^{-8} s. If a hydrogen atom is in the $n = 2$ state, about how many orbits will it undergo before the atom returns to the ground state? Assume Bohr's model for hydrogen.

INTERPRET and ANTICIPATE

Using the information in the problem statement, we must determine the period of the orbit of the electron. Once we have the period, we can divide the total time of existence in the $n = 2$ excited state by the period to get the number of orbits of the electron. Assuming a circular orbit from the Bohr model, we can find the radius (and then the circumference) using Eq. 42.7. The speed of an electron in an excited state can be expressed using Eq. 42.5. By using the circumference (distance traveled in the period), and the speed, we can find the period. Given the strong bond between the electron and the nucleus, we expect there to be many revolutions of the electron, even in the short amount of time the electron is in the excited state.

459

SOLVE First, we use Eq. 42.7 to express the radius of the electron's orbit.	$r = n^2 r_B = (2)^2 r_B = 4 r_B$
Then, use Eq. 42.5 to express the speed of the electron in this state.	$v = \dfrac{nh}{2\pi m_e r} = \dfrac{2h}{2\pi m_e (4 r_B)} = \dfrac{h}{4\pi m_e r_B}$
The period is given by the ratio of the circumference of the path (distance traveled, d) and the speed, as found above.	$T = \dfrac{d}{v} = \dfrac{2\pi r}{v} = \dfrac{2\pi (4 r_B)}{v} = \dfrac{8\pi r_B}{v}$ $T = (8\pi r_B) \cdot \left(\dfrac{4\pi m_e r_B}{h} \right) = \dfrac{32\pi^2 m_e r_B^2}{h}$
We can then divide the total time by the period to find the number of orbits that will occur during the allotted time.	$n = \dfrac{t_{total}}{T} = (1 \times 10^{-8}\ \text{s}) \cdot \left(\dfrac{h}{32\pi^2 m_e r_B^2} \right)$ $n = (1 \times 10^{-8}\ \text{s}) \cdot \left(\dfrac{6.626 \times 10^{-34}\ \text{J} \cdot \text{s}}{32\pi^2 (9.109 \times 10^{-31}\ \text{kg})(5.29 \times 10^{-11}\ \text{m})^2} \right)$ $n = \boxed{8 \times 10^6}$

CHECK and THINK

The electron undergoes many orbits in a very short amount of time in this model of the atom. Of course, the existence of the electron is a bit different than that described by the Bohr model, but the model is quite useful in making predictions about the necessary absorbed energy, or the energy released as the electron makes transitions from one state to another in the hydrogen atom.

20. (N) The Balmer series consists of the spectral lines from hydrogen for an electron making a transition from an excited state to the $m = 2$ state. The Lyman series consists of the spectral lines from hydrogen for an electron making a transition from an excited state to the $m = 1$ state. Determine the wavelengths of the first four spectral lines of the Lyman series ($n = 2, 3, 4,$ and 5).

INTERPRET and ANTICIPATE

In order to find the first four wavelengths of the emitted photons in the Lyman series, we use Eq. 42.3 with $m = 1$. The wavelength emitted should decrease as we examine transitions from greater and greater states back to the $m = 1$ state.

SOLVE	
We begin by writing Eq. 42.3 for transitions from the state n to the state $m = 1$.	$$\frac{1}{\lambda_n} = R\left(\frac{1}{m^2} - \frac{1}{n^2}\right)$$ $$\frac{1}{\lambda_n} = R\left(\frac{1}{1^2} - \frac{1}{n^2}\right)$$
Find the wavelength for the transition with $n = 2$.	$$\frac{1}{\lambda_2} = \left(1.097 \times 10^7 \text{ m}^{-1}\right)\left(\frac{1}{1^2} - \frac{1}{2^2}\right)$$ $$\lambda_2 = \frac{1}{\left(1.097 \times 10^7 \text{ m}^{-1}\right)(1 - 1/4)} = \boxed{1.22 \times 10^{-7} \text{ m}}$$
Find the wavelength for the transition with $n = 3$.	$$\frac{1}{\lambda_3} = \left(1.097 \times 10^7 \text{ m}^{-1}\right)\left(\frac{1}{1^2} - \frac{1}{3^2}\right)$$ $$\lambda_3 = \frac{1}{\left(1.097 \times 10^7 \text{ m}^{-1}\right)(1 - 1/9)} = \boxed{1.03 \times 10^{-7} \text{ m}}$$
Find the wavelength for the transition with $n = 4$.	$$\frac{1}{\lambda_4} = \left(1.097 \times 10^7 \text{ m}^{-1}\right)\left(\frac{1}{1^2} - \frac{1}{4^2}\right)$$ $$\lambda_4 = \frac{1}{\left(1.097 \times 10^7 \text{ m}^{-1}\right)(1 - 1/16)} = \boxed{9.72 \times 10^{-8} \text{ m}}$$
Find the wavelength for the transition with $n = 5$.	$$\frac{1}{\lambda_5} = \left(1.097 \times 10^7 \text{ m}^{-1}\right)\left(\frac{1}{1^2} - \frac{1}{5^2}\right)$$ $$\lambda_5 = \frac{1}{\left(1.097 \times 10^7 \text{ m}^{-1}\right)(1 - 1/25)} = \boxed{9.50 \times 10^{-8} \text{ m}}$$

CHECK and THINK

Notice that the wavelength decreases as we consider transitions from greater initial states. Remember that if the wavelength of the photon is less, the energy of the photon is greater. This makes sense since the emitted photon should have more energy when the electron transitions from a higher energy state to the $m = 1$ state.

Chapter 42 – Atoms

24. (N) The Bohr model for the hydrogen atom can be extended to cover other atoms when they are stripped free of all but one electron. When this occurs, the energy levels for the single electron in an atom with atomic number, Z, are given by

$E_n = (-13.6 \text{ eV}) Z^2/n^2$ (see Example 42.4). Calculate the electron energy for the first five energy levels ($n = 1$ to $n = 5$) of ionized lithium (Li^{++}).

INTERPRET and ANTICIPATE

In examining Example 42.4, we find an equation that we can use to help model the energy levels of the ionized lithium atom. We expect that the bound state energies will be negative and will get closer to zero as we calculate the energy of higher and higher excited states.

SOLVE

Use the equation from Example 42.4 to calculate the bound state energies for the ground state and the first four excited states.

$E_1 = (-13.6 \text{ eV})(3)^2/(1)^2 = \boxed{-122 \text{ eV}}$

$E_2 = (-13.6 \text{ eV})(3)^2/(2)^2 = \boxed{-30.6 \text{ eV}}$

$E_3 = (-13.6 \text{ eV})(3)^2/(3)^2 = \boxed{-13.6 \text{ eV}}$

$E_4 = (-13.6 \text{ eV})(3)^2/(4)^2 = \boxed{-7.65 \text{ eV}}$

$E_5 = (-13.6 \text{ eV})(3)^2/(5)^2 = \boxed{-4.90 \text{ eV}}$

CHECK and THINK

If we were to continue to increase n and calculate the bound sate energy for greater excited states, the result would get closer to 0. Once there is no bound state energy, the electron is free.

---

29. (N) A hydrogen atom is in its $n = 6$ state. Find the de Broglie wavelength of its electron.

INTERPRET and ANTICIPATE

We can express the radius of the orbit of the electron in the $n = 6$ state using Eq. 42.7. Then, we can use Eq. 42.13 to find the wavelength of the electron while it is in this excited state.

SOLVE

First, use Eq. 42.7 to express the radius of the electron orbit in the $n = 6$ excited state.

$r = (6)^2 r_B$

Then, use this result with Eq. 42.13 to find the wavelength of the electron.	$\lambda_n = \dfrac{2\pi r}{n}$
	$\lambda_6 = \dfrac{2\pi(6)^2 r_B}{6}$
	$\lambda_6 = 2\pi(6)(5.29 \times 10^{-11} \text{ m}) = \boxed{1.99 \times 10^{-9} \text{ m}}$

CHECK and THINK

This wavelength is not the wavelength of a photon absorbed or emitted during a transition of the electron, but rather the wavelength of the electron itself, while it is in a particular state (in this case $n = 6$). In this, we model the electron as a standing wave, somewhat similar to how we modeled the standing wave modes of vibration of a string fixed at both ends much earlier in the book.

31. (N) If the state of hydrogen is given by $n = 3$ and $\ell = 2$, find the possible values of m, L and L_z.

INTERPRET and ANTICIPATE

We can use the principal quantum number and the orbital quantum number to find the allowed values of m, L, and L_z. By using Eq. 42.14, we can find the value of L. Also, by knowing the orbital quantum number, we can determine the possible values of m, the magnetic quantum number, based on the rules discussed in the chapter. Lastly, Eq. 42.15 can be used to find the value of L_z for each value of m allowed.

SOLVE First, find the value of L using Eq. 42.14.	$L = \sqrt{\ell(\ell+1)}\hbar = \sqrt{2(2+1)}\hbar = \boxed{\sqrt{6}\hbar}$
Now, as discussed in the chapter, the allowed values for m are $-\ell, -\ell+1, \ldots, 0, \ldots, \ell-1, \ell$. In our case, $\ell = 2$.	$m = \boxed{0, \ \pm 1, \ \pm 2}$
Finally, we use Eq. 42.15 to find the possible values of L_z which are dictated by the allowed values of m.	$L_z = m\hbar$ $L_z = \boxed{0, \pm\hbar, \pm 2\hbar}$

CHECK and THINK

Note that both L and L_z are multiples of $\hbar$. The quantity $\hbar$ has units of angular momentum. The fact that the allowed values are not continuous means that they are quantized.

35. (N) If hydrogen is in its $n = 3$ state, what are the possible values of $\left(\mu_{orbit} \right)_z$ in terms of the Bohr magneton?

INTERPRET and ANTICIPATE	
Since we know the value of the principal quantum number, we can determine the maximum allowed orbital quantum number. By knowing this, we can determine the possible values of m when the electron is in the state $n = 3$. Once we know the allowed values of m, we can use Eq. 42.23 to find the possible values of $\left(\mu_{orbit} \right)_z$. The results can be expressed as multiples of the Bohr magneton and they should be quantized.	
SOLVE If in the $n = 3$ state, the maximum allowed value of ℓ is 2.	$\ell_{max} = n - 1 = 3 - 1 = 2$
As discussed in the chapter, the allowed values for m are $-\ell, -\ell+1, \ldots, 0, \ldots, \ell-1, \ell$.	$m = 0, \pm 1, \pm 2$
Now, use Eq. 42.23 to find the possible values of $\left(\mu_{orbit} \right)_z$.	$\left(\mu_{orbit} \right)_z = -m \mu_{Bohr} = \boxed{0, \ \pm \mu_{Bohr}, \ \pm 2 \mu_{Bohr}}$

CHECK and THINK
Our result is quantized. Note that the possible values are directly tied to the value of the magnetic quantum number. Thus, there are the same number of possible orbital magnetic moments as there are possible values of the magnetic quantum number.

36. (N) An electron in a hydrogen atom is in a state with a principle quantum number of 3 and orbital quantum number of 2. What is the number of possible states for the electron? Take into account the possible values of m, and m_s.

INTERPRET and ANTICIPATE	
An electron with an orbital quantum number of 2 is in the d subshell. We can determine the possible values of m, the magnetic quantum number, by using the information provided in the chapter. Each of the allowed values of m defines two possible states of the electron. This is because the spin magnetic quantum number has two possible values, 1/2 and −1/2. Thus, there are two possible states for each value of m.	
SOLVE As discussed in the chapter, the allowed values for m are $-\ell, -\ell+1, \ldots, 0, \ldots, \ell-1, \ell$.	$m = 0, \pm 1, \pm 2$

The total number of possible values of m is 5. If the electron is in anyone of these possible states, there are two possible values of m_s.	$m_s = \pm 1/2$
Because each value of m defines two states, one for each possible value of the spin magnetic quantum number, the total number of possible states is 2 times the number of possible values of m.	$N_{states} = 2(5) = \boxed{10}$

CHECK and THINK

Because the principal quantum number and orbital quantum number were both defined, this limited the number of possible states of the electron. If the problem had only defined the principal quantum number, we would have had to consider each of the possible values of the orbital angular momentum, and the possible values of the magnetic quantum number for each case.

---

42. (N, C) Is it possible for the spin magnetic moment of an electron to cancel its orbital magnetic moment? If so, what value(s) of m make this possible? If not, explain why not.

INTERPRET and ANTICIPATE	

It would be possible for the spin magnetic moment to cancel the orbital magnetic moment if they are equal but opposite. We can test this possibility by setting Eq. 42.23 equal to the opposite of Eq. 42.26. Set the first equation equal to the negative of the second equation and observe any requirements that must be met for this to be possible. If it is possible, we expect to find a relationship involving m and m_s that is possibly true.

SOLVE Set Eq. 42.23 equal to the opposite of Eq. 42.26 and solve for m.	$\left(\mu_{orbit} \right)_z = - \left(\mu_{spin} \right)_z$ $-m\mu_{Bohr} = 2m_s \mu_{Bohr}$ $m = -2m_s$

Thus, the requirement would be that the magnetic quantum number be double the spin magnetic quantum number and opposite in sign. This is possible, because we could have $m = 1$ and $m_s = -1/2$, or $m = -1$ and $m_s = 1/2$.

CHECK and THINK

Here, we found that it is possible for the spin magnetic moment to cancel the orbital magnetic moment if the electron is in a state with either $m = 1$ or $m = -1$. And even then, the electron must have the right spin magnetic quantum number for this to be so, as described in the solution to this problem.

47. (N) What is the ground state electron configuration of iodine (I)?

INTERPRET and ANTICIPATE
Using the periodic table, we see that neutral iodine has 53 electrons. We proceed to fill the appropriate shells and subshells, following the rules and order for filling the subshells. Recall that the $4s$ subshell fills before $3d$ subshell, and that the $5s$ subshell fills before the $4d$ subshell.
SOLVE
$$\boxed{1s^2 2s^2 2p^6 3s^2 3p^6 4s^2 3d^{10} 4p^6 5s^2 4d^{10} 5p^5}$$
CHECK and THINK
Each s subshell can hold two electrons, each p 6 electrons, and each d 10 electrons. The total number of electrons in this case is 53. If we were to ionize the iodine atom by freeing electrons, we would start removing electrons from the outermost subshell.

51. (N) What atom's ground state electron configuration is given by $1s^2 2s^2 2p^6 3s^2 3p^6 4s^2 3d^8$?

INTERPRET and ANTICIPATE
We can figure out the total number of electrons in this ground state configuration by summing the superscripts on each subshell. Then, we can use the periodic table to determine the neutral atom that corresponds to this arrangement.
SOLVE
There are 28 electrons in the atom, found by summing the superscripts in the configuration. $$N_{electrons} = 2 + 2 + 6 + 2 + 6 + 2 + 8 = 28$$ Using the periodic table, the atom must be nickel, $\boxed{\text{Ni}}$.
CHECK and THINK
We assumed that the configuration did not represent an ion, in answering this question. If the atom is in an excited state, we would find at least one electron in a higher subshell, not represented here in this list.

54. (A) Calculate the energy levels for hydrogen's $3d$ subshell when hydrogen is in an external magnetic field B pointing in the z direction. Express your answer in terms of the Bohr magneton.

INTERPRET and ANTICIPATE

When the hydrogen atom is in a magnetic field, the energy degeneracy of the orbitals is removed. We can use Eq. 42.29 to express the new energy of each state, based on the magnetic quantum number. For the $3d$ subshell of hydrogen, we know that $n = 3$, $\ell = 2$, so the allowed values of m are $-2, -1, 0, 1$, and 2.

SOLVE

Use Eq. 42.29 for each possible value of m to find the energy of each state with $n = 3$, and $\ell = 2$.

$$E_{nm} = -\frac{13.6 \text{ eV}}{n^2} + m\mu_{Bohr}B$$

$$E_{3,-2} = -\frac{13.6 \text{ eV}}{3^2} + (-2)\mu_{Bohr}B = \boxed{-1.51 \text{ eV} - 2\mu_{Bohr}B}$$

$$E_{3,-1} = -\frac{13.6 \text{ eV}}{3^2} + (-1)\mu_{Bohr}B = \boxed{-1.51 \text{ eV} - \mu_{Bohr}B}$$

$$E_{3,0} = -\frac{13.6 \text{ eV}}{3^2} + (0)\mu_{Bohr}B = \boxed{-1.51 \text{ eV}}$$

$$E_{3,1} = -\frac{13.6 \text{ eV}}{3^2} + (1)\mu_{Bohr}B = \boxed{-1.51 \text{ eV} + \mu_{Bohr}B}$$

$$E_{3,2} = -\frac{13.6 \text{ eV}}{3^2} + (2)\mu_{Bohr}B = \boxed{-1.51 \text{ eV} + 2\mu_{Bohr}B}$$

CHECK and TIIINK

We can see that the state with $m = 0$ has the energy we would normally associate with an electron in the $n = 3$ state and in any subshell and orbital combination. However, due to the magnetic field, the energy of the electron is different if it is in a particular orbital other than $m = 0$.

57. (N) An electron in a hydrogen atom is in the $n = 5$ shell and $\ell = 4$ subshell. What is the difference between the highest and lowest possible energies of the electron when the atom is exposed to a 4.000 T magnetic field?

INTERPRET and ANTICIPATE

When the hydrogen atom is in a magnetic field, the energy degeneracy of the orbitals is removed. We can use Eq. 42.29 to express the new energy of each state, based on the magnetic quantum number. In this case, with $n = 5$ and $\ell = 4$, the allowed values of m are $-4, -3, -2, -1, 0, 1, 2, 3$ and 4. The highest and lowest possible energies of the electron would be for the states with $m = -4$ and $m = 4$. Once we find the energy for these two states, we can take the difference between the results.

SOLVE Use Eq. 42.29 to compute the energies of the states with $m = -4$ and $m = 4$.	$E_{nm} = -\dfrac{13.6 \text{ eV}}{n^2} + m\mu_{\text{Bohr}}B$ $E_{5,-4} = -\dfrac{13.6 \text{ eV}}{5^2} + (-4)(5.789\times10^{-5} \text{ eV/T})(4.000 \text{ T})$ $E_{5,4} = -\dfrac{13.6 \text{ eV}}{5^2} + (4)(5.789\times10^{-5} \text{ eV/T})(4.000 \text{ T})$

Now, take the difference between the two results.

$$\Delta E = \left| E_{5,4} - E_{5,-4} \right|$$

$$\Delta E = \left| \left[-\frac{13.6 \text{ eV}}{5^2} + (4)(5.789\times10^{-5} \text{ eV/T})(4.000 \text{ T}) \right] - \left[-\frac{13.6 \text{ eV}}{5^2} + (-4)(5.789\times10^{-5} \text{ eV/T})(4.000 \text{ T}) \right] \right|$$

$$\Delta E = (8)(5.789\times10^{-5} \text{ eV/T})(4.000 \text{ T}) = \boxed{1.852\times10^{-3} \text{ eV}}$$

CHECK and THINK

This result gives us some perspective on the spread in energies caused by a strong magnetic field. The difference between the largest and smallest energy values within the subshell is only about 2 meV. That being said, though, this difference is detectable!

62. (A) The magnitude of the orbital angular momentum is $L = \sqrt{30}\hbar$. What are the possible values of m?

INTERPRET and ANTICIPATE

We need to use the given information and Eq. 42.14 to find the orbital angular momentum. Once we know the orbital angular momentum, we can use the information in the chapter to determine the possible values of the magnetic quantum number, m. As discussed in the chapter, the allowed values for m are $-\ell, -\ell+1, \ldots, 0, \ldots, \ell-1, \ell$.

SOLVE Use Eq. 42.14 to find the orbital angular momentum.	$L = \sqrt{\ell(\ell+1)}\hbar = \sqrt{30}\hbar$ $\ell(\ell+1) = 30$ $\ell^2 + \ell - 30 = 0$ $(\ell+6)(\ell-5) = 0$ So either: $\ell+6=0$ or $\ell-5=0$ $\ell = -6$ or $\ell = 5$

The orbital angular momentum cannot be negative, so the result is 5.	$\ell = 5$
The allowed values for m are $-\ell, -\ell+1, \ldots, 0, \ldots, \ell-1, \ell$.	$m = \boxed{0, \pm1, \pm2, \pm3, \pm4, \pm5}$

CHECK and THINK

There are eleven possible values of m given that the orbital quantum number is 5. This means there are lots of states that have the value of the angular momentum given. Each electron in a $\ell = 5$ subshell would have this angular momentum.

65. (N) An electron in a hydrogen atom makes a transition from the ground state to the $n = 4$ state after absorbing a photon

a. What is the minimum energy of the photon?

b. What is the frequency of the photon with this minimum energy?

INTERPRET and ANTICIPATE

To model this transition, we must use Eq. 42.10 to find the bound state energy of the ground state and the $n = 4$ excited state. We can then find the magnitude of the difference between these two energies to find the minimum energy a photon must have to make this transition possible. Once we know the energy, we can use Eq. 40.17 to find the frequency of this photon. The energy should be on the order of 1 or 10 eV, as this is representative of the energy levels in hydrogen.

SOLVE **a.** First, use Eq. 42.10 to express the ground state energy and the energy of the $n = 4$ excited state.	$E_1 = \left(-13.6 \text{ eV}\right)/\left(1\right)^2$ $E_4 = \left(-13.6 \text{ eV}\right)/\left(4\right)^2$
Now, we find the magnitude of the difference between these two energies. This is the minimum energy a photon must have to make this transition possible.	$\Delta E = \left| E_4 - E_1 \right| = \left| \left(-13.6 \text{ eV}\right)/\left(4\right)^2 - \left(-13.6 \text{ eV}\right)/\left(1\right)^2 \right|$ $\Delta E = \boxed{12.8 \text{ eV}}$

Chapter 42 – Atoms

b. Use Eq. 40.17 to find the frequency of this photon.	$f = \Delta E / h$ $f = \left\| (-13.6 \text{ eV}) / (4)^2 - (-13.6 \text{ eV}) / (1)^2 \right\| / (4.136 \times 10^{-15} \text{ eV} \cdot \text{s})$ $f = \boxed{3.08 \times 10^{15} \text{ Hz}}$

CHECK and THINK

Note that the transition is still possible, even if the photon has an energy slightly larger than the 12.8 eV found here, as long as the energy is also less than the difference between the ground state and the $n = 5$ excited state. In that case, the small amount of excess energy would be released as low-energy, low-frequency radiation.

70. **(N)** An imaginary atom has only one electron, and its energy levels are given by $E_n = (-19.5 \text{ eV}) / n^2$ where $n = 1, 2, 3 \ldots$. If the atom goes from the $n = 3$ state to the $n = 1$ state, what is the frequency of the emitted photon?

INTERPRET and ANTICIPATE

In order to answer this question, we must use the given model for the energy levels of this atom to find the energy of the $n = 3$ and the $n = 1$ state. When the electron decays, it will emit a photon equal to the magnitude of the energy difference between these two levels. Once we know the energy of the photon, we can use Eq. 40.17 to find the frequency of the photon.

SOLVE First, use the model to find the energy of the $n = 3$ and the $n = 1$ state.	$E_1 = (-19.5 \text{ eV}) / 1^2$ $E_3 = (-19.5 \text{ eV}) / 3^2$
Now, we express the magnitude of the difference in energy between these two levels. This is the energy of the emitted photon.	$\Delta E = \left\| E_1 - E_3 \right\| = \left\| (-19.5 \text{ eV}) / 1^2 - (-19.5 \text{ eV}) / 3^2 \right\|$
Finally, use Eq. 40.17 to find the frequency of this photon.	$f = E / h$ $f = \left\| (-19.5 \text{ eV}) / 1^2 - (-19.5 \text{ eV}) / 3^2 \right\| / (4.136 \times 10^{-15} \text{ J} \cdot \text{s})$ $f = \boxed{4.19 \times 10^{15} \text{ Hz}}$

CHECK and THINK

When an electron decays directly from a particular higher energy state to a particular lower energy state, the photon emitted will always have the same energy. Because each atom will have a distinct set of energy levels, the energies of emitted photons act like a fingerprint for the atom.

73. Consider an electron in a hydrogen atom in the $n = 2$, $\ell = 1$, $m = 1$ excited state.

a. (N) What magnitude of an applied magnetic field would cause the energy of this electron to be equal to zero?

b. (C) Given that the energy of this electron would be zero under these conditions (if the electron does not undergo a transition to another state), what would be true about the electron?

INTERPRET and ANTICIPATE

Given the particular state the electron occupies, we can use Eq. 42.29 for the case when the, $n = 2$, and $m = 1$, while setting the energy, $E_{nm} = 0$. We expect the field strength to be rather large. After all, what is being described here, is an applied magnetic field that is so strong it is causing the bound state energy to change from its typical value,

$E_2 = (-13.6 \text{ eV})/2^2 = -3.4 \text{ eV}$, to 0.

SOLVE

a. Set up Eq. 42.29, and set the energy equal to 0 to solve for the magnetic field strength.

$$E_{n,m} = -\frac{13.6 \text{ eV}}{n^2} + m\mu_{\text{Bohr}}B$$

$$0 = -\frac{13.6 \text{ eV}}{2^2} + (1)\mu_{\text{Bohr}}B$$

$$B = \frac{13.6 \text{ eV}}{2^2 \mu_{\text{Bohr}}} = \frac{13.6 \text{ eV}}{4(5.789 \times 10^{-5} \text{ eV/T})} = \boxed{5.87 \times 10^4 \text{ T}}$$

---

b. If the state has a bound state energy of 0, then the electron would no longer be bound to the atom! It has been freed.

CHECK and THINK

What we have seen here is that some of the orbitals of our normal hierarchy of bound states can vanish when the atom is in the presence of a strong magnetic field. By "vanish" what we mean is that those particular orbitals are no longer bound states, because if the electron were to try to occupy that state, it would have enough energy to be freed under these conditions.

---

76. (A) Suppose an electron in a hydrogen atom transitions from its current state, n, to the state, $n/2$. Use Eq. 42.2 to express the wavelength of the resulting emitted photon in terms of R, and n.

INTERPRET and ANTICIPATE

As the questions requests, we will set up Eq. 42.2 with the initial state equal to n and the final state equal to $n/2$, and solve for the wavelength of the emitted photon. The result should be in terms of the Rydberg constant, R and the initial state, n.

Chapter 42 – Atoms

SOLVE Express Eq. 42.2 with the appropriate states and solve for the wavelength.	$\dfrac{1}{\lambda} = R\left(\dfrac{1}{m^2} - \dfrac{1}{n^2}\right)$
	$\dfrac{1}{\lambda} = R\left(\dfrac{1}{(n/2)^2} - \dfrac{1}{n^2}\right) = R\left(\dfrac{4}{n^2} - \dfrac{1}{n^2}\right) = R\left(\dfrac{3}{n^2}\right)$
	$\lambda = \boxed{\dfrac{n^2}{3R}}$

CHECK and THINK

This is a special case where the electron decays from its current state to a state with a principal quantum number equal to half the initial value. Note that this means the initial state must have an even principal quantum number (i.e., 2, 4, 6, …).

43

Nuclear and Particle Physics

4. (N) The element nitrogen (N) has two stable isotopes—one with an atomic mass number of 14 and the other with an atomic mass number of 15.
a. How many neutrons are in each of these nuclei?
b. If the natural abundance of ^{15}N is 0.37%, what is the natural abundance of ^{14}N?

INTERPRET and ANTICIPATE

We can find the number of neutrons in each isotope by using the periodic table to find the atomic number (number of protons) of nitrogen. This can be subtracted from each of the atomic mass numbers to find the number of neutrons. Then, the sum of the natural abundances of the isotopes of an atom must equal 100%, so we can subtract the natural abundance of ^{15}N from 100% to find the natural abundance of ^{14}N.

SOLVE	
a. Using the periodic table, the atomic number of nitrogen is 7. Subtract this from the atomic mass number of each isotope to find the number of neutrons in each case.	$N = A - Z$ $N_{14} = 14 - 7 = \boxed{7}$ $N_{15} = 15 - 7 = \boxed{8}$
b. Subtract the 0.37% from 100% to find the natural abundance of ^{14}N.	natural abundance $^{14}N = 100\% - 0.37\% = \boxed{99.63\%}$

CHECK and THINK

If there were more isotopes to consider for an element, then the sum of all the natural abundances would still need to be 100%. Often, one isotope of an element is far more likely to be found than any of the other isotopes of that element.

---

7. (N) Find the nuclear radius of ^{40}Ca in fermis.

INTERPRET and ANTICIPATE

The atomic mass number of calcium is 40. We can use Eq. 43.2 to find the nuclear radius. We expect that the radius will be a few fermis.

SOLVE	
Use Eq. 43.2 to find the radius. Then convert the result to the unit of fermis (1 fermi $= 1 \times 10^{-15}$ m).	$R = r_0 A^{1/3} = \left(1.2 \times 10^{-15} \text{ m}\right)(40)^{1/3}$ $R = 4.1 \times 10^{-15} \text{ m} = \boxed{4.1 \text{ fm}}$

CHECK and THINK
Most nuclei are similarly-sized – on the order of a few fermis. The size of the nucleus increases with the addition of more nucleons.

8. (N, C) Consider two protons that are separated by 1.5 fm. What is the magnitude of the Coulomb repulsive force between them? Assume that these protons are one another's nearest neighbors, and that the strong force between them is about 2000 N. Will these protons hold together or fly apart?

INTERPRET and ANTICIPATE
We can use Eq. 23.3 to find the Coulomb repulsion between the two protons. In considering whether or not these protons will hold together or fly apart, we can compare the magnitude of the repulsive force to that of the strong force between the protons.

SOLVE	
Use Eq. 23.3 to find the magnitude of the Coulomb repulsion between the protons.	$F_E = \dfrac{k\lvert q_1 q_2 \rvert}{r^2} = \dfrac{\left(8.99 \times 10^9 \text{ N} \cdot \text{m}^2/\text{C}^2\right)\left(1.60 \times 10^{-19} \text{ C}\right)^2}{\left(1.5 \times 10^{-15} \text{ m}\right)^2}$ $F_E = \boxed{1.0 \times 10^2 \text{ N}}$

The strong nuclear force, being 2000 N, is much greater than this repulsive force, so the protons $\boxed{\text{will not fly apart}}$.

CHECK and THINK
Here, we only considered the Coulomb repulsion from one proton. In a nucleus with many protons, each one would exert a repulsive force on the subject proton, and the nucleus will not hold together if the net repulsive force is greater than the attractive strong force between the proton and its nearest neighbors.

12. (G, C) Fill in an energy level diagram (Fig. P43.12) for $^{16}_{8}\text{O}$. Based on your diagram, do you expect $^{16}_{8}\text{O}$ to be stable? Explain.

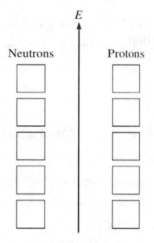

Figure P43.12

INTERPRET and ANTICIPATE

By subtracting the atomic number from the atomic mass number of this isotope, we find the number of neutrons is also 8 – the same as the number of protons. We then fill in the energy level diagram by obeying the Pauli exclusion principle (2 nucleons in each orbital – one spin up and one spin down, as indicated by arrows)

SOLVE

The number of protons is equal to the number of neutrons (8) and so both sides of the energy level diagram will appear identical.

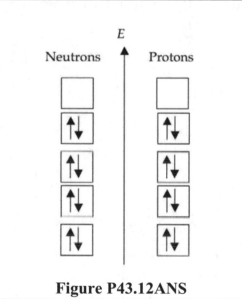

Figure P43.12ANS

CHECK and THINK

This isotope is called a *doubly magic* isotope since both the number of protons and the number of neutrons completely fill a shell in the energy model of the nucleus. In this case, the $1s$, $2s$, and $2p$ subshells are filled, which means the first 2 shells are full.

18. (N) The half-life of ^{14}C is 5730 yr, and a constant ratio of ^{14}C$/^{12}$C $= 1.3 \times 10^{-12}$ is maintained in all living tissues. A fossil is found to have ^{14}C$/^{12}$C $= 2.65 \times 10^{-13}$. How old is the fossil?

INTERPRET and ANTICIPATE

First, we must recognize that the ratio ^{14}C$/^{12}$C will decrease over time. Thus, it behaves very similar to the number of ^{14}C isotopes, which decrease over time. It is just that in this case, their number is referenced to the number of ^{12}C isotopes. This means we can use Eq. 43.8, but replace the number of nuclei with the ratio given here. We can set up this equation and solve for the time that must have passed, or the age of the fossil. It looks like the ratio is about $1/4^{\text{th}}$ of the initial value from the living tissue, so the time that must have passed is about 2 half-lives or roughly 12000 yr.

SOLVE	
Setup Eq. 43.8 for the ratio, as discussed above, and solve for the age of the fossil. Note that in this solution, we do make use of some rules for log functions in solving for t.	$$\left[\frac{N\left(^{14}\text{C}\right)}{N\left(^{12}\text{C}\right)} \right] = \left[\frac{N\left(^{14}\text{C}\right)}{N\left(^{12}\text{C}\right)} \right]_0 \left(\frac{1}{2} \right)^{t/T_{1/2}}$$ $$2.65 \times 10^{-13} = 1.3 \times 10^{-12} \left(\frac{1}{2} \right)^{t/5730 \text{ yr}}$$ $$\left(\frac{1}{2} \right)^{t/5730 \text{ yr}} = \frac{2.65 \times 10^{-13}}{1.3 \times 10^{-12}}$$ $$\ln \left[\left(\frac{1}{2} \right)^{t/5730 \text{ yr}} \right] = \ln \left[\frac{2.65 \times 10^{-13}}{1.3 \times 10^{-12}} \right]$$ $$\left(t/5730 \text{ yr} \right)\left(-\ln(2) \right) = \ln \left[\frac{2.65 \times 10^{-13}}{1.3 \times 10^{-12}} \right]$$ $$t = -\left(5730 \text{ yr} \right) \ln \left[\frac{2.65 \times 10^{-13}}{1.3 \times 10^{-12}} \right] / \ln(2) = \boxed{1.3 \times 10^4 \text{ yr}}$$

CHECK and THINK

As expected, the number of years that have passed is about 12000 (it turned out to be a little more). The half-life of an isotope is how much time must pass for the initial population to be halved in number.

21. (N) The number of radioactive isotopes in a sample is found to drop to 22.50% of its original value in 924.5 s. What is the half-life of the isotope?

INTERPRET and ANTICIPATE

We must first use Eq. 43.5 to find the decay constant. Note that the ratio of the number of nuclei currently to the initial number of nuclei must be 0.2250 because 22.50% of the original number remain in the sample. Once we have the decay constant, we can use Eq. 43.7 to find the half-life.

SOLVE Write Eq. 43.5 and find an expression for the decay constant.	$N = N_0 e^{-\eta t}$ $N/N_0 = e^{-\eta t}$ $0.2250 = e^{-\eta(924.5\ \text{s})}$ $\ln(0.2250) = -\eta(924.5\ \text{s})$ $\eta = -\dfrac{\ln(0.2250)}{(924.5\ \text{s})}$
Now, use Eq. 43.7 to find the half-life.	$T_{1/2} = \dfrac{\ln 2}{\eta} = \dfrac{\ln 2}{-\ln(0.2250)/(924.5\ \text{s})} = \boxed{429.6\ \text{s}}$

CHECK and THINK

Note that we also could have used Eq. 43.8 to find the half-life. Because there is about 25% of the original amount of isotopes remaining, this means that the sample has gone through about 2 half-lives. Thus, we could have guessed that the half-life was about half the time that had passed in the problem statement.

28. (A) Show that the half-life is given by Equation 43.7, $T_{1/2} = (\ln 2)/\eta$.

INTERPRET and ANTICIPATE

We are looking to find the time when the number of nuclei is half the original amount. Use Eq. 43.5 under these conditions to find the time, or the half-life, in terms of the time constant. Note that the ratio of the number of nuclei currently to the initial number of nuclei must be ½ because 50% of the original number remain in the sample after one half-life has passed.

SOLVE Write Eq. 43.5.	$N = N_0 e^{-\eta t}$
Now, divide both sides by N_0 and note that the ratio is equal to ½. Also, we note here that the time we are looking for is the half-life since the original number of isotopes is halved.	$N/N_0 = e^{-\eta t}$ $1/2 = e^{-\eta T_{1/2}}$

Take the natural log of both sides of the equation and solve for the half-life.	$\ln(1/2) = \ln\left(e^{-\eta T_{1/2}}\right)$ $\ln(1/2) = -\eta T_{1/2}$ $\ln(2^{-1}) = -\eta T_{1/2}$ $-\ln(2) = -\eta T_{1/2}$ $T_{1/2} = \dfrac{\ln(2)}{\eta}$

CHECK and THINK

We have found Eq. 43.7 by applying Eq. 43.5 at the particular moment in time when half the nuclei have decayed. Imagine beginning this process again, but after half the original number of isotopes have decayed. Would the half-life turn out to be any different?

30. (E) CASE STUDY Before free neutrons in the early Universe had much time to beta decay, the ratio of neutrons to protons was $N_n/N_p \approx 1/5$. At this point, what was the abundance of helium $N_{He}/(N_H + N_{He})$? Express your answer as a percentage.

INTERPRET and ANTICIPATE

In helium, the number of protons is equal to the number of neutrons, while in hydrogen, there is only one proton. We begin by assuming all of the neutrons are part of the helium nuclei. Then, an equal number of protons are part of the helium nuclei. The number of helium nuclei that can form is equal to half the number of neutrons. This is assumed because the most common (and stable) isotope of helium contains two neutrons.

We assume the remaining protons make up the hydrogen nuclei (subtract the number of neutrons from the number of protons). At this point, we can move to expressing the number of helium and hydrogen nuclei in terms of the number of protons and neutrons. The goal is to create an expression that depends on the ratio of neutrons to protons because we have a value for that quantity. In the end, we expect a fraction, or percentage less than 100% as a final answer.

SOLVE

Let's begin with the ratio we are asked to resolve by incorporating the idea that the number of helium nuclei must be half the number of neutrons.	$\dfrac{N_{He}}{N_{He} + N_H} = \dfrac{\frac{1}{2}N_n}{\frac{1}{2}N_n + N_H}$

Now, as we stated, since there must be the same number of protons in the helium as neutrons, the number of hydrogen nuclei is the number of protons left over, or the original number of protons minus the number of neutrons.	$$\frac{N_{He}}{N_{He} + N_H} = \frac{\frac{1}{2}N_n}{\frac{1}{2}N_n + \left(N_p - N_n\right)}$$
We now use the ratio given in the problem statement by solving for the number of neutrons and substituting this into our expression. We can then factor out and cancel the number of protons from the relationship and solve.	$N_n \approx \frac{1}{5}N_p$ $$\frac{N_{He}}{N_{He} + N_H} = \frac{\frac{1}{2}\left(\frac{1}{5}N_p\right)}{\frac{1}{2}\left(\frac{1}{5}N_p\right) + \left(N_p - \left(\frac{1}{5}N_p\right)\right)}$$ $$\frac{N_{He}}{N_{He} + N_H} = \frac{\frac{1}{2}\left(\frac{1}{5}\right)}{\frac{1}{2}\left(\frac{1}{5}\right) + \left(1 - \frac{1}{5}\right)}$$ $$\frac{N_{He}}{N_{He} + N_H} = \frac{\frac{1}{10}}{\frac{1}{10} + \frac{4}{5}} = \frac{\frac{1}{10}}{\frac{1}{10} + \frac{8}{10}} = \frac{\frac{1}{10}}{\frac{9}{10}}$$ $$\frac{N_{He}}{N_{He} + N_H} = \frac{1}{9} = \boxed{11\%}$$

CHECK and THINK

This accounting is possible due to the few possible states of matter in the early universe. It is quite amazing that the myriad elements of the periodic table emerge from these initial building blocks, and allow for the more complicated structures we find in nature today.

33. (C) Use Equation 43.13 to express the electron capture process of each of the following parent isotopes: **a.** $^{188}_{77}\text{Ir}$, **b.** $^{127}_{54}\text{Xe}$ **c.** $^{57}_{27}\text{Co}$.

INTERPRET and ANTICIPATE

In each case, the daughter nucleus should have an atomic mass number that is the same as the parent and an atomic number that is less than the parent atomic number by 1. Use the periodic table to aide in identifying the nuclei, and then use Eq. 43.13 to express the process.

SOLVE	
a. Use Eq. 43.13. The daughter is $^{188}_{76}\text{Os}$.	$^{188}_{77}\text{Ir} + \text{e}^- \rightarrow {}^{188}_{76}\text{Os} + \nu$

b. Use Eq. 43.13. The daughter is $^{127}_{53}\text{I}$.	
	$\boxed{^{127}_{54}\text{Xe} + e^- \rightarrow\ ^{127}_{53}\text{I} + v}$
c. Use Eq. 43.13. The daughter is $^{57}_{26}\text{Fe}$.	
	$\boxed{^{57}_{27}\text{Co} + e^- \rightarrow\ ^{57}_{26}\text{Fe} + v}$

CHECK and THINK

The electron capture process is a very specific process involving the absorption of an electron and the emission of a neutrino. The neutrino is produced by the proton decay (part of this process) that occurs in the nucleus of the parent isotope.

---

37. (C) Oxygen undergoes the following reaction: $^{19}_{?}\text{O} \rightarrow\ ^{19}_{?}\text{F} + ? + ?$. Rewrite this reaction, filling in all the missing information. What sort of reaction is this?

INTERPRET and ANTICIPATE

When we compare this reaction to the various types of reactions we saw in this chapter (beta decay, inverse beta decay, electron capture, …) we can determine that this is a $\boxed{\text{beta decay}}$ reaction. We know this because fluorine has a greater atomic number than oxygen and the atomic mass number of the isotope on each side of the reaction is the same (19). This means that a neutron has decayed to a proton, which involves a beat decay process.

The reaction can be completed by using Eq. 43.10 and the periodic table to fill in the missing information.

SOLVE	
Use Eq. 43.10 to complete the reaction. The atomic number of oxygen is 8 and the atomic number of fluorine is 9.	$\boxed{^{19}_{8}\text{O} \rightarrow\ ^{19}_{9}\text{F} + e^- + \bar{v}}$

CHECK and THINK

A neutron decay is involved in the beta decay process. The daughter nucleus will always have an atomic number that is 1 greater than the parent's atomic number in this process.

---

42. (N) According to Figure 43.10, uranium-238 alpha decays into thorium-234. Calculate the energy released in this alpha decay. The rest masses of $^{238}_{92}\text{U}$, $^{234}_{90}\text{Th}$ and $^{4}_{2}\text{He}$ are 238.05079 u, 234.04363 u and 4.00260 u respectively.

INTERPRET and ANTICIPATE

We can compute the difference in mass between the products and reactants in this process, which we will also call the mass deficit (though this is different than the mass deficit of a particular isotope, Eq. 43.15). Once we know this change in mass, we can use Eq. 43.18 to express the energy that must have been released (that is no longer part of the rest-mass energy!). The rest masses are given in atomic mass units, which we will convert to MeV/c^2.

SOLVE	
First, we compute the mass deficit and convert the result to MeV/c^2.	$\Delta M = m_U - m_{Th} - m_{He}$ $\Delta M = \left(238.05079\ u - 234.04363\ u - 4.00260\ u\right)$ $\Delta M = 0.00456\ u$ $\Delta M = \left(0.00456\ u\right)\left(\dfrac{931.494\ MeV/c^2}{1\ u}\right)$
Now, use Eq. 43.18 to find the energy released.	$E_{released} = \Delta Mc^2 = \left(0.00456\ u\right)\left(\dfrac{931.494\ MeV/c^2}{1\ u}\right)c^2$ $E_{released} = \boxed{4.25\ MeV}$

CHECK and THINK

We assume we can ignore the kinetic energies of the isotopes because the energy would be so little compared to the difference in rest-mass energy between the products and the reactants. Thus, we have really solved a conservation of energy problem for this reaction, where most of the energy is still rest-mass energy after the reaction, and the difference is released as radiation (EM waves).

43. (N) Use the binding energy curve (Fig. 43.11) to find the binding energy per nucleon for ^4_2He. Then find the binding energy for the whole nucleus.

INTERPRET and ANTICIPATE

We can use the binding energy curve to find the value, ε, for ^4_2He. The binding energy for the whole nucleus can then be found by using $\varepsilon = E_B/A$, where we recognize that the atomic mass number of our nucleus is 4.

SOLVE

According to the binding energy curve, the binding energy per nucleon is

$\varepsilon \sim \boxed{7\ MeV\ per\ nucleon}$.

Now, we can use the atomic mass number, 4, to find the binding energy with the relationship, $\varepsilon = E_B/A$.	$\varepsilon = E_B/A$ $E_B = \varepsilon A \sim (7 \text{ MeV})(4)$ $E_B \sim \boxed{28 \text{ MeV}}$

CHECK and THINK

Note that the binding energy per nucleon curve has a peak. Fusion processes (with no extra energy required) allow for the lighter elements to create heavier elements up to Fe, while heavier elements create lighter elements via fission (without extra energy required) down to Fe.

48. (N) Consider the fission reaction from the previous problem (Problem 43.47), where $^{239}_{94}\text{Pu}$ (239.05216 u) undergoes fission into $^{148}_{58}\text{Ce}$ (147.9242 u) and $^{A}_{Z}\text{X}$ (88.91764 u). Compute the binding energy of **a.** $^{239}_{94}\text{Pu}$, **b.** $^{148}_{58}\text{Ce}$, and **c.** $^{A}_{Z}\text{X}$.

INTERPRET and ANTICIPATE

The method of solution for each part will be the same in this problem. We must first express the mass deficit of the nucleus, using Eq. 43.15, where we consider the number of protons and neutrons. Note that the number of neutrons in the difference between the atomic mass number and the atomic number for each isotope. Once we have the mass deficit, we can use Eq. 43.16 to find the binding energy of the nucleus. As the atomic number decreases, we expect the binding energy to decrease.

SOLVE

a. Use Eq. 43.15 to express the mass deficit of $^{239}_{94}\text{Pu}$. The number of neutrons is $145 = 239 - 94$.

$$\Delta m = \sum m_i - M = N_p m_p + N_n m_n - M$$

$$\Delta m = (94)(938.27 \text{ MeV}/c^2) + (145)(939.57 \text{ MeV}/c^2) - (239.05216 \text{ u})\left(\frac{931.494 \text{ MeV}/c^2}{1 \text{ u}}\right)$$

Now, use Eq. 43.16 to find the binding energy.

$$E_B = \Delta m c^2$$

$$E_B = \left[(94)(938.27 \text{ MeV}/c^2) + (145)(939.57 \text{ MeV}/c^2) - (239.05216 \text{ u})\left(\frac{931.494 \text{ MeV}/c^2}{1 \text{ u}}\right)\right]c^2$$

$$E_B = \boxed{1759.4 \text{ MeV}}$$

b. Use Eq. 43.15 to express the mass deficit of $^{148}_{58}\text{Ce}$. The number of neutrons is $90 = 148 - 58$.

$$\Delta m = \sum m_i - M = N_p m_p + N_n m_n - M$$

$$\Delta m = (58)(938.27 \text{ MeV}/c^2) + (90)(939.57 \text{ MeV}/c^2) - (147.9242 \text{ u})\left(\frac{931.494 \text{ MeV}/c^2}{1 \text{ u}}\right)$$

Now, use Eq. 43.16 to find the binding energy.

$$E_B = \Delta mc^2$$

$$E_B = \left[(58)(938.27 \text{ MeV}/c^2) + (90)(939.57 \text{ MeV}/c^2) - (147.9242 \text{ u})\left(\frac{931.494 \text{ MeV}/c^2}{1 \text{ u}}\right)\right]c^2$$

$$E_B = \boxed{1190.5 \text{ MeV}}$$

c. In order to solve this part, we must actually solve P43.47 first! In looking at that problem, we are to complete the given reaction to find the unknown isotope. Observe that the total number of protons and neutrons must be the same on both sides of the reaction. We can then use the periodic table to identify the missing nucleus.

$$^{239}_{94}\text{Pu} + ^{1}_{0}\text{n} \rightarrow ^{148}_{58}\text{Ce} + 3\left(^{1}_{0}\text{n}\right) + ^{89}_{36}\text{Kr}$$

This means the unknown isotope is $^{89}_{36}\text{Kr}$.

Use Eq. 43.15 to express the mass deficit of $^{89}_{36}\text{Kr}$. The number of neutrons is $53 = 89 - 36$.

$$\Delta m = \sum m_i - M = N_p m_p + N_n m_n - M$$

$$\Delta m = (36)(938.27 \text{ MeV}/c^2) + (53)(939.57 \text{ MeV}/c^2) - (88.91764 \text{ u})\left(\frac{931.494 \text{ MeV}/c^2}{1 \text{ u}}\right)$$

Now, use Eq. 43.16 to find the binding energy.

$$E_B = \Delta mc^2$$

$$E_B = \left[(36)(938.27 \text{ MeV}/c^2) + (53)(939.57 \text{ MeV}/c^2) - (88.91764 \text{ u})\left(\frac{931.494 \text{ MeV}/c^2}{1 \text{ u}}\right)\right]c^2$$

$$E_B = \boxed{748.68 \text{ MeV}}$$

> **CHECK and THINK**
> While the binding energy per nucleon actually increases as we climb up the curve shown in Fig 43.11, the actual binding energy of these nuclei decreases.

53. (N) The mass of tritium is 2808.9261 MeV/c^2. Use this and Table 43.2 to verify that the energy released in the fusion reaction depicted in Eq. 43.21, $^2_1H + ^3_1H \rightarrow ^4_2He + ^1_0n$ is 17.59 MeV.

> **INTERPRET and ANTICIPATE**
> Here, we want to verify the energy released (17.59 MeV). We can compute the difference in mass between the products and reactants in this process, which we will also call the mass deficit (though this is different than the mass deficit of a particular isotope, Eq. 43.15). Once we know this change in mass, we can use Eq. 43.18 to express the energy that must have been released (that is no longer part of the rest-mass energy!).

> **SOLVE**
> We use the mass of each product and reactant to express the difference in mass for the reaction.
>
> $$\Delta M = 2808.9261 \text{ MeV}/c^2 + 1875.612859 \text{ MeV}/c^2 - 3727.379 \text{ MeV}/c^2 - 939.57 \text{ MeV}/c^2$$
>
> Now, use Eq. 43.18 to find the energy released.
>
> $$E_{released} = \Delta Mc^2$$
> $$E_{released} = \left[2808.9261 \text{ MeV}/c^2 + 1875.612859 \text{ MeV}/c^2 - 3727.379 \text{ MeV}/c^2 - 939.57 \text{ MeV}/c^2 \right] c^2$$
> $$E_{released} = \boxed{17.59 \text{ MeV}}$$

> **CHECK and THINK**
> This energy from the fusion of hydrogen may not seem like much, but when you consider that there are about 10^{38} of these fusion events per second in the Sun, the total energy released each second is quite substantial in that instance. How many reactions would we need to produce per second to be comparable to the energy output of a typical power plant?

57. (N) During a chest X-ray a patient is exposed to a dose equivalent of 0.3 mSv. The X-ray has an energy of 25 keV. If the portion of the chest that absorbed the radiation has a mass of 2.5 kg, how many X-ray photons were absorbed?

> **INTERPRET and ANTICIPATE**
> To determine the number of X-ray photons absorbed, we need to first find the total energy absorbed. This is because we know how much energy is carried by each X-ray

photon ($25 \text{ keV} = 25 \times 10^3 \text{ eV}$). We can divide the total energy absorbed by the energy of each photon to find their number. In order to find the total energy absorbed, we can use Eq. 43.22, after we have used Eq. 43.23 to express the dose, D. For X-rays, the quality factor, Q, should be equal to 1.0 Sv/Gy.

SOLVE First, use Eq. 43.23 to express the dose. For X-rays, the quality factor is 1.0 Sv/Gy.	$D = H/Q = (0.3 \times 10^{-3} \text{ Sv})/(1.0 \text{ Sv/Gy})$ $D = 0.3 \times 10^{-3} \text{ Gy}$
Then, the total energy absorbed can be found using Eq. 43.22.	$D = \dfrac{E}{m}$ $E = mD = (2.5 \text{ kg})(0.3 \times 10^{-3} \text{ Gy})$ $E = 7.5 \times 10^{-4} \text{ J}$ $E = (7.5 \times 10^{-4} \text{ J})(1 \text{ eV}/1.60 \times 10^{-19} \text{ J})$
Now, divide the total energy by the energy of each photon to find the number of X-ray photons.	$N = E/E_{\text{X-ray}}$ $N = (7.5 \times 10^{-4} \text{ J})(1 \text{ eV}/1.60 \times 10^{-19} \text{ J})/(25 \times 10^3 \text{ eV})$ $N = \boxed{1.9 \times 10^{11} \text{ photons}}$

CHECK and THINK

It takes a good number of X-ray photons to produce a single X-ray image! While occasional exposure at this level may not be harmful, it is best to avoid continued exposure, and exposure to sensitive organs.

---

59. (N) While walking outside of New Vegas, you pick up a piece of metal and are exposed to alpha radiation with a quality factor of 20. The dose absorbed by your body is 30.0 rad.

a. What was the absorbed dose expressed in the units of grays (Gy)?

b. What was the dose equivalent expressed in units of sieverts (Sv)?

c. What was the dose equivalent expressed in units of rem?

INTERPRET and ANTICIPATE

In this problem, we convert the absorbed dose into several other unit expressions. In the first case, the absorbed dose is expressed in grays, while the other two parts of the problem are expressions of dose equivalent, which requires a quality factor, Q. In the problem statement, we are told the quality factor of this radiation is 20. Use proper unit conversions and Eq. 43.23 to solve the problem.

<table>
<tr><td colspan="2">SOLVE</td></tr>
<tr>
<td>a. Convert the absorbed dose to Gy.</td>
<td>$$D = 30.0 \text{ rad} \times \frac{0.01 \text{ Gy}}{1 \text{ rad}} = \boxed{0.300 \text{ Gy}}$$</td>
</tr>
<tr>
<td>b. Use Eq. 43.23 to find the dose equivalent, where $Q = 20$.</td>
<td>$$H = QD = (20 \text{ Sv/Gy})(0.300 \text{ Gy}) = \boxed{6.00 \text{ Sv}}$$</td>
</tr>
<tr>
<td>c. Convert the dose equivalent to rem.</td>
<td>$$H = 6.00 \text{ Sv} \times \frac{1 \text{ rem}}{0.01 \text{ Sv}} = \boxed{6.00 \times 10^2 \text{ rem}}$$</td>
</tr>
</table>

CHECK and THINK

The absorbed dose does not consider the type of radiation that is being absorbed. It is fundamentally related to the amount of energy absorbed. The dose equivalent does depend on the type of radiation and, thus, the effect it might have on the human body.

64. (C) a. Prior to the discovery of the electron antineutrino, its existence was predicted because of the varied amount of kinetic energy of the electron resulting from neutron decay. Given what you have learned about lepton number in this chapter, why is the following reaction dissatisfying: ${}_0^1 n \rightarrow {}_1^1 p + e$? **b.** Include an electron antineutrino in this reaction such that the issue you identified in part (a) is resolved. Write the resulting reaction.

INTERPRET and ANTICIPATE

a. The reaction is not satisfying because the lepton number is not conserved. Neither the proton or the neutron are leptons, but the electron is! A lepton particle has a lepton number of 1 and a lepton antiparticle has a lepton number of -1. There must be a lepton that is an antiparticle appearing as a product, so that the total lepton number of the products is 0 (the same lepton number of the reactants).

SOLVE

b. We include an antineutrino in the products and write the resulting reaction, where the lepton number is now conserved.

$$\boxed{{}_0^1 n \rightarrow {}_1^1 p + e + \bar{\nu}}$$

CHECK and THINK

There are 6 lepton particles and 6 corresponding lepton antiparticles. The total lepton number of the reactants must be the same as that of the products.

66. (N) A bone found in a crypt is being dated. If 25% of the original amount of carbon-14 has decayed, what is the age of the bone? The half-life of ^{14}C is 5730 yr.

INTERPRET and ANTICIPATE	
We can use Eq. 43.7 to express the decay constant in terms of the half-life, and then use Eq. 43.5 to find the time that has passed. Note that if 25% of the original isotope has decayed, then 75% of the isotope remains. This means that the ratio $N/N_0 = 0.75$. Since only $1/4^{th}$ of the original number of nuclei have decayed, we expect that about 2365 yr have passed, or half the half-life.	
SOLVE Substitute Eq. 43.7 into Eq. 43.5.	$N = N_0 e^{-\eta t} = N_0 e^{-t\ln(2)/T_{1/2}}$
Divide both sides by the initial number of nuclei and solve using the information in the problem statement.	$N/N_0 = e^{-t\ln(2)/T_{1/2}}$ $0.75 = e^{-t\ln(2)/5730 \text{ yr}}$ $\ln(0.75) = \ln\left(e^{-t\ln(2)/5730 \text{ yr}}\right)$ $\ln(0.75) = -t\ln(2)/5730 \text{ yr}$ $t = -(5730 \text{ yr})\ln(0.75)/\ln(2) = \boxed{2378 \text{ yr}}$
CHECK and THINK	
Every 5730 yr, the number of remaining carbon-14 is halved. Eventually, the number of remaining nuclei is so few that we cannot continue to use carbon-14 for dating purposes reliably. It is a great technique for dating fossils extending back about 50,000 yr, but can be difficult to use beyond that point. There are other dating techniques, however!	

73. (N) A sample material is observed to have an activity of 1.45×10^9 decays/s. Express the activity in **a.** Bq, and **b.** Ci. **c.** If the material has a half-life of 30.07 yr, how long will it take for the activity to decrease by 10.0%? **d.** How long will it take for the activity to decrease by 20.0%?

INTERPRET and ANTICIPATE	
In the first two parts of this problem, we need only convert the given activity to the appropriate unit expressions. For parts (c) and (d), we can use Eq. 43.7 to express the decay constant in terms of the given half-life and then use Eq. 43.6 to solve for the new activity. Note that the final activity in part (c) would be 90.0% of the original activity, since it has decreased by 10.0%. Likewise, the final activity in part (d) would be 80.0% of the original activity, since it had decreased by 20.0%.	
SOLVE	
a. A Bq is the same as a decay/s, so no formal conversion is necessary.	$a = \boxed{1.45 \times 10^9 \text{ Bq}}$

b. Convert the answer from part (a) to Ci.	$a = \left(1.45 \times 10^9 \text{ Bq}\right) \times \left(\dfrac{1 \text{ Ci}}{3.7 \times 10^{10} \text{ Bq}}\right) = \boxed{0.0392 \text{ Ci}}$
c. Use Eq. 43.7 to express the decay constant.	$\eta = \dfrac{\ln 2}{T_{1/2}} = \dfrac{\ln 2}{30.07 \text{ yr}}$
If the activity decreases by 10.0% from its current value, this would mean the new activity would be 90.0% of the initial value. Substitute our expression of Eq. 43.7 into Eq. 43.6 and solve to find the time required.	$a = a_0 e^{-\eta t}$ $a/a_0 = e^{-\eta t}$ $0.900 = e^{-\left(\frac{\ln 2}{30.07 \text{ yr}}\right)t}$ $\ln(0.900) = \ln\left(e^{-\left(\frac{\ln 2}{30.07 \text{ yr}}\right)t}\right)$ $\ln(0.900) = -\left(\dfrac{\ln 2}{30.07 \text{ yr}}\right)t$ $t = -\dfrac{\ln(0.900)(30.07 \text{ yr})}{\ln 2} = \boxed{4.57 \text{ yr}}$
d. If the activity decreases by 20.0% from its current value, this would mean the new activity would be 80.0% of the initial value. Substitute our expression of Eq. 43.7 (from the beginning of part (c)) into Eq. 43.6 and solve to find the time required.	$a = a_0 e^{-\eta t}$ $a/a_0 = e^{-\eta t}$ $0.800 = e^{-\left(\frac{\ln 2}{30.07 \text{ yr}}\right)t}$ $\ln(0.800) = \ln\left(e^{-\left(\frac{\ln 2}{30.07 \text{ yr}}\right)t}\right)$ $\ln(0.800) = -\left(\dfrac{\ln 2}{30.07 \text{ yr}}\right)t$ $t = -\dfrac{\ln(0.800)(30.07 \text{ yr})}{\ln 2} = \boxed{9.68 \text{ yr}}$

CHECK and THINK

The activity is modeled as behaving exponentially as is the number of nuclei remaining. Both decrease exponentially as time goes on. Also, because the decay of each quantity depends on the same decay constant for that particular type of nuclei, the relationship between the number of current nuclei and the amount of activity is linear. We can see this in the relationship between the initial activity and the initial number of nuclei, $a_0 = \eta N_0$.

78. (N) The Chernobyl power station was capable of producing 4.00×10^3 MW of electrical power, prior to the disaster at one of its four reactors in 1986. Assuming that each fission reaction produces about 200 MeV, and that the power plant is about 30.0% efficient, determine the mass of $^{235}_{92}$U that must be used in operating the plant for 1.00 hours. Retain three significant figures in your answer.

INTERPRET and ANTICIPATE
To begin, we actually must solve P43.77 to find the number of fission events that occur per second. Once we have that information, we can multiply that result by the number of seconds in 1.00 hr (3600 s) to determine the total number of fission events. The number of fission events is the number if uranium nuclei required. Thus, we will need to multiply the number of fission events by the mass of one uranium nucleus to get the total mass required.

SOLVE	
First, we express the usable energy generated by each fission event, using the 30.0% (0.300) given in the problem statement. The conversion is there to express the energy in Joules.	$E_{fission} = (0.300)(200 \text{ MeV})\left(\dfrac{1 \times 10^6 \text{ eV}}{1 \text{ MeV}}\right)\left(\dfrac{1.60 \times 10^{-19} \text{ J}}{1 \text{ eV}}\right)$
Then, we can take the power and divide by this energy per fission event to find the number of fission events per second, N, required. Remember that a W is a J/s.	$N = \dfrac{P}{E_{fission}}$ $N = \dfrac{(4.00 \times 10^3 \text{ MW})\left(\dfrac{1 \times 10^6 \text{ W}}{1 \text{ MW}}\right)}{(0.300)(200 \text{ MeV})\left(\dfrac{1 \times 10^6 \text{ eV}}{1 \text{ MeV}}\right)\left(\dfrac{1.60 \times 10^{-19} \text{ J}}{1 \text{ eV}}\right)}$ $N = \boxed{4.17 \times 10^{20} \text{ events/s}}$
Using this result, we can find the number of nuclei, N_U, that must decay in 1.00 hr = 3600 s by multiplying by this time.	$N_U = (4.17 \times 10^{20} \text{ events/s})(3600 \text{ s}) = 1.50 \times 10^{24} \text{ events}$ $N_U = 1.50 \times 10^{24} \text{ nuclei}$

Multiply the number of nuclei by the mass of one Uranium nucleus to find the total mass used in 1.00 hr.

$$m_{total} = m_U \left(1.50 \times 10^{24}\right)$$

$$m_{total} = (221696.64 \text{ MeV}/c^2)\left(\frac{1 \text{ u}}{931.494 \text{ MeV}/c^2}\right)\left(\frac{1.66 \times 10^{-27} \text{ kg}}{1 \text{ u}}\right)\left(1.50 \times 10^{24}\right)$$

$$m_{total} = \boxed{0.593 \text{ kg}}$$

CHECK and THINK
Nuclear power through fission does not necessarily require a large amount of material, as we can see here, but there is not a lot readily available, so that is a good thing. Consider the type a fuel that would be needed for nuclear power through fusion. Is there a lot of that available? Is it worth pushing forward in researching the design of a fusion power generator?